Controllability of Dynamical Systems

Mathematics and Its Applications (*East European Series*)

Volume 48

Controllability of Dynamical Systems

by

Jerzy Klamka

Silesian Technical University
Faculty of Automatics
Gliwice, Poland

KLUWER ACADEMIC PUBLISHERS
DORDRECHT / BOSTON / LONDON

PWN – POLISH SCIENTIFIC PUBLISHERS
WARSZAWA

Published by PWN—Polish Scientific Publishers,
Miodowa 10, 00-251 Warszawa, Poland
in co-edition with Kluwer Academic Publishers,
P.O. Box 17, 3300 AA Dordrecht, The Netherlands

Distributors for the U.S.A. and Canada:
Kluwer Academic Publishers,
101 Philip Drive, Norwell, MA 02061, U.S.A.

Distributors for Albania, Bulgaria, Cuba, Czechoslovakia, Hungary, Korean People's Republic, Mongolia, People's Republic of China, Poland, Romania, the U.S.S.R., Vietnam, and Yugoslavia:
ARS POLONA, Krakowskie Przedmieście 7, 00-068 Warszawa, Poland

Distributors for all remaining countries:
Kluwer Academic Publishers Group,
P.O. Box 322, 3300 AH Dordrecht, The Netherlands

Library of Congress Cataloging-in-Publication Data
Klamka, Jerzy.
Controllability of dynamical systems / Jerzy Klamka.
p. cm.—(Mathematics and its applications (Kluwer Academic Publishers). East Europan series: 48)
Includes bibliographical references (p.) and index.
ISBN 0-7923-0822-0
1, Control theory I. Title. II, Series.
QA402.3.K53 1990
629.8'312—dc20 90-4753

Printed in Poland by D.N.T.

To my Parents

Author

Series Editor's Preface

'Et moi, ..., si j'avait su comment en revenir, je n'y serais point allé.'

Jules Verne

The series is divergent; therefore we may be able to do something with it.

O. Heaviside

One service mathematics has rendered the human race. It has put common sense back where it belongs, on the topmost shelf next to the dusty canister labelled 'discarded nonsense'.

Eric T. Bell

Mathematics is a tool for thought. A highly necessary tool in a world where both feedback and nonlinearities abound. Similarly, all kinds of parts of mathematics serve as tools for other parts and for other sciences.

Applying a simple rewriting rule to the quote on the right above one finds such statements as: 'One service topology has rendered mathematical physics ...'; 'One service logic has rendered computer science ...'; One service category theory has rendered mathematics ...'. All arguably true. And all statements obtainable this way form part of the raison d'être of this series.

This series, *Mathematics and Its Applications*, started in 1977. Now that over one hundred volumes have appeared is seems opportune to reexamine its scope. At the time I wrote

> "Growing specialization and diversification have brought a host of monographs and textbooks on increasingly specialized topics. However, the 'tree' of knowledge of mathematics and related fields does not grow only by putting forth new branches. It also happens, quite often in fact, that branches which were thought to be completely disparate are suddenly seen to be related. Further, the kind and level of sophistication of mathematics applied in various sciences has changed drastically in recent years: measure theory is used (non-trivially) in regional and theoretical economics; algebraic geometry interacts with physics; the Minkowsky lemma, coding theory and the structure of water meet one another in packing and covering theory; quantum fields, crystal defects and mathematical programming profit from homotopy theory; Lie algebras are relevant to

> filtering; and prediction and electrical engineering can use Stein spaces. And in addition to this there are such new emerging subdisciplines as 'experimental mathematics', 'CFD', 'completely integrable systems', 'chaos, synergetics and large-scale order', which are almost impossible to fit into the existing classification schemes. They draw upon widely different sections of mathematics."

By and large, all this still applies today. It is still true that at first sight mathematics seems rather fragmented and that to find, see, and exploit the deeper underlying interrelations more effort is needed and so are books that can help mathematicians and scientists do so. Accordingly MIA will continue to try to make such books available.

If anything, the description I have in 1977 is now an understatement. To the examples of interaction areas one should add string theory where Riemann surfaces, algebraic geometry, modular functions, knots, quantum field theory, Kac-Moody algebras, monstrous moonshine (and more) all come together. And to the examples of things which can be usefully applied let me add the topic 'finite geometry'; a combination of words which sounds like it might not even exist, let alone be applicable. And yet it is being applied: to statistics via designs, to radar/sonar detection arrays (via finite projective planes), and to bus connections of VLSI chips (via difference sets). There seems to be no part of (so-called pure) mathematics that is not in immediate danger of being applied. And, accordingly, the applied mathematician needs to be aware of much more. Besides analysis and numerics, the traditional workhorses, he may need all kinds of combinatorics, algebra, probability, and so on.

In addition, the applied scientist needs to cope increasingly with the nonlinear world and the extra mathematical sophistication that this requires. For that is where the rewards are. Linear models are honest and a bit sad and depressing: proportional efforts and results. It is in the nonlinear world that infinitesimal inputs may result in macroscopic outputs (or vice versa). To appreciate what I am hinting at: if electronics were linear we would have no fun with transistors and computers; we would have no TV; in fact you would not be reading these lines.

There is also no safety in ignoring such outlandish things as nonstandard analysis, superspace and anticommuting integration, p-adic and ultrametric space. All three have applications in both electrical engineering and physics. Once, complex numbers were equally outlandish, but they frequently proved the shortest path between 'real' results. Similarly, the first two topics named have already provided a number of 'wormhole' paths. There is no telling where all this is leading—fortunately.

Thus the original scope of the series, which for various (sound) reasons now comprises five subseries: white (Japan), yellow (China), red (USSR), blue (Eastern Europe), and green (everything else), still applies. It has been enlarged a bit to include books treating of the tools from one subdiscipline which are used in others. Thus the series still aims at books dealing with:

— a central concept which plays an important role in several different mathematical and/or scientific specialization areas;
— new applications of the results and ideas from one area of scientific endeavour into another;
— influences which the results, problems and concepts of one field of enquiry have, and have had, on the development of another.

Control theory (sytems theory) is concerned with equations of the form $\dot{x} = f(x, u)$ (or suitable discrete analogues), where x denotes the state of the system under consideration and u denotes the input or control parameters. Thus, probably, the subject is concerned with 'redoing' all analysis 'with parameters' and, in addition, vast collections of new conceptual ideas and problems arise. So far, in the field, the last part has dominated, and even for linear systems $\dot{x} = Ax+Bu$, particularly in the infinite-dimensional case, but also in the finite-dimensional case, there is a great deal to be said.

Two of the dominants which have emerged are stabilizability and controllability (and the two are very much related). This book is about controllability in almost all its aspects by an author who has made substantial contributions. Wisely, the author does not claim completeness and indeed it is not in that it does not treat the somewhat specialized and technically difficult topic of what happens to controllability and related concepts, as the system parameters vary. Still, this is an almost encyclopaedic and exhaustive treatment and, by a wide margin, it is the most complete treatment of controllability of linear finite, and especially infinite-dimensional linear systems (such as control systems with delays) that I know of.

The shortest path between two truths in the real domain passes through the complex domain.

J. Hadamard

La physique ne nous donne pas seulement l'occasion de résoudre des problèmes ... elle nous fait pressentir la solution.

H. Poincaré

Never lend books, for no one ever returns them; the only books I have in my library are books that other folk have lent me.

Anatole France

The function of an expert is not to be more right than other people, but to be wrong for more sophisticated reasons.

David Butler

Bussum, 25 April 1990

Michiel Hazewinkel

Contents

Preface

In recent decades modern control theory of linear dynamical systems has been the subject of considerable interest of many research scientists. It has been motivated, on the one hand, by the wide range of applications of linear models in various areas of science and engineering and, on the other hand, by the difficult and stimulating theoretical poblems posed by such systems. This book is intended to provide information about various fundamental research results obtained in the field of controllability of linear dynamical systems and discusses their significance and consequences.

The book is mainly concerned with the analysis of various kinds of controllability for linear dynamical systems. Several controllability conditions for continuous-time, finite-dimensional systems, discrete-time finite-dimensional systems and continuous-time, infinite-dimensional systems will be presented and proved. Moreover, dynamical systems with delays in the state variables and in control will be considered and some controllability criteria for them will be given. The book contains several illustrative numerical examples.

In order to attain a high level of precision and clarity we have freely used modern mathematical notation and terminology and have stated all significant results and conclusions in the form of theorems, lemmas or corollaries. To facilitate the understanding of the subject matter and the use of the book as a reference, we have given numbers to all the theorems, lemmas, corollaries and examples. The book contains a long list of recent references. Since much of the material presented is new, our exposition of it has no claim to completeness. This applies in particular to the discussion of controllability of infinite-dimensional dynamical systems and dynamical systems with delays. For the convenience of the reader, each chapter includes a short introduction, remarks and comments.

The text is intended for university courses at the senior-graduate level, of analysis and design of linear control systems. It is also intended for engineers and applied mathematicians as material for independent study and reference. The mathematical background expected from the reader is a working knowledge of matrix manipulation and an elementary knowledge of differential equations, difference equations and partial differential equations. Moreover, some fundamental knowledge of the theory of functional differential equations and linear operators in infinite-dimensional spaces is also required.

J. Klamka

Katowice, February 1990

Index of Notations

I. General conventions

1. Lower case bold-face letters denote vectors, e.g., $\mathbf{u}$, $\mathbf{x}$, $\mathbf{y}$
2. Capital bold-face letters denote matrices, e.g., $\mathbf{A}$, $\mathbf{B}$, $\mathbf{C}$
3. Greek and italic type are used for scalar-valued variables, functions and operators, e.g., α, $u(t)$, P
4. f or $f(\cdot)$ denotes a function, with the dot standing for an undesigned variable; $f(t)$ denotes the value of f at time t. The latter convention is not adhered to strictly, so that in many instances $f(t)$ is used in the same sense as f or $f(\cdot)$
5. A dot over a time function indicates its time derivative, e.g., $\dot{x}$. $x^{(n)}$ or $x^{(n)}(t)$ stands for the n-th derivative of $x(t)$
6. Superscript asterisk (*) denotes the adjoint of an operator, e.g., P^*, or the conjugate transpose of a matrix, e.g., $\mathbf{A}^*$
7. Superscript capital ($^\mathrm{T}$) denotes the transposition of a vector, e.g., $\mathbf{x}^\mathrm{T}$, or transposition of a matrix, e.g., $\mathbf{A}^\mathrm{T}$
8. Superscript minus one ($^{-1}$) denotes the inverse of an operator, e.g., P^{-1}, or the inverse of a matrix, e.g., $\mathbf{A}^{-1}$
9. Superscript perpendicular ($^\perp$) denotes the orthogonal complement of a subspace, e.g., $Q^\perp$
10. Superscript dagger ($^+$) denotes the pseudo inverse of a matrix, e.g., $\mathbf{A}^+$.

II. General symbols and abbreviations

$\Rightarrow$ implication
$\wedge$ for all
$\vee$ there exists
$\supset$ contains

$\cup$ union
$\cap$ intersection
$\times$ Cartesian product
(t_0, t_1) open interval
$(t_0, t_1]$ semiclosed interval
$[t_0, t_1]$ closed interval
$\langle \cdot, \cdot \rangle$ scalar product (dots stand for undesignated variables)
$\| \cdot \|_X$ norm in the space X
$\mathscr{L}$ Laplace transform
det determinant
tr trace
diag diagonal matrix
■ end of proof

III. Symbols with special meaning

$L^p_{\text{loc}}([t_0, \infty), R^q)$, space of locally p-integrable functions with values in R^q

$\mathbf{x}(t, \mathbf{x}(t_0), \mathbf{u})$, state at time t given that the state at t_0 is $\mathbf{x}(t_0)$ and the input $\mathbf{u}_n$ is applied in time interval $[t_0, t)$

$\mathbf{F}(t, t_0)$, state transition matrix

$\mathbf{x}(t)$, state at time t of a dynamical system

$\exp(\mathbf{A}t)$, exponential of the matrix $\mathbf{A}t$

$[\mathbf{A}_1 \vdots \mathbf{A}_2 \vdots \ldots \vdots \mathbf{A}_i \vdots \ldots \vdots \mathbf{A}_n]$, block matrix with components matrices $\mathbf{A}_1, \mathbf{A}_2, \ldots, \mathbf{A}_i, \ldots, \mathbf{A}$

$\text{sp}\{\mathbf{x}_i, i = 1, 2, \ldots, n\}$, linear subspace spanned by the vectors $\mathbf{x}_i$, $i = 1, 2, \ldots, n$

$\overline{\text{sp}}\{\mathbf{x}_i, i = 1, 2, \ldots, n\}$, closure of the $\text{sp}\{\mathbf{x}_i, i = 1, 2, \ldots, n\}$

$\sigma(P)$, spectrum of the operator P

$\varrho(P)$, resolvent set of the operator P

rank $\mathbf{A}$, rank of the matrix $\mathbf{A}$

$D(P)$, domain of the operator P

$W_2^{(1)}([-h, 0], R^n)$, the Sobolev space of R^n-valued functions, defined in $[-h, 0]$ with a square integrable first derivative

$\overline{Q}$, closure of the set Q

I, identity operator or identity matrix

$(i_1, i_2, \ldots, i_j, \ldots, i_n)$, n-tuple of scalars $i_1, i_2, \ldots, i_j, \ldots, i_n$

s_i, i-th eigenvalue of the matrix $\mathbf{A}$

$S(t)$, semigroup of linear bounded operators

$\text{int}Q$, interior of the set Q

CHAPTER 1

Controllability of Continuous-Time Dynamical Systems

1.1. Introduction

The most popular and most fully investigated of the theories of different kinds of dynamical systems is the controllability theory for linear, continuous-time, finite-dimensional dynamical systems. This follows both from the simple form of the mathematical model and from the great number of applications.

In this chapter we shall formulate numerous necessary and sufficient conditions for various kinds of controllability of linear, continuous-time, finite-dimensional dynamical systems. We shall consider the nonstationary (time-varying) and stationary (time-invariant) dynamical systems. The relations between different types of controllability will also be given and explained.

The notion of controllability of a dynamical system has also proved very useful in the investigation of the so-called "minimum energy control problem" for linear dynamical systems. This problem will be solved by using only the fundamental properties of the norm and of the scalar product in Euclidean spaces, without resorting to any additional theorems from modern control theory.

Considerable space in this chapter will be devoted to the study of the so-called "constrained controllability" of dynamical systems, i.e., the controllability of dynamical systems with constrained control. Several different criteria for constrained controllability will be presented, especially for time-invariant, finite-dimensional dynamical systems.

One section of this chapter deals with the dependence of controllability on the perturbations of system parameters. From the practical point of view this is a very important problem but it is not completely solved yet.

The duality principle for controllability and observability will also be presented and some comments will be given. Moreover, we shall analyse and investigate the so-called "structural controllability" of time-invarinat dynamical system. Structural controllability is one of the most important and difficult problems in controllability theory for linear dynamical systems.

Moreover, we shall consider the numerical algorithm related to computing controllability and some canonical forms related to controllability, using the so-called "singular value decomposition" (SVD).

The chapter contains a large number of illustrative examples and extensive comments.

1.2 System description and the basic definitions

In this section we shall give the fundamental definitions of various kinds of controllability for linear time-varying continuous finite-dimensional dynamical systems, described by the following ordinary differential equation:

$$\dot{\mathbf{x}}(t) = \mathbf{A}(t)\mathbf{x}(t)+\mathbf{B}(t)\mathbf{u}(t), \qquad t \geqslant t_0, \tag{1.2.1}$$

where $\mathbf{x}(t) \in \boldsymbol{R}^n$ is the state vector, $\mathbf{u} \in L^2_{\text{loc}}([t_0, \infty), \boldsymbol{R}^m)$ is an admissible control, $\mathbf{A}(t)$ is the $(n \times n)$-dimensional matrix with elements $a_{ij} \in L^1_{\text{loc}}([t_0, \infty), \boldsymbol{R})$ for $i = 1, 2, \ldots, n$, $j = 1, 2, \ldots, n$, $\mathbf{B}(t)$ is the $(n \times m)$-dimensional matrix with elements $b_{ij} \in L^2_{\text{loc}}([t_0, \infty), \boldsymbol{R})$ for $i = 1, 2, \ldots, n, j = 1, 2, \ldots, m$.

For a given initial condition $\mathbf{x}(t_0) \in \boldsymbol{R}^n$ and admissible control $\mathbf{u}$, there exists a unique solution of equation (1.2.1), $\mathbf{x}(t, \mathbf{x}(t_0), \mathbf{u})$ which is absolutely continuous in $[t_0, \infty)$ (Chen (1970, Ch. 4.2), Lee and Markus (1968, Ch.2.1), Zadeh and Desoer (1963, Ch. 11.3)). This solution has the form

$$\mathbf{x}(t, \mathbf{x}(t_0), \mathbf{u}) = \mathbf{F}(t, t_0)\mathbf{x}(t_0)+ \int_{t_0}^{t} \mathbf{F}(t, \tau)\mathbf{B}(\tau)\mathbf{u}(\tau)d\tau, \qquad t \geqslant t_0, \tag{1.2.2}$$

where $\mathbf{F}(t, \tau)$ is the $(n \times n)$-dimensional *transition matrix* for the dynamical system (1.2.1). The transition matrix $\mathbf{F}(t, \tau)$ is defined for each t and τ in $(-\infty, +\infty)$ and has the following properties:

$$\begin{aligned} &\mathbf{F}(t, t) = \mathbf{I}_{n\times n} && \text{for } t \in (-\infty, +\infty), \\ &\mathbf{F}^{-1}(t, \tau) = \mathbf{F}(\tau, t) && \text{for } t, \tau \in (-\infty, +\infty), \\ &\mathbf{F}(t_2, t_1)\mathbf{F}(t_1, t_0) = \mathbf{F}(t_2, t_0) && \text{for } t_0, t_1, t_2 \in (-\infty, +\infty). \end{aligned}$$

DEFINITION 1.2.1. The dynamical system (1.2.1) is said to be *controllable in the time interval* $[t_0, t_1]$ if, for any initial state $\mathbf{x}(t_0) \in \boldsymbol{R}^n$ and any vector $\mathbf{x}_1 \in \boldsymbol{R}^n$, there exists a control $\mathbf{u} \in L^2([t_0, t_1], \boldsymbol{R}^m)$ such that the corresponding trajectory of the dynamical system (1.2.1), $\mathbf{x}(t, \mathbf{x}(t_0, \mathbf{u})$ satisfies the following condition:

$$\mathbf{x}(t_1, \mathbf{x}(t_0), \mathbf{u}) = \mathbf{x}_1 .$$

DEFINITION 1.2.2. The dynamical system (1.2.1) is said to be *controllable at time* t_0 if there exists a $t_1 \in (t_0, \infty)$ such that it is controllable in $[t_0, t_1]$.

DEFINITION 1.2.3. The dynamical system (1.2.1) is said to be *controllable* if it is controllable at any time $t_0 \in (-\infty, +\infty)$.

DEFINITION 1.2.4. The dynamical system (1.2.1) is said to be *uniformly controllable* if it is controllable in any time interval $[t_0, t_1]$ $(t_0 < t_1)$.

Controllability is strongly related to the notion of the attainable set $K(t_0, t_1)$ for the dynamical system (1.2.1). The attainable set from the zero initial state $K(t_0, t_1)$ is defined as follows:

$$K(t_0, t_1) = \left\{\mathbf{x} \in \mathbf{R}^n \colon \mathbf{x} = \int_{t_0}^{t_1} \mathbf{F}(t_1, \tau)\mathbf{B}(\tau)\mathbf{u}(\tau)d\tau \colon \mathbf{u} \in L^2([t_0, t_1], \mathbf{R}^m)\right\}. \tag{1.2.3}$$

Therefore controllability of the dynamical system (1.2.1) in time interval $[t_0, t_1]$ is equivalent to the condition $K(t_0, t_1) = \mathbf{R}^n$. In a similar way we can express the other types of controllability through the notions of the attainable sets.

Between the various kinds of controllability given by Definitions 1.2.1–1.2.4, there exist the following implications:

$$\begin{array}{ccccc} & \nearrow & \begin{array}{c}\text{controllability}\\ \text{in } [t_0,\ t_1]\end{array} & \searrow & \\ \begin{array}{c}\text{uniform}\\ \text{controllability}\end{array} & & & & \begin{array}{c}\text{controllability}\\ \text{at } t_0\end{array} \\ & \searrow & \text{controllability} & \nearrow & \end{array} \tag{1.2.4}$$

Besides the types of controllability given above, we may also introduce the notions of regular controllability and selective controllability (Zunde (1971)). Selective controllability is extensively used in interaction investigations (Zunde (1971)).

DEFINITION 1.2.5. The dynamical system (1.2.1) is said to be *regularly controllable in time interval* $[t_0, t_1]$ if every dynamical system of the form

$$\dot{\mathbf{x}}(t) = \mathbf{A}(t)\mathbf{x}(t) + \mathbf{b}_j(t)u_j(t), \quad t \geqslant t_0,\ j = 1, 2, \ldots, m, \tag{1.2.5}$$

where $\mathbf{b}_j(t) \in \mathbf{R}^n$ is the j-th column of the matrix $\mathbf{B}(t)$, $j = 1, 2, \ldots, m$, $u_j(t) \in L^2_{\text{loc}}([t_0, \infty), \mathbf{R})$, $j = 1, 2, 1\ldots, m$, is the scalar admissible control, is controllable in the time interval $[t_0, t_1]$.

In other words, the dynamical system (1.2.1) is regularly controllable in the time interval $[t_0, t_1]$ if it is controllable in this time interval with respect to each scalar input separately.

Definition 1.2.5 immediately implies that the dynamical system (1.2.1) regularly controllable in $[t_0, t_1]$ is of course also controllable in this interval. The inverse implication is in general false: there are dynamical systems of the form (1.2.1) which are controllable in $[t_0, t_1]$, but not regularly controllable in this time interval (for example one of the columns of the matrix $\mathbf{B}(t)$ may be equal to zero). Similarly we can define uniform regular controllability, regular controllability and regular controllability at t_0 of the dynamical system (1.2.1).

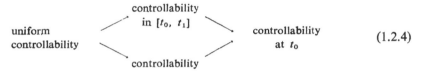

Next we shall define so-called selective controllability for the dynamical system (1.2.1).

Definition 1.2.6. The dynamical system (1.2.1) is said to be *selectively controllable with respect to the i-th* ($i = 1, 2, \dots, n$) *state variable by the j-th* ($j = 1, 2, \dots, m$) *control variable* $u_j(t)$ in $[t_0, t_1]$ if the dynamical system of the form (1.2.5) has the i-th state variable $x_i(t) \in \boldsymbol{R}$ controllable in $[t_0, t_1]$.

In other words, the dynamical system (1.2.1) is selectively controllable with respect to the i-th state variable by the j-th control variable in $[t_0, t_1]$ if, for any initial condition $x_i(t_0) \in \boldsymbol{R}$ and any real number x_{i1}, there exists a scalar control $u_j \in L^2([t_0, t_1], \boldsymbol{R})$ such that $x_i(t_1, x_i(t_0), u_j) = x_{i1}$.

Similarly to the state controllabilities defined with the aid of the differential state equation (1.2.1), we may define various kinds of so-called output controllability. In order to do that, let us introduce the algebraic linear output equation of the dynamical system (1.2.1) given by

$$\mathbf{y}(t) = \mathbf{C}(t)\mathbf{x}(t) + \mathbf{D}(t)\mathbf{u}(t), \qquad t \geqslant t_0, \tag{1.2.6}$$

where $\mathbf{y}(t) \in \boldsymbol{R}^p$ is the output vector, $\mathbf{C}(t)$ is the $(p \times n)$-dimensional matrix with elements $c_{ij} \in L^1_{\text{loc}}([t_0, \infty), \boldsymbol{R})$, $i = 1, 2, \dots, p$, $j = 1, 2, \dots, n$, $\mathbf{D}(t)$ is the $(p \times m)$-dimensional matrix with elements $d_{ij} \in L^2_{\text{loc}}([t_0, \infty), \boldsymbol{R})$, $i = 1, 2, \dots, p$, $j = 1, 2, \dots, m$

Definition 1.2.7. The dynamical system with state equation (1.2.1) and output equation (1.2.6) is said to be *output controllable in the time interval* $[t_0, t_1]$ if for any output $\mathbf{y}(t_0) \in \boldsymbol{R}^p$ and any vector $\mathbf{y}_1 \in \boldsymbol{R}^p$, there exists an admissible control $\mathbf{u} \in L^2([t_0, t_1], \boldsymbol{R}^m)$ such that the corresponding trajectory of the output $\mathbf{y}(t, \mathbf{y}(t_0), \mathbf{u})$ satisfies the following condition:

$$\mathbf{y}(t_1, \mathbf{y}(t_0), \mathbf{u}) = \mathbf{y}_1. \tag{1.2.7}$$

In practice the dimension m of the control vector and the dimension p of the output vector are less than the dimension n of the state vector. Hence, taking into account the equality (1.2.6), we conclude that there are no relations between state and output controllabilities. Indeed, there exist dynamical systems which are controllable in $[t_0, t_1]$ and not output controllable in the same time interval. Conversely, there exist dynamical systems output controllable in $[t_0, t_1]$ which are not controllable in this time interval.

Similarly, we can define other types of output controllability, i.e., uniform output controllability, output controllability and output controllability at time t_0. It should be noted that although the concepts of controllability and output controllability are similar, the controllability is a property of a differential state equation (1.2.1), whereas output controllability is a property of the state equation and algebraic output equation (1.2.6).

Now let us consider the special case of the dynamical system (1.2.1), namely linear time-invariant finite-dimensional dynamical system, described by the equations with constant coefficients. Therefore, the time-invariant dynamical system is given by the following set of equations:

$$\dot{\mathbf{x}}(t) = \mathbf{A}\mathbf{x}(t)+\mathbf{B}\mathbf{u}(t), \quad t \geqslant 0, \tag{1.2.8}$$

$$\mathbf{y}(t) = \mathbf{C}\mathbf{x}(t)+\mathbf{D}\mathbf{u}(t), \quad t \geqslant 0, \tag{1.2.9}$$

where **A**, **B**, **C**, **D** are constant matrices respectively of dimensions $n \times n$, $n \times m$, $p \times n$, and $p \times m$.

In the case of time-invariant dynamical systems, the $(n \times n)$-diemensional transition matrix $\mathbf{F}(t, \tau)$ can easily be computed. Indeed, by Zadeh and Desoer (1963, Ch. 6.2) we have

$$\mathbf{F}(t, \tau) = \exp\big(\mathbf{A}(t-\tau)\big) \quad \text{for } t, \tau \in \boldsymbol{R}. \tag{1.2.10}$$

For time-invariant dynamical systems some of the definitions of various kinds of controllability become simplified. In this case the time interval is no longer substantial for the concept of controllability. Therefore for time-invariant dynamical system uniform controllability, controllability, controllability in $[t_0, t_1]$ and controllability at time t_0 are equivalent concepts (this will be proved in Lemma 1.4.1). As will be illustrated in the next sections, controllability conditions for time-invariant dynamical systems are substantially simpler than the controllability conditions for time-varying dynamical systems. It should also be pointed out, that in the time-invariant case the characteristics of the equations do not change with time; hence there is no loss of generality in choosing the initial time t_0 to be 0. Finally, it should be stressed that almost all physical systems are time-varying. Thus, the notion of a time invariant dynamical systems is essentially an idealization of physical systems whose characteristics change very slowly in time in relation to variation in the input.

1.3 Controllability conditions for time-varying dynamical systems

At the beginning of this section we shall introduce the concept of linear independence of a set of functions of a real variable and give the conditions for linear independence. Next we shall apply these conditions to obtain controllability criteria for dynamical system (1.2.1).

Let $f_i \in L^2([t_0, t_1], \boldsymbol{R}^m)$, $i = 1, 2, \ldots, n$.

DEFINITION 1.3.1. The set of vector-valued functions $f_1, f_2, \ldots, f_n$ is said to be *linearly dependent on the time interval* $[t_0, t_1]$ *over the field of complex numbers* if there exists a nonzero complex vector $\mathbf{a} = [a_1\ a_2\ \ldots\ a_n]^{\mathrm{T}} \in \boldsymbol{C}^n$ such that

$$a_1 f_1(t)+a_2 f_2(t)+\ \ldots\ +a_n f_n(t) = 0 \quad \text{for a.e. } t \text{ in } [t_0, t_1]. \tag{1.3.1}$$

Otherwise, the set of functions $f_1, f_2, \ldots, f_m$ is said to be *linearly independent in the time interval* $[t_0, t_1]$ *over the field of complex numbers.*

The length of the time interval $[t_0, t_1]$ plays an important role in Definition 1.3.1. The set of functions which are linearly independent over the field of complex numbers on $[t_0, t_1]$ need not be linearly independent on each subinterval of $[t_0, t_1]$. However, there exists a subinterval with nonzero length, on which this set is linearly independent. On the other hand, if a set of functions is linearly independent on an interval $[t_0, t_1]$, then the set of functions is linearly independent on any interval that contains $[t_0, t_1]$.

Writing $[f_1(t)\ f_2(t)\ \ldots\ f_n(t)]^T = \mathbf{F}(t)$, we can express the relation (1.3.1) in a compact form: $\mathbf{a}^T\mathbf{F}(t) = 0$ for a.e. t in $[t_0, t_1]$.

LEMMA 1.3.1. *The functions* $f_1, f_2, \ldots, f_n$ *are linearly independent on* $[t_0, t_1]$ *if and only if the* $(n \times n)$*-dimensional constant matrix*

$$\mathbf{G}(t_0, t_1) = \int_{t_0}^{t_1} \mathbf{F}(t)\mathbf{F}^T(t)\,dt \tag{1.3.2}$$

is nonsingular.

Proof:

Necessity. By contradiction. Assume that the set of functions $f_1, f_2, \ldots, f_n$ is linearly independent on $[t_0, t_1]$ but the matrix $\mathbf{G}(t_0, t_1)$ is singular. Then there exists a nonzero vector $\mathbf{a} \in \boldsymbol{R}^n$ such that $\mathbf{a}^T\mathbf{G}(t_0, t_1) = 0$. Consequently

$$\mathbf{a}^T\mathbf{G}(t_0, t_1)\mathbf{a} = \int_{t_0}^{t_1} (\mathbf{a}^T\mathbf{F}(t))(\mathbf{a}^T\mathbf{F}(t))^T dt = 0. \tag{1.3.3}$$

The relation (1.3.3) implies that $\mathbf{a}^T\mathbf{F}(t) = 0$ a.e. in $[t_0, t_1]$. This contradicts the linear independence assumption of the set of functions $f_1, f_2, \ldots, f_n$. Hence if the functions $f_1, f_2, \ldots, f_n$ are linearly independent on $[t_0, t_1]$ then the matrix $\mathbf{G}(t_0, t_1)$ is nonsingular.

We next prove the sufficiency of the theorem. By contradiction. Suppose that the matrix $\mathbf{G}(t_0, t_1)$ is nonsingular, but the functions $f_1, f_2, \ldots, f_n$ are linearly dependent on $[t_0, t_1]$. Then, by Definition 1.3.1, there exists a nonzero constant vector $\mathbf{a} \in \boldsymbol{R}^n$ such that $\mathbf{a}^T\mathbf{F}(t) = 0$ a.e. in $[t_0, t_1]$. Consequently, by (1.3.2) we have $\mathbf{a}^T\mathbf{G}(t_0, t_1) = 0$, which contradicts the assumption that the matrix $\mathbf{G}(t_0, t_1)$ is nonsingular. Hence our theorem follows. ■

The determinant of $\mathbf{G}(t_0, t_1)$ is called the *Gram determinant* of the set of functions $f_1, f_2, \ldots, f_n$. If the functions f_i, $i = 1, 2, \ldots, n$ have continuous derivatives up to order $n-1$, then we may use the following lemma.

LEMMA 1.3.2. *Assume that $f_i \in C^{n-1}([t_0, t_1], R^m)$, $i = 1, 2, \ldots, n$. Then the set of functions $f_1, f_2, \ldots, f_n$ is linearly independent on $[t_0, t_1]$ if there exists a certain $t_2 \in [t_0, t_1]$ such that the $(n \times nm)$-dimensional constant matrix*

$$[\mathbf{F}(t_2) \vdots \mathbf{F}^{(1)}(t_2) \vdots \mathbf{F}^{(2)}(t_2) \vdots \ \ldots \ \vdots \mathbf{F}^{(n-1)}(t_2)] \tag{1.3.4}$$

has rank n.

Proof:

We prove the lemma by contradiction. Suppose that there exists a certain $t_2 \in [t_0, t_1]$ such that

$$\text{rank}[\mathbf{F}(t_2) \vdots \mathbf{F}^{(1)}(t_2) \vdots \mathbf{F}^{(2)}(t_2) \vdots \ \ldots \ \vdots \mathbf{F}^{(n-1)}(t_2)] = n$$

but the set of functions f_i, $i = 1, 2, \ldots, n$ is linearly dependent on $[t_0, t_1]$. Then by Definition 1.3.1 there exists a nonzero vector $\mathbf{a} \in R^n$ such that

$$\mathbf{a}^T\mathbf{F}(t) = 0 \quad \text{for all } t \in [t_0, t_1].$$

Differentiating k times the above equality, we have

$$\mathbf{a}^T\mathbf{F}^{(k)}(t) = 0 \quad \text{for } k = 1, 2, \ldots, n-1 \text{ and all } t \in [t_0, t_1].$$

This imples that

$$\mathbf{a}^T[\mathbf{F}(t) \vdots \mathbf{F}^{(1)}(t) \vdots \mathbf{F}^{(2)}(t) \vdots \ \ldots \ \vdots \mathbf{F}^{(n-1)}(t)] = 0 \quad \text{for all } t \in [t_0, t_1].$$

In particular, the above equality holds for $t = t_2$, which implies that all the rows of the matrix (1.3.4) are linearly dependent. This contradicts the hypothesis that this matrix has rank n. Hence the set of functions f_i, $i = 1, 2, \ldots, n$ is linearly independent on $[t_0, t_1]$. ■

Lemma 1.3.2 is only a sufficient but not a necessary condition for a set of functions to be linearly independent on $[t_0, t_1]$.

In order to check the linear independence of a set of functions, if the functions are generally discontinuous we can employ Lemma 1.3.1, which requires an integration over an interval, and if the functions are $n-1$ times differentiable we can employ Lemma 1.3.2, which requires a differentiation in the time interval $[t_0, t_1]$. It is clear that Lemma 1.3.2 is easier to use than Lemma 1.3.1; however, it gives only sufficient conditions for linear independence. If the functions are analytic in $[t_0, t_1]$, then we can use Lemma 1.3.3, which is based on the fact that if a function is analytic on $[t_0, t_1]$, then the function is completely determinable from a point in $[t_0, t_1]$, provided all the derivatives of the function at that point are known.

LEMMA 1.3.3. *Assume that for each $i = 1, 2, \ldots, n$ the function f_i is analytic on $[t_0, t_1]$. Let t_2 be any fixed point in $[t_0, t_1]$. Then the functions f_i, $i = 1, 2, \ldots, n$ are linearly independent on $[t_0, t_1]$ if and only if*

$$\text{rank}[\mathbf{F}(t_2) \vdots \mathbf{F}^{(1)}(t_2) \vdots \mathbf{F}^{(2)}(t_2) \vdots \ \ldots \ \vdots F^{(n-1)}(t_2) \vdots \ \ldots] = n. \tag{1.3.5}$$

Proof:

Sufficiency. The sufficiency of Lemma 1.3.3 can be proved as in Lemma 1.3.2.

Necessity. Now we prove by contradiction the necessity of Lemma 1.3.3. Suppose that condition (1.3.2) is not satisfied. Then there exists a nonzero vector $\mathbf{a} \in \boldsymbol{R}^n$ such that

$$\mathbf{a}^{\mathrm{T}}[\mathbf{F}(t_2) \vdots \mathbf{F}^{(1)}(t_2) \vdots \mathbf{F}^{(2)}(t_2) \vdots \ \ldots \ \vdots \mathbf{F}^{(n-1)}(t_2) \vdots \ \ldots] = 0. \tag{1.3.6}$$

Expanding the matrix function $\mathbf{F}(t)$ in a Taylor series in the neighbourhood of the point t_2 and using the analyticity assumption of $\mathbf{F}(t)$, we have by (1.3.6)

$$\mathbf{a}^{\mathrm{T}}\mathbf{F}(t) = 0 \quad \text{for all } t \text{ in } [t_0, t_1],$$

which is equivalent to the linear dependence of the functions f_i, $i = 1, 2, \ldots, n$ on $[t_0, t_1]$. This is a contradiction, and hence our lemma follows. ∎

A direct consequence of Lemma 1.3.3 are the following corollaries.

COROLLARY 1.3.1. *If a set of analytic functions is linearly independent on* $[t_0, t_1]$, *then*

$$\operatorname{rank}[\mathbf{F}(t) \vdots \mathbf{F}^{(1)}(t) \vdots \mathbf{F}^{(2)}(t) \vdots \ \ldots \ \vdots \mathbf{F}^{(n-1)}(t) \vdots \ \ldots] = n \quad \textit{for all } t \textit{ in } [t_0, t_1].$$

COROLLARY 1.3.2. *If a set of analytic functions is linearly independent on* $[t_0, t_1]$, *then it is also linearly independent in each subinterval of* $[t_0, t_1]$.

COROLLARY 1.3.3. *Assume that the functions* f_i, $i = 1, 2, \ldots, n$ *are analytic on* $[t_0, t_1]$. *Then* $f_1, f_2, \ldots, f_n$ *are linearly independent on* $[t_0, t_1]$ *if and only if*

$$\operatorname{rank}[\mathbf{F}(t) \vdots \mathbf{F}^{(1)}(t) \vdots \mathbf{F}^{(2)}(t) \vdots \ \ldots \ \vdots \mathbf{F}^{(n-1)}(t)] = n \quad \textit{for almost all } t \textit{ in } [t_0, t_1]. \tag{1.3.7}$$

Using the preceding lemmas and a well-known theorem from functional analysis we shall formulate several conditions for the controllability of time-varying dynamical systems. First we shall give the fundamental necessary and sufficient conditions of controllability in $[t_0, t_1]$ for time-varying dynamical systems of the form (1.2.1).

THEOREM 1.3.1. *The dynamical system* (1.2.1) *is controllable on* $[t_0, t_1]$ *if and only if the* $(n \times n)$*-dimensional controllability matrix*

$$\mathbf{W}(t_0, t_1) = \int_{t_0}^{t_1} \mathbf{F}(t_1, t)\mathbf{B}(t)\mathbf{B}^{\mathrm{T}}(t)\mathbf{F}^{\mathrm{T}}(t_1, t)\,dt \tag{1.3.8}$$

is nonsingular.

Proof:

Sufficiency. Let $\mathbf{x}(t_0) \in \boldsymbol{R}^n$ be any initial state of the dynamical system (1.2.1) and $\mathbf{x}_1 \in \boldsymbol{R}^n$ an arbitrary vector. We shall show that the control $\mathbf{u} \in L^2([t_0, t_1], \boldsymbol{R}^m)$ given by

$$\mathbf{u}(t) = \mathbf{B}^{\mathrm{T}}(t)\mathbf{F}^{\mathrm{T}}(t_1, t)\mathbf{W}^{-1}(t_0, t_1)\big(\mathbf{x}_1 - \mathbf{F}(t_1, t_0)\mathbf{x}(t_0)\big), \quad t \in [t_0, t_1] \tag{1.3.9}$$

steers the dynamical system (1.2.1) from an initial state $\mathbf{x}(t_0)$ to a final state $\mathbf{x}_1$. Substituting (1.3.9) in (1.2.2) and using the properties of the transition matrix $\mathbf{F}(t, s)$, we obtain after some calculations the following equalities:

$$\begin{aligned} &\mathbf{x}(t_1, \mathbf{x}(t_0), \mathbf{u}) \\ &= \mathbf{F}(t_1, t_0)\mathbf{x}(t_0)+ \\ &\quad + \int_{t_0}^{t_1} \mathbf{F}(t_1, t)\mathbf{B}(t)\mathbf{B}^{\mathrm{T}}(t)\mathbf{F}^{\mathrm{T}}(t_1, t)\,dt\,\mathbf{W}^{-1}(t_0, t_1)\big(\mathbf{x}_1-\mathbf{F}(t_1, t_0)\mathbf{x}(t_0)\big) \\ &= \mathbf{F}(t_1, t_0)\mathbf{x}(t_0)+\mathbf{W}(t_0, t_1)\mathbf{W}^{-1}(t_0, t_1)\big(\mathbf{x}_1-\mathbf{F}(t_1, t_0)\mathbf{x}(t_0)\big) = \mathbf{x}_1. \end{aligned}$$

Since $\mathbf{x}(t_0)$ and $\mathbf{x}_1$ were arbitrary, then by Definition 1.2.1 the dynamical system (1.2.1) is controllable on $[t_0, t_1]$.

Necessity. By contradiction. Suppose that the dynamical system (1.2.1) is controllable on $[t_0, t_1]$, but the matrix $\mathbf{W}(t_0, t_1)$ is singular. Then by Lemma 1.3.1 the rows of the matrix $\mathbf{F}(t_1, t)\mathbf{B}(t)$ are linearly dependent on $[t_0, t_1]$. Therefore there exists a nonzero vector $\mathbf{a} \in \boldsymbol{R}^n$ such that

$$\mathbf{a}^{\mathrm{T}}\mathbf{F}(t_1, t)\mathbf{B}(t) = 0 \quad \text{for all } t \text{ in } [t_0, t_1]. \tag{1.3.10}$$

Multiplying the equality (1.2.2) by $\mathbf{a}^{\mathrm{T}}$ and using (1.3.10) for $t = t_1$, we obtain

$$\mathbf{a}^{\mathrm{T}}\mathbf{x}(t_1, \mathbf{x}(t_0), \mathbf{u}) = \mathbf{a}^{\mathrm{T}}\mathbf{F}(t_1, t_0)\mathbf{x}(t_0). \tag{1.3.11}$$

Since by assumption the dynamical system (1.2.1) is controllable on $[t_0, t_1]$, the equality (1.3.11) must be satisfied for any $\mathbf{x}(t_0)$ and $\mathbf{x}_1 = \mathbf{x}(t_1, \mathbf{x}(t_0), \mathbf{u})$. However, taking $\mathbf{x}_1 = 0$ and $\mathbf{x}(t_0) = \mathbf{F}(t_0, t_1)\mathbf{a} \neq 0$, we obtain $\mathbf{a}^{\mathrm{T}}\mathbf{a} = 0$. Since by assumption the vector $\mathbf{a}$ is nonzero, this is a contradiction. Hence our theorem follows. ■

The proof of Theorem 1.3.1 presented above is a constructive one, because it gives the formula for admissible control, which steers the dynamical system (1.2.1) from a given initial state $\mathbf{x}(t_0)$ to the desired final state $\mathbf{x}_1$ at time t_1.

Theorem 1.3.1 can also be proved in a simple way by using well-known theorems from linear functional analysis (Balakrishnan (1977), Sastry and Desoer (1982)). However, it should be pointed out that this proof is not constructive, i.e. does not give the formula for steering control.

The second version of the proof of Theorem 1.3.1. First of all let us define the so-called *controllability operator* $P(t_0, t_1)\colon L^2([t_0, t_1], \boldsymbol{R}^m) \to \boldsymbol{R}^n$ as follows:

$$P(t_0, t_1)\mathbf{u} = \int_{t_0}^{t_1} \mathbf{F}(t_1, t)\mathbf{B}(t)\mathbf{u}(t)\,dt. \tag{1.3.12}$$

The controllability operator $P(t_0, t_1)$ is linear and continuous from the Hilbert space $L^2([t_0, t_1], \boldsymbol{R}^m)$ to the Hilbert space $\boldsymbol{R}^n$. Therefore the adjoint operator $P^*(t_0, t_1)$: $\boldsymbol{R}^n \to L^2([t_0, t_1], \boldsymbol{R}^m)$ is of the following form (Sastry and Desoer (1982)):

$$P^*(t_0, t_1)\mathbf{x} = \mathbf{B}^{\mathrm{T}}(t)\mathbf{F}^{\mathrm{T}}(t_1, t)\mathbf{x} \quad \text{for } t \in [t_0, t_1]. \tag{1.3.13}$$

By (1.2.3) the range of the controllability operator $P(t_0, t_1)$ is equal to the attainable set $K(t_0, t_1)$. Therefore the controllability on $[t_0, t_1]$ of the dynamical system (1.2.1) is equivalent to the statement that the range of the controllability operator is the whole space $\boldsymbol{R}^n$. Consequently, this is equivalent to the invertibility of the following linear bounded operator: $P(t_0, t_1)P^*(t_0, t_1)$ (Sastry and Desoer (1982)). Operator $P(t_0, t_1) \cdot P^*(t_0, t_1)\colon \boldsymbol{R}^n \to \boldsymbol{R}^n$ is by (1.3.12) and (1.3.13) equal to the controllability matrix $\mathbf{W}(t_0, t_1)$. Hence our theorem follows. ∎

The second version of the proof of Theorem 1.3.1 is entirely based on the theory of linear operators in Hilbert spaces. This method of proof can be extended to cover a wider class of dynamical systems, for instance dynamical systems described by the linear partial differential equations with mixed boundary conditions.

In order to apply Theorem 1.3.1 we must compute a transition state matrix $\mathbf{F}(t, s)$.

As we mentioned earlier, this is generally a difficult task. Hence, Theorem 1.3.1 is not readily applicable. In the following we shall give controllability criteria based entirely on the matrices $\mathbf{A}(t)$ and $\mathbf{B}(t)$ without solving the state equation. However, in order to do so, we need some additional assumptions on $\mathbf{A}(t)$ and $\mathbf{B}(t)$. These assumptions concern the regularity of the matrices $\mathbf{A}(t)$ and $\mathbf{B}(t)$ as functions of time $t \in [t_0, t_1[$. These controllability criteria will be formulated in detail in the next theorems and corollaries. Finally, it should be stressed that some of these conditions are only sufficient conditions for controllability, while others are also necessary conditions.

Theorem 1.3.2. *Assume that the matrices* $\mathbf{A}(t)$ *and* $\mathbf{B}(t)$ *are* $n-1$ *times continuously differentiable on* $[t_0, t_1]$. *Then the dynamical system* (1.2.1) *is controllable on* $[t_0, t_1]$ *if there exists a* $t_2 \in [t_0, t_1]$ *such that*

$$\operatorname{rank}[\mathbf{M}_0(t_2) \vdots \mathbf{M}_1(t_2) \vdots \mathbf{M}_2(t_2) \vdots \ \ldots \ \vdots \mathbf{M}_{n-1}(t_2)] = n, \tag{1.3.14}$$

where the $(n \times m)$*-dimensional matrices* $\mathbf{M}_k(t)$, $k = 0, 1, 2, \ldots, n-1$ *are defined as follows*:

$$\mathbf{M}_0(t) = \mathbf{B}(t), \tag{1.3.15}$$

$$\mathbf{M}_{k+1}(t) = -\mathbf{A}(t)\mathbf{M}_k(t) + \frac{d}{dt}\mathbf{M}_k(t) \quad \textit{for } k = 0, 1, 2, \ldots, n-1. \tag{1.3.16}$$

Proof:

Using the properties of the transition matrix $\mathbf{F}(t, s)$ and formulas (1.3.15), (1.3.16), we observe that

$$\frac{d^k}{dt^k}\mathbf{F}(t_0, t)\mathbf{B}(t) = \mathbf{F}(t_0, t)\mathbf{M}_k(t) \quad \text{for } k = 0, 1, 2, \ldots, n-1. \tag{1.3.17}$$

Hence

$$\mathbf{F}(t_0, t_2)[\mathbf{M}_0(t_2) \vdots \mathbf{M}_1(t_2) \vdots \mathbf{M}_2(t_2) \vdots \ \cdots \ \vdots \mathbf{M}_{n-1}(t_2)]$$

$$= \left[\mathbf{F}(t_0, t_2)\mathbf{B}(t_2) \vdots \frac{d}{dt}\mathbf{F}(t_0, t)\mathbf{B}(t)\Big|_{t=t_2} \vdots \ \cdots \ \vdots \frac{d^{n-1}}{dt^{n-1}}\mathbf{F}(t_0, t)\mathbf{B}(t)\Big|_{t=t_2}\right]. \tag{1.3.18}$$

Since the matrix $\mathbf{F}(t_0, t_2)$ is nonsingular, equality (1.3.14) together with (1.3.18) implies that

$$\operatorname{rank}\left[\mathbf{F}(t_0, t_2)\mathbf{B}(t_2) \vdots \frac{d}{dt}\mathbf{F}(t_0, t)\mathbf{B}(t)\Big|_{t=t_2} \vdots \ \cdots \ \vdots \frac{d^{n-1}}{dt^{n-1}}\mathbf{F}(t_0, t)\mathbf{B}(t)\Big|_{t=t_2}\right] = n.$$

Thus by Lemma 1.3.2 the rows of the matrix $\mathbf{F}(t_0, t_2)\mathbf{B}(t_2)$ are linearly independent on $[t_0, t_2]$, and since $[t_0, t_2] \subset [t_0, t_1]$ they are of course linearly independent on $[t_0, t_1]$. Therefore by Lemma 1.3.1 and Theorem 1.3.1 the dynamical system (1.2.1) is controllable on $[t_0, t_1]$. Hence our theorem follows. ■

Theorem 1.3.2 is only a sufficient but not necessary condition for controllability on $[t_0, t_1]$ of the dynamical system (1.2.1). Under some additional assumptions on the matrices $\mathbf{A}(t)$ and $\mathbf{B}(t)$ concerning analyticity we can derive the necessary and sufficient conditions for controllability on $[t_0, t_1]$ of the dynamical system (1.2.1).

THEOREM 1.3.3. *Assume that the matricres* $\mathbf{A}(t)$ *and* $\mathbf{B}(t)$ *are analytic on* $[t_0, t_1]$. *Then the dynamical system* (1.2.1) *is controllable on* $[t_0, t_1]$ *if and only if, for any fixed* $t_2 \in [t_0, t_1]$,

$$\operatorname{rank}[\mathbf{F}_0(t_2) \vdots \mathbf{M}_1(t_2) \vdots \mathbf{M}_2(t_2) \vdots \ \cdots \ \vdots \mathbf{M}_{n-1}(t_2) \vdots \ \cdots\ ] = n. \tag{1.3.19}$$

Proof:

If the matrix $\mathbf{A}(t)$ is analytic on $[t_0, t_1]$, it can be shown that the transition matrix $\mathbf{F}(t_1, t)$ is analytic on $[t_0, t_1]$ and hence, since $\mathbf{B}(t)$ is also analytic on $[t_0, t_1]$, this implies that the matrix $\mathbf{F}(t_1, t)\mathbf{B}(t)$ is analytic on $[t_0, t_1]$. Next we use Lemma 1.3.3 and the remainder of the proof is quite similar to the proof of Theorem 1.3.2. Hence it will be omitted here. ■

From Definitions 1.2.2, 1.2.3 and 1.2.4, by using Lemmas 1.3.1, 1.3.2 and 1.3.3 we can obtain the necessary and sufficient conditions or only the sufficient conditions for various kinds of controllability, i.e. for uniform controllability, controllability at t_0, or for controllabilty. These controllability criteria are similar in form to the criteria given in Theorems 1.3.1, 1.3.2 and 1.3.3. They differ only in the length of the time interval involved.

Now, we shall list some popular controllability criteria for the dynamical system (1.2.1).

COROLLARY 1.3.4. *The dynamical system* (1.2.1) *is controllable at time* t_0 *if and only if there exists a time* $t_1 > t_0$ *such that the matrix* $\mathbf{W}(t_0, t_1)$ *is nonsingular.*

COROLLARY 1.3.5. *If the matrices* $\mathbf{A}(t)$ *and* $\mathbf{B}(t)$ *are analytic on* $(-\infty, +\infty)$, *then the dynamical system* (1.2.1) *is uniformly controllable if and only if for each* $t \in (-\infty, +\infty)$ *we have*

$$\operatorname{rank}[\mathbf{M}_0(t) \vdots \mathbf{M}_1(t) \vdots \mathbf{M}_2(t) \vdots \ldots \vdots \mathbf{M}_{n-1}(t) \vdots \ldots] = n. \tag{1.3.20}$$

COROLLARY 1.3.6. *If the matrices* $\mathbf{A}(t)$ *and* $\mathbf{B}(t)$ *are analytic on* $[t_0, \infty)$, *then the dynamical system* (1.2.1) *is controllable at time* t_0 *if and only if it is controllable.*

Taking into account Definition 1.2.5 and Theorem 1.3.1 we can easily obtain the necessary and sufficient conditions for the regular controllability on $[t_0, t_1]$ of the dynamical system (1.2.1).

COROLLARY 1.3.7. *The dynamical system* (1.2.1) *is regularly controllable on* $[t_0, t_1]$ *if and only if for every* $j = 1, 2, \ldots, m$ *we have*

$$\det \int_{t_0}^{t_1} \mathbf{F}(t_1, t)\mathbf{b}_j(t)\mathbf{b}_j^{\mathrm{T}}(t)\mathbf{F}^{\mathrm{T}}(t_1, t)\,dt \neq 0. \tag{1.3.21}$$

Proof:

Since $\mathbf{B}(t) = [\mathbf{b}_1(t) \vdots \mathbf{b}_2(t) \vdots \ldots \vdots \mathbf{b}_m(t)]$, by Definition 1.2.5 the dynamical system (1.2.1) is regularly controllable on $[t_0, t_1]$ if and only if every dynamical system (1.2.5) is controllable on $[t_0, t_1]$. Thus, by Theorem 1.3.1, this is equivalent to the statement that condition (1.3.21) holds for every $j = 1, 2, \ldots, m$. Hence our corollary follows. ∎

Similarly, we can obtain the necessary and sufficient conditions for other types of regular controllability of the dynamical system (1.2.1).

Now, using Definition 1.2.7 and Theorem 1.3.1, we shall formulate the necessary and sufficient conditions for the output controllability on $[t_0, t_1]$ of the dynamical system (1.2.1) together with output equation (1.2.6).

THEOREM 1.3.4 (Albrecht, Grasse and Wax (1986)). *The dynamical system* (1.2.1), (1.2.6) *is output controllable on* $[t_0, t_1]$ *if and only if the* $(p \times p)$*-dimensional matrix*

$$\begin{aligned}\mathbf{V}_{\mathbf{D}}(t_0, t_1) &= \int_{t_0}^{t_1} \mathbf{C}(t_1)\mathbf{F}(t_1, t)\mathbf{B}(t)\mathbf{B}^{\mathrm{T}}(t)\mathbf{F}^{\mathrm{T}}(t_1, t)\mathbf{C}^{\mathrm{T}}(t_1)\,dt + \mathbf{D}(t_1)\mathbf{D}^{\mathrm{T}}(t_1) \\ &= \mathbf{V}(t_0, t_1) + \mathbf{D}(t_1)\mathbf{D}^{\mathrm{T}}(t_1)\end{aligned} \tag{1.3.22}$$

is nonsingular.

Proof:

Sufficiency. Let $\mathbf{y}(t_0) \in \boldsymbol{R}^p$ be any initial output vector of the dynamical system (1.2.1), and let $\mathbf{y}_1 \in \boldsymbol{R}^p$ be any vector. We shall show that the admissible control $\mathbf{u} \in L^2([t_0, t_1], \boldsymbol{R}^m)$ of the form

$$\mathbf{u}(t) = \mathbf{B}^{\mathrm{T}}(t)\mathbf{F}^{\mathrm{T}}(t_1, t)\mathbf{C}^{\mathrm{T}}(t_1)\mathbf{V}^{-1}(t_0, t_1)(\mathbf{y}_1 - \mathbf{C}(t_1)\mathbf{F}(t_1, t_0)\mathbf{x}(t_0) - \\ -\mathbf{D}(t_1)\mathbf{u}(t_1)), \quad t \in [t_0, t_1] \tag{1.3.23}$$

steers the output $\mathbf{y}(t)$ from $\mathbf{y}(t_0) \in \boldsymbol{R}^p$ to $\mathbf{y}_1 \in \boldsymbol{R}^p$ at time t_1, i.e. $\mathbf{y}(t_1) = \mathbf{y}_1$. Indeed, substituting the control (1.3.23) in (1.2.2) and taking into account the output equation (1.2.6) and relation (1.3.22) for $t = t_1$ we obtain

$$\begin{aligned}\mathbf{y}(t_1) &= \mathbf{C}(t_1)\mathbf{F}(t_1, t_0)\mathbf{x}(t_0) + \mathbf{C}(t_1)\int_{t_0}^{t_1} \mathbf{F}(t_1, t)\mathbf{B}(t)\mathbf{u}(t)\,dt + \mathbf{D}(t_1)\mathbf{u}(t_1) \\ &= \mathbf{C}(t_1)\mathbf{F}(t_1, t_0)\mathbf{x}(t_0) + \mathbf{C}(t_1)\int_{t_0}^{t_1} \mathbf{F}(t_1, t)\mathbf{B}(t)\mathbf{B}^{\mathrm{T}}(t)\mathbf{F}^{\mathrm{T}}(t_1, t)\mathbf{C}^{\mathrm{T}}(t_1) \times \\ &\quad \times \mathbf{V}^{-1}(t_0, t_1)(\mathbf{y}_1 - \mathbf{C}(t_1)\mathbf{F}(t_1, t_0)\mathbf{x}(t_0) - \mathbf{D}(t_1)\mathbf{u}(t_1))\,dt + \mathbf{D}(t_1)\mathbf{u}(t_1) \\ &= \mathbf{C}(t_1)\mathbf{F}(t_1, t_0)\mathbf{x}(t_0) + \mathbf{V}(t_0, t_1)\mathbf{V}^{-1}(t_0, t_1)(\mathbf{y}_1 - \mathbf{C}(t_1)\mathbf{F}(t_1, t_0)\mathbf{x}(t_0) - \\ &\quad - \mathbf{D}(t_1)\mathbf{u}(t_1)) + \mathbf{D}(t_1)\mathbf{u}(t_1) = \mathbf{y}_1,\end{aligned} \tag{1.3.24}$$

where

$$\mathbf{V}(t_0, t_1) = \int_{t_0}^{t_1} \mathbf{C}(t_1)\mathbf{F}(t_1, t)\mathbf{B}(t)\mathbf{B}^{\mathrm{T}}(t)\mathbf{F}^{\mathrm{T}}(t_1, t)\mathbf{C}^{\mathrm{T}}(t_1)\,dt. \tag{1.3.25}$$

Hence the control $\mathbf{u}(t)$ given by formula (1.3.23) steers the output of the dynamical system (1.2.1), (1.2.6) from $\mathbf{y}(t_0)$ to $\mathbf{y}_1$ at time t_1.

In the case where the matrix $\mathbf{V}(t_0, t_1)$ is singular, but $\mathbf{V}_{\mathbf{D}}(t_0, t_1)$ is nonsingular (which is equivalent to the statement that $\operatorname{rank} \mathbf{D}(t_1) = p$) we cannot use the control given by formula (1.3.23) since the matrix $\mathbf{V}^{-1}(t_0, t_1)$ does not exist. Then we can apply the following admissible control:

$$\mathbf{u}(t) = \begin{cases} 0 & \text{for } t \in [t_0, t_1), \\ (\mathbf{D}^{\mathrm{T}}(t_1)\mathbf{D}(t_1))^{-1}\mathbf{D}^{\mathrm{T}}(t_1)\mathbf{y}_1 & \text{for } t = t_1, \end{cases} \tag{1.3.26}$$

for which we have $\mathbf{y}(t_1) = \mathbf{y}_1$. It should be also stressed that the control (1.3.23) is not the only control which steers the output of the dynamical system (1.2.1), (1.2.6) from $\mathbf{y}(t_0)$ to $\mathbf{y}_1$ at time t_1 (Albrecht, Grasse and Wax (1986)).

Necessity. The necessity of the theorem can be proved as in Theorem 1.3.1. ■

COROLLARY 1.3.8 (Chen (1967)). *If* $\mathbf{D}(t_1) = 0$, *then the dynamical system* (1.2.1) (1.2.6) *is output controllable on* $[t_0, t_1]$ *if and only if the matrix* $\mathbf{V}(t_0, t_1)$ *given by formula* (1.2.5) *is nonsingular.*

Proof:

If $\mathbf{D}(t_1) = 0$, then by (1.3.22) the matrix $\mathbf{V}_{\mathbf{D}}(t_0, t_1)$ is equal to $\mathbf{V}(t_0, t_1)$. Hence Corollary 1.3.8 follows directly from Theorem 1.3.4. ■

Under some additional assumptions concerning the regularity of the matrices $\mathbf{A}(t)$ and $\mathbf{B}(t)$ we can formulate the conditions of output controllability without computing the state transition matrix $\mathbf{F}(t, s)$.

THEOREM 1.3.5 (Albrecht, Grasse and Wax (1986)). *Assume that the matrices* $\mathbf{A}(t)$ *and* $\mathbf{B}(t)$ *are* $n-1$ *times continuously differentiable. Then the dynamical system* (1.2.1), (1.2.6) *is output controllable on* $[t_0, t_1]$ *if there exists a* $t_2 \in [t_0, t_1]$ *such that*

$$\operatorname{rank}[\mathbf{C}(t_1)\mathbf{M}_0(t_2) \vdots \mathbf{C}(t_1)\mathbf{M}_{k1}(t_2) \vdots \mathbf{C}(t_1)\mathbf{M}_2(t_2) \vdots \ldots \vdots \mathbf{C}(t_1)\mathbf{M}_{n-1}(t_2) \vdots \mathbf{D}(t_1)] = p, \tag{1.3.27}$$

where $(n \times m)$*-dimensional matrices* $\mathbf{M}_k(t)$, $k = 0, 1, 2, \ldots, n-1$ *are defined by* (1.3.15) *and* (1.3.16).

Proof:

From Definition 1.2.7 it follows that if the dynamical system (1.2.1) is controllable on $[t_0, t_1]$, then the necessary and sufficient condition for output controllability on $[t_0, t_1]$ is that $\operatorname{rank}[\mathbf{C}(t_1) \vdots \mathbf{D}(t_1)] = p$. Therefore Theorem 1.3.5 immediately follows from Theorem 1.3.2. ∎

COROLLARY 1.3.9. *If the matrices* $\mathbf{A}(t)$ *and* $\mathbf{B}(t)$ *are analytic on* $[t_0, t_1]$, *then the dynamical system* (1.2.1), (1.2.6) *is output controllable on* $[t_0, t_1]$ *if and only if there exists a* $t_2 \in [t_0, t_1]$ *such that*

$$\operatorname{rank}[\mathbf{C}(t_1)\mathbf{M}_0(t_2) \vdots \mathbf{C}(t_1)\mathbf{M}_1(t_2) \vdots \ldots \vdots \mathbf{C}(t_1)\mathbf{M}_{n-1}(t_2) \vdots \ldots \vdots \mathbf{D}(t_1)] = p. \tag{1.3.28}$$

Proof:

Corollary 1.3.9 immediately follows from Theorem 1.3.3 and the remark given in the proof of Corollary 1.3.8. ∎

COROLLARY 1.3.10. *If* $\mathbf{D}(t_1) = 0$ *and the dynamical system* (1.2.1) *is controllable on* $[t_0, t_1]$, *then it is output controllable on* $[t_0, t_1]$ *if and only if*

$$\operatorname{rank}\mathbf{C}(t_1) = p. \tag{1.3.29}$$

Proof:

Corollary 1.3.10 follows from Theorem 1.3.5 and the form of the output equation. ∎

Of course, if $\mathbf{C}(t_1) = 0$, then the dynamical system (1.2.1), (1.2.6) is output controlable on $[t_0, t_1]$ if and only if

$$\operatorname{rank}\mathbf{D}(t_1) = p. \tag{1.3.30}$$

Example 1.3.1

Let us consider the dynamical system (1.2.1), (1.2.6) with the matrices $\mathbf{A}(t)$, $\mathbf{B}(t)$ and $\mathbf{C}(t)$:

$$\mathbf{A}(t) = \begin{bmatrix} t^2 & 1 \\ t & t \end{bmatrix}, \quad \mathbf{B}(t) = \begin{bmatrix} t & t^2 \\ 1 & t \end{bmatrix}, \quad \mathbf{C}(t) = \begin{bmatrix} t & 1 \\ 1 & t \end{bmatrix} \tag{1.3.31}$$

defined for $t \geqslant t_0 = 0$.

Using (1.3.15) and (1.3.16) we obtain

$$\mathbf{M}_0(t) = \mathbf{B}(t) = \begin{bmatrix} t & t^2 \\ 1 & t \end{bmatrix},$$

$$\mathbf{M}_1(t) = -\mathbf{A}(t)\mathbf{M}_0(t) + \frac{d}{dt}\mathbf{M}_0(t) = -\mathbf{A}(t)\mathbf{B}(t) + \frac{d}{dt}\mathbf{B}(t)$$

$$= \begin{bmatrix} -t^3 & -t^4+t \\ -t^2-t & -t^3-t^2+1 \end{bmatrix}.$$

Since the matrix $[\mathbf{M}_0(t) \vdots \mathbf{M}_1(t)]$ has rank equal to 2 for all $t > 0 = t_0$, by Theorem 1.3.3 the dynamical system (1.3.31) is controllable on any interval $[t_0, t_1]$, $t_1 > t_0$. The output controllability of the dynamical system (1.3.31) can be checked by using Corollary 1.3.10. It is easily verified that the assumptions of Corollary 1.3.10 are satisfied. Since $\operatorname{rank}\mathbf{C}(t) = 2$ for all $t > t_0 = 0$ except the point $t = 1$, then by Corollary 1.3.10 the dynamical system (1.3.31) is output controllable on $[t_0, t_1]$ for all $t_1 > t_0$ except $t_1 = 1$.

Example 1.3.2 (Chen (1970), p. 173)

Let us consider the dynamical system (1.2.1) with the matrices $\mathbf{A}(t)$ and $\mathbf{B}(t)$:

$$\mathbf{A}(t) = \begin{bmatrix} t & 1 & 0 \\ 0 & t & 0 \\ 0 & 0 & t^2 \end{bmatrix}, \quad \mathbf{B}(t) = \mathbf{B} = \begin{bmatrix} 0 \\ 1 \\ 1 \end{bmatrix} \tag{1.3.23}$$

defined for $t \geqslant t_0 = 0$.

Using (1.3.15) and (1.3.16) we obtain

$$\mathbf{M}_0(t) = \mathbf{B}(t) = \mathbf{B} = \begin{bmatrix} 0 \\ 1 \\ 1 \end{bmatrix},$$

$$\mathbf{M}_1(t) = -\mathbf{A}(t)\mathbf{M}_0(t) + \frac{d}{dt}\mathbf{M}_0(t) = \begin{bmatrix} -1 \\ -t \\ -t^2 \end{bmatrix},$$

$$\mathbf{M}_2(t) = -\mathbf{A}(t)\mathbf{M}_1(t) + \frac{d}{dt}\mathbf{M}_1(t) = \begin{bmatrix} 2t \\ t^2 \\ t^4 \end{bmatrix} + \begin{bmatrix} 0 \\ -1 \\ -2t \end{bmatrix} = \begin{bmatrix} 2t \\ t^2-1 \\ t^4-2t \end{bmatrix}.$$

Since

$$\operatorname{rank}[\mathbf{M}_0(t) \vdots \mathbf{M}_1(t) \vdots \mathbf{M}_2(t)] = \operatorname{rank}\begin{bmatrix} 0 & -1 & 2t \\ 1 & -t & t^2-1 \\ 1 & -t^2 & t^4-2t \end{bmatrix} = 3 = n$$

for all $t > t_0 = 0$, by Theorem 1.3.3 the dynamical system (1.3.32) is controllable on any time interval $[t_0, t_1]$, $t_1 > t_0$. Since in fact the matrices $\mathbf{A}(t)$ and $\mathbf{B}(t)$ are analytic on $(-\infty, +\infty)$, it is easy to deduce that the dynamical system (1.3.32) is controllable on any time interval with a positive measure.

1.4 Controllability conditions for time-invariant dynamical systems

In this section, we shall study the controllability of the linear time-invariant dynamical system (1.2.8). In this case the controllability conditions are simpler than for the time-varying dynamical system (1.2.1). It follows from the fact, that for time-invariant case the transition matrix can be easily computed by the formula (1.2.10). Moreover, for time-invariant dynamical systems the notions of uniform controllability, controllability in $[t_0, t_1]$, controllability at t_0 and controllability are equivalent (see Lemma 1.4.1).

LEMMA 1.4.1. *For the time-invariant dynamical system* (1.2.8) *the notions of uniform controllability, controllability in* $[t_0, t_1]$, *controllability at* t_0 *and controllability are equivalent.*

Proof:
For the time-invariant dynamical system (1.2.8) the matrix

$$\mathbf{F}(t_1, t)\mathbf{B}(t) = \exp\big(\mathbf{A}(t_1 - t)\big)\mathbf{B}(t)$$

is an analytic function for $t \in (-\infty, +\infty)$. Thus the linear independence of the rows of the matrix $\exp\big(\mathbf{A}(t_1 - t)\big)\mathbf{B}(t)$ in a certain fixed time interval is equivalent to the linear independence in each time interval. Hence the notions of the various kinds of controllability are equivalent. ∎

Since the notions of the various kinds of controllability are equivalent, then for simplicity we shall use the term "controllability".

Using the results presented in Section 1.3 and specially Theorem 1.3.3 we can formulate the fundamental necessary and sufficient condition for the controllability of time-invariant dynamical systems.

THEOREM 1.4.1. *The dynamical system* (1.2.8) *is controllable if and only if*

$$\operatorname{rank}[\mathbf{B} \vdots \mathbf{AB} \vdots \mathbf{A}^2\mathbf{B} \vdots \ldots \vdots \mathbf{A}^{n-1}\mathbf{B}] = n. \tag{1.4.1}$$

Proof:

From (1.3.15) and (1.3.16) and by substituting $\mathbf{A}(t) = \mathbf{A}$, $\mathbf{B}(t) = \mathbf{B}$ we have

$$\mathbf{M}_k(t) = (-1)^k \mathbf{A}^k \mathbf{B} \quad \text{for } k = 0, 1, 2, \ldots$$

Substituting the above matrices $\mathbf{M}_k(t)$ in formula (1.3.19) and using Theorem 1.3.3, we conclude, that the dynamical system (1.2.8) is controllable if and only if

$$\operatorname{rank}[\mathbf{B} \,\vdots\, -\mathbf{AB} \,\vdots\, \mathbf{A}^2\mathbf{B} \,\vdots\, \ldots \,\vdots\, (-1)^{n-1}\mathbf{A}^{n-1}\mathbf{B} \,\vdots\, \ldots] = n. \tag{1.4.2}$$

From the Cayley–Hamilton theorem it follows (Chen (1970 p. 53)) that the powers $\mathbf{A}^k$, $k \geqslant n$, can be expressed as linear combinations of the matrices $\mathbf{I}, \mathbf{A}, \mathbf{A}^2, \ldots, \mathbf{A}^{n-1}$. Hence the columns of the matrices $\mathbf{A}^k\mathbf{B}$ for $k \geqslant n$ are linear combinations of the columns of the matrices $\mathbf{B}, \mathbf{AB}, \mathbf{A}^2\mathbf{B}, \ldots, \mathbf{A}^{n-1}\mathbf{B}$. Hence relation (1.4.2) is equivalent to condition (1.4.1). Therefore the time-invariant dynamical system (1.2.8) is controllable if and only if equality (1.4.1) holds. ■

It should be noticed that for the investigation of the controllability of the time-invariant dynamical system (1.2.8) we shall also use the Laplace transform of the matrix $\exp(\mathbf{A}t)\mathbf{B}$. Taking the Laplace transform of $\exp(\mathbf{A}t)\mathbf{B}$, we have

$$\mathscr{L}\{\exp(\mathbf{A}t)\mathbf{B}\} = (s\mathbf{I} - \mathbf{A})^{-1}\mathbf{B}. \tag{1.4.3}$$

THEOREM 1.4.2. *The time-invariant dynamical system* (1.2.8) *is controllable if and only if the rows of the matrix* $(s\mathbf{I} - \mathbf{A})^{-1}\mathbf{B}$ *are linearly independent on* $[0, \infty)$ *over the field of complex numbers.*

Proof:

Since the Laplace transform is a one-to-one linear operator, if the rows of $\exp(\mathbf{A}t)\mathbf{B}$ are linearly independent on $[0, \infty)$ over the field of complex numbers, then so are the rows of $(s\mathbf{I} - \mathbf{A})^{-1}\mathbf{B}$ and vice versa. ■

From Theorem 1.4.1 it follows that in order to investigate the controllability of the dynamical system (1.2.8) we must compute the rank of the $(n \times nm)$-dimensional matrix given on the right-hand side of formula (1.4.1). It should be pointed out that in some special cases we can reduce the dimensionality of this matrix. This leads to simpler controllability conditions for the dynamical system (1.2.8). Let us denote

$$\mathbf{W}_k = [\mathbf{B} \,\vdots\, \mathbf{AB} \,\vdots\, \mathbf{A}^2\mathbf{B} \,\vdots\, \ldots \,\vdots\, \mathbf{A}^{k-1}\mathbf{B} \,\vdots\, \mathbf{A}^k\mathbf{B}], \quad k = 0, 1, 2, \ldots, \tag{1.4.4}$$

$\mathbf{W}_k$ is the $(n \times km)$-dimensional matrix, $r = \operatorname{rank}\mathbf{B}$, n_0—the degree of the minimal polynomial of the matrix $\mathbf{A}$, $n_0 \leqslant n$ (Chen (1970, p. 50)).

THEOREM 1.4.3. *If j is the least integer such that*

$$\operatorname{rank}\mathbf{W}_j = \operatorname{rank}\mathbf{W}_{j+1}$$

then

$$\operatorname{rank}\mathbf{W}_k = \operatorname{rank}\mathbf{W}_j \quad \textit{for all integers } k > j$$

and

$$j \leqslant \min(n-r, n_0-1).$$

Proof:

Since the columns of $\mathbf{W}_j$ are in $\mathbf{W}_{j+1}$, the condition $\operatorname{rank}\mathbf{W}_j = \operatorname{rank}\mathbf{W}_{j+1}$ implies that every column of the matrix $\mathbf{A}^{j+1}\mathbf{B}$ is linearly dependent on the columns of the matrices $\mathbf{B}$, $\mathbf{AB}$, $\mathbf{A}^2\mathbf{B}$, ..., $\mathbf{A}^j\mathbf{B}$. Consequently, proceeding further, we conclude that every column of $\mathbf{A}^k\mathbf{B}$ with $k \geqslant j$ is linearly dependent on the columns of $\mathbf{B}$, $\mathbf{AB}$, $\mathbf{A}^2\mathbf{B}$, ..., $\mathbf{A}^j\mathbf{B}$. This proves that $\operatorname{rank}\mathbf{W}_k = \operatorname{rank}\mathbf{W}_j$ for all integers $k > j$. The rank of the matrix $[\mathbf{B} \vdots \mathbf{AB} \vdots \mathbf{A}^2\mathbf{B} \vdots \ldots]$ must increase by at least 1 as each submatrix of the form $\mathbf{A}^k\mathbf{B}$ is added; otherwise its rank will cease to increase. The maximum rank of $[\mathbf{B} \vdots \mathbf{AB} \vdots \mathbf{A}^2\mathbf{B} \vdots \ldots]$ is n, and hence if the rank of $\mathbf{B}$ is r, in checking the maximum rank of $[\mathbf{B} \vdots \mathbf{AB} \vdots \mathbf{A}^2\mathbf{B} \vdots \ldots]$ it is sufficient to add at most $n-r$ submatrices of the form $\mathbf{A}^k\mathbf{B}$. Thus we have $j \leqslant n-r$. That $j \leqslant n_0-1$ follows from the definition of the minimal polynomial of the matrix $\mathbf{A}$. Hence we conclude that $j \leqslant \min(n-r, n_0-1)$. ∎

THEOREM 1.4.4. *If the rank of* $\mathbf{B}$ *is r, then the dynamical system* (1.2.8) *is controllable if and only if*

$$\operatorname{rank}\mathbf{W}_{n-r} = \operatorname{rank}[\mathbf{B} \vdots \mathbf{AB} \vdots \mathbf{A}^2\mathbf{B} \vdots \ldots \vdots \mathbf{A}^{n-r}\mathbf{B}] = n. \tag{1.4.5}$$

Proof:

From Theorem 1.4.3 we have $\operatorname{rank}\mathbf{W}_{n-1} = \operatorname{rank}\mathbf{W}_{n-r}$. Thus by (1.4.5) and Theorem 1.4.1 we obtain the necessary and sufficient condition for the controllability of the dynamical system (1.2.8). ∎

COROLLARY 1.4.1. *Time-invariant dynamical system* (1.2.8) *is controllable if and only if the* $(n \times n)$*-dimensional matrix* $\mathbf{W}_{n-r}\mathbf{W}_{n-r}^{\mathrm{T}}$ *is nonsingular.*

Proof:

Corollary 1.4.1 is an immediate consequence of Theorem 1.4.4 and well-known inequalities concerning the rank of the product of matrices (Chen (1970), p. 33). ∎

It should be stressed that in practice to compute the degree n_0 of the minimal polynomial is a rather difficult task.

In the case where the eigenvalues of the matrix $\mathbf{A}$, s_i, $i = 1, 2, \ldots, n$ are known, we can check controllability, using the condition formulated in Theorem 1.4.5.

THEOREM 1.4.5. *The dynamical system* (1.2.8) *is controllable if and only if*

$$\operatorname{rank}[s_i\mathbf{I} - \mathbf{A} \vdots \mathbf{B}] = n \quad \textit{for all } s_i,\ i = 1, 2, \ldots, n. \tag{1.4.6}$$

Proof:

Necessity. By contradiction. Let us suppose that the dynamical system (1.2.8). is controllable, i.e., that condition (1.4.1) holds but condition (1.4.6) is not satisfied Then there exists an eigenvalue, say s_k, $1 \leqslant k \leqslant n$, and a nonzero vector $\mathbf{d} \in R^n$ such that $\mathbf{d}^T[s_k\mathbf{I}-\mathbf{A}\vdots\mathbf{B}] = 0$, i.e., $s_k\mathbf{d}^T\mathbf{I} = \mathbf{d}^T\mathbf{A}$ and $\mathbf{d}^T\mathbf{B} = 0$. Thus we conclude that for any natural number j we have $\mathbf{d}^T\mathbf{A}^j\mathbf{B} = 0$, which contradicts the condition (1.4.1). Hence our necessary condition follows.

Sufficiency. By contradiction. Let us suppose that condition (1.4.6) is satisfied, but the dynamical system (1.2.8) is not controllable, i.e., condition (1.4.1) does not hold. Thus there exists a nonzero vector $\mathbf{d} \in R^n$ such that $\mathbf{d}^T\mathbf{A}^j\mathbf{B} = 0$ for all $j = 0, 1, 2, \ldots$ (by the Cayley–Hamilton theorem (Chen (1970, p. 53)). Therefore $\mathbf{d}^T\mathbf{B} = 0$ and $\mathbf{d}^T\mathbf{A} = 0 = \mathbf{d}^T s_i$ (by the definition of the eigenvalue). Hence we have $\mathbf{d}^T[s_i\,\mathbf{I}-\mathbf{A}\vdots\mathbf{B}] = 0$, which is a contradiction. ∎

Example 1.4.1

Let us consider the dynamical system (1.2.8) with matrices **A** and **B** of the following form:

$$\mathbf{A} = \begin{bmatrix} -9 & 1 & 5 \\ -6 & -1 & 4 \\ -8 & 1 & 4 \end{bmatrix}, \quad \mathbf{B} = \begin{bmatrix} 1 & 0 & 3 \\ 3 & 3 & 0 \\ 2 & -2 & 4 \end{bmatrix}. \tag{1.4.7}$$

Hence we have

$$\operatorname{rank}[\mathbf{B}\vdots\mathbf{AB}\vdots\mathbf{A}^2\mathbf{B}] = \operatorname{rank}\left[\begin{array}{ccc:ccc:ccc} 1 & 0 & 3 & 4 & -7 & -7 & 8 & 57 & 21 \\ 3 & 3 & 0 & 1 & -11 & -2 & 26 & -13 & 77 \\ 2 & -2 & 4 & 3 & -5 & -8 & -21 & 25 & 22 \end{array}\right]$$
$$= 3 = n\,.$$

Thus by Theorem 1.4.1 the dynamical system (1.4.7) is controllable. Since the characteristic polynomial $\varphi(s) = (s+1)(s+2)(s+3)$, the eigenvalues of the matrix **A** are, rspectively, $s_1 = -1$, $s_2 = -2$, and $s_3 = -3$. Moreover, all these eigenvalues are singular, whence the minimal polynomial of the matrix **A**, $\psi(s)$, is the same as the characteristic polynomial $\varphi(s)$. By (1.4.7) we have $\operatorname{rank}\mathbf{B} = 3 = r$, and hence from Theorem 1.4.3 it follows that $j \leqslant \min(n-r, n_0-1) = \min(0, 2) = 0$. Therefore we can use Theorem 1.4.4 to calculate the controllability of the dynamical system (1.4.7).

Now let us investigate the controllability of the dynamical system (1.4.7), using Theorem 1.4.5. Taking into account the values of the eigenvalues s_1, s_2 and s_3, by condition (1.4.6) we have

$$\operatorname{rank}\begin{bmatrix} 8 & -1 & -5 & 1 & 0 & 3 \\ 6 & 0 & -4 & 3 & 3 & 0 \\ 8 & -1 & -5 & 2 & -2 & 4 \end{bmatrix} = 3 = n \qquad \text{for } s_1 = -1,$$

$$\operatorname{rank}\begin{bmatrix} 7 & -1 & -5 & 1 & 0 & 3 \\ 6 & -1 & -4 & 3 & 3 & 0 \\ 8 & -1 & -6 & 2 & -2 & 4 \end{bmatrix} = 3 = n \quad \text{for } s_2 = -2,$$

$$\operatorname{rank}\begin{bmatrix} 6 & -1 & -5 & 1 & 0 & 3 \\ 6 & -2 & -4 & 3 & 3 & 0 \\ 8 & -1 & -7 & 2 & -2 & 4 \end{bmatrix} = 3 = n \quad \text{for } s_3 = -3.$$

Thus by Theorem 1.4.5 the dynamical system (1.4.7) is controllable. Example 1.4.1 will also be considered in Section 1.5, where the Jordan canonical form will be extensively used to check the controllability of the time-invariant dynamical systems (1.2.8).

Example 1.4.2

Let us consider the dynamical system (1.2.8) with the matrices **A** and **B** of the following form:

$$\mathbf{A} = \begin{bmatrix} -3 & 0 \\ 0 & -5 \end{bmatrix}, \quad \mathbf{B} = \begin{bmatrix} 1 & a \\ 1 & 2 \end{bmatrix} = [b_1 \vdots b_2], \tag{1.4.8}$$

where a is a given constant parameter. Since

$$\operatorname{rank}[\mathbf{B} \vdots \mathbf{AB}] = \operatorname{rank}\begin{bmatrix} 1 & a & -3 & -3a \\ 1 & 2 & -5 & -10 \end{bmatrix} = 2 = n \quad \text{for each constant } a,$$

by Theorem 1.4.1 the dynamical system (1.4.8) is controllable for each a.

In the next part of Example 1.4.2 we shall consider the regular controllability of the dynamical system (1.2.8). We shall use Definition 1.2.5 and Theorem 1.4.1. In order to do that let us introduce the following matrices:

$$[\mathbf{b}_1 \vdots \mathbf{Ab}_1] = \begin{bmatrix} 1 & -3 \\ 1 & -5 \end{bmatrix} \quad \text{and} \quad [\mathbf{b}_2 \vdots \mathbf{Ab}_2] = \begin{bmatrix} a & -3a \\ 2 & -10 \end{bmatrix}.$$

The matrix $[\mathbf{b}_1 \vdots \mathbf{Ab}_1]$ is nonsingular, while the matrix $[\mathbf{b}_2 \vdots \mathbf{Ab}_2]$ is nonsingular if and only if $a \neq 0$. Hence the dynamical system (1.4.8) is regularly controllable if and only if the parameter $a \neq 0$. It is quite easy to check that if $a = 0$ and $\mathbf{u}_1(t) = 0$ then the dynamical system (1.4.8) reduces to the following system:

$$\dot{x}_1(t) = -3x_1(t),$$
$$\dot{x}_2(t) = -5x_2(t) + 2u_2(t).$$

Therefore the coordinate $x_1(t)$ is not steered by the control $\mathbf{u}_2(t)$. This explains the nature of the lack of regular controllability for the case $a = 0$.

Example 1.4.3

Let us consider the dynamical system (1.2.8) with the matrices **A** and **B** of the following form:

$$\mathbf{A} = \begin{bmatrix} -4 & 1 \\ -3 & 0 \end{bmatrix}, \qquad \mathbf{B} = \begin{bmatrix} 1 & b \\ 1 & b \end{bmatrix}, \tag{1.4.9}$$

where b is a given constant parameter. Since

$$\operatorname{rank}[\mathbf{B} \vdots \mathbf{AB}] = \operatorname{rank}\begin{bmatrix} 1 & b & \vdots & -3 & -3b \\ 1 & b & \vdots & -3 & -3b \end{bmatrix} = 1 < 2 = n \quad \text{for each parameter } b,$$

by Theorem 1.4.1 the dynamical system (1.4.9) is not controllable for each parameter b. Moreover, the dynamical system (1.4.9) is of course not regularly controllable for each parameter b. It should also be stressed that we have $\operatorname{rank}\mathbf{B} = 1$ for each value of the parameter b.

A controllability of the linear stationary dynamical system (1.2.8) is strongly related to its stabilizability by linear state feedback of the form

$$\mathbf{u}(t) = \mathbf{F}\mathbf{x}(t) + \mathbf{v}(t), \qquad t \geqslant 0, \tag{1.4.10}$$

where $\mathbf{v}(t) \in \boldsymbol{R}^n$ is a new control, $\mathbf{F}$ is the $(m \times n)$-dimensional state feedback matrix. Introducing linear state feedback of the form (1.4.10) we obtain the linear state-feedback dynamical system of the following form:

$$\dot{\mathbf{x}}(t) = (\mathbf{A} + \mathbf{BF})\mathbf{x}(t) + \mathbf{B}\mathbf{v}(t), \tag{1.4.11}$$

which is characterized by the pair of matrices $(\mathbf{A}+\mathbf{BF}, \mathbf{B})$.

An interesting result is the equivalence between controllability of the dynamical systems (1.2.8) and (1.4.11). This fact is illustrated by Lemma 1.4.2.

LEMMA 1.4.2 (Chen (1970, Ch. 7.3)). *The dynamical system* (1.2.8) *is controllable if and only if for arbitrary matrix* $\mathbf{F}$ *the dynamical system* (1.4.11) *is controllable.*

From Lemma 1.4.2 it follows that under the controllability assumption we can arbitrarily form the spectrum of the dynamical system (1.2.8) by the introduction of state feedback.

THEOREM 1.4.6 (Chen (1970, Ch. 7.3)). *The pair od matrices* $(\mathbf{A}, \mathbf{B})$ *is controllable if and only if for each set* $\Lambda \subset \boldsymbol{C}$ *consisting of n complex numbers and symmetric with respect to real axis, there exists state feedback matrix* $\mathbf{F}$ *such that*

$$\sigma(\mathbf{A}+\mathbf{BF}) = \Lambda. \tag{1.4.12}$$

Practically, in the design of a system, sometimes it is required only to change unstable eigenvalues (the eigenvalues with nonnegative real parts) to stable eigenvalues (the eigenvalues with negative real parts). This is called *stabilization* of the dynamical system (1.2.8).

DEFINITION 1.4.1. The dynamical system (1.2.8) is said to be *stabilizable* if there exists state feedback matrix $\mathbf{F}$ such that the spectrum of the matrix $\mathbf{A}+\mathbf{BF}$ entirely lies in the left-hand side of the complex plane.

An immediate relation between controllability and stabilizability of the dynamical system (1.2.8) gives Theorem 1.4.7.

THEOREM 1.4.7 (Chen (1970, Ch. 7.3)). *The pair of matrices* (**A**, **B**) *is stabilizable if and only if all unstable modes of the dynamical system* (1.2.8) *are controllable.*

Since in practice the number of state variables is generally larger that the number of output variables, the distinction between state feedback and output feedback should be made. In fact, what can be achieved by output feedback can always be achieved by state feedback, but the converse is not true (see e.g. Chen (1970, Ch. 7.7)).

1.5 Controllability of Jordan-form dynamical systems

In this section we shall study the linear, time-invariant dynamical systems (1.2.8). Suppose the dynamical system (1.2.8) is controllable. It is natural to ask whether it remains controllable after the equivalence transformation of the state variables. The answer is yes. This is intuitively clear because an equivalence transformation changes only the basis of the state space. Therefore, the controllability of the dynamical system (1.2.8) should not be affected.

In this section we shall use the so called Jordan canonical form of the dynamical system (1.2.8) to investigate controllability (Chen (1970), Ch. 2.6 and 5.5, Chen, Desoer and Niederliński (1966), Chen and Desoer (1967, 1968), Klamka (1972, 1973, 1974a, 1974b, 1975d); Zadeh and Desoer (1963)). The *Jordan canonical form* of the dynamical system (1.2.8) is obtained by using a special equivalence transformation represented by the $(n \times n)$-dimensional, nonsingular matrix **T**. Since the controllability of the dynamical system (1.2.8) is preserved under any equivalence transformations, it is conceivable that we might obtain a simpler controllability criterion by transforming the dynamical equation into a special form. If we transform linear, time-invariant dynamical systems (1.2.8) into Jordan canonical forms, their controllabilities can be determined almost by inspection. It should also be stressed that a dynamical system (1.2.8) given in the Jordan canonical form has an extremely simple structure, which immediately shows the connections between the inputs, state variables and outputs of a dynamical system.

Now let us consider the Jordan canonical form in a more rigorous way.

LEMMA 1.5.1 (Chen (1970, Ch. 5.5)). *The controllability of the dynamical system* (1.2.8) *is invariant under any linear equivalence transformation* $\mathbf{x} = \mathbf{T}\mathbf{z}$, *where* $\mathbf{x} \in \mathbf{R}^n$ $\mathbf{z} \in \mathbf{R}^n$ *and* **T** *is an* $(n \times n)$*-dimensional, nonsingular transformation matrix.*

Proof:

Using the transformation $\mathbf{x} = \mathbf{T}\mathbf{z}$ and taking into account that the inverse matrix $\mathbf{T}^{-1}$ exists, we transform the state equation of the dynamical system (1.2.8) into the following equivalent form:

$$\dot{\mathbf{z}}(t) = \mathbf{J}\mathbf{z}(t) + \mathbf{G}\mathbf{u}(t), \quad t \geqslant 0, \tag{1.5.1}$$

where

$$\mathbf{J} = \mathbf{T}^{-1}\mathbf{A}\mathbf{T}, \tag{1.5.2}$$

$$\mathbf{G} = \mathbf{T}^{-1}\mathbf{B}. \tag{1.5.3}$$

By Theorem 1.4.1 the dynamical system (1.2.8) is controllable if and only if $\operatorname{rank}\mathbf{W}_{n-1} = n$. Thus by (1.5.2) and (1.5.3) we have the equality:

$$\begin{aligned}\mathbf{W}_{n-1} &= [\mathbf{B} \vdots \mathbf{AB} \vdots \mathbf{A}^2\mathbf{B} \vdots \dots \vdots \mathbf{A}^{n-1}\mathbf{B}] = [\mathbf{TG} \vdots \mathbf{TJG} \vdots \mathbf{TJ}^2\mathbf{G} \vdots \dots \vdots \mathbf{TJ}^{n-1}\mathbf{G}] \\ &= \mathbf{T}[\mathbf{G} \vdots \mathbf{JG} \vdots \mathbf{J}^2\mathbf{G} \vdots \dots \mathbf{J}^{n-1}\mathbf{G}].\end{aligned}$$

Since $\mathbf{T}$ is a nonsingular matrix, we have $\operatorname{rank}\mathbf{W}_{n-1} = \operatorname{rank}[\mathbf{G} \vdots \mathbf{JG} \vdots \mathbf{J}^2\mathbf{G} \vdots \dots \vdots \mathbf{J}^{n-1}\mathbf{G}]$. Thus controllability is invariant under the equivalence transformation. ■

It should be pointed out that Lemma 1.5.1 also holds for the linear, time-varying equivalence transformation represented by the matrix $\mathbf{T}(t)$, nonsingular for all $t \in (-\infty, +\infty)$. Choosing appropriately the transformation matrix $\mathbf{T}$, we can obtain the *Jordan canonical form* of the dynamical system (1.2.8) represented by the matrices $\mathbf{J}$ and $\mathbf{G}$, which are given by the following equalities:

$$\mathbf{J} = \begin{bmatrix} \mathbf{J}_1 & & & 0 \\ & \mathbf{J}_2 & & \\ & & \ddots & \\ 0 & & & \mathbf{J}_k \end{bmatrix}, \quad \mathbf{G} = \begin{bmatrix} \mathbf{G}_1 \\ \mathbf{G}_2 \\ \vdots \\ \mathbf{G}_k \end{bmatrix}, \tag{1.5.4}$$

$$\mathbf{J}_i = \begin{bmatrix} \mathbf{J}_{i1} & & & 0 \\ & \mathbf{J}_{i2} & & \\ & & \ddots & \\ 0 & & & \mathbf{J}_{ir(i)} \end{bmatrix}, \quad \mathbf{G}_i = \begin{bmatrix} \mathbf{G}_{i1} \\ \mathbf{G}_{i2} \\ \vdots \\ \mathbf{G}_{ir(i)} \end{bmatrix}, \quad i = 1, 2, \dots, k, \tag{1.5.5}$$

$$\mathbf{J}_{ij} = \begin{bmatrix} s_i & 1 & & & 0 \\ & s_i & 1 & & \\ & & \ddots & \ddots & \\ & & & & 1 \\ 0 & & & & s_i \end{bmatrix}, \quad \mathbf{G}_{ij} = \begin{bmatrix} \mathbf{g}_{ij1} \\ \mathbf{g}_{ij2} \\ \vdots \\ \mathbf{g}_{ijn(ij)} \end{bmatrix}, \tag{1.5.6}$$

$$i = 1, 2, \dots, k, j = 1, 2, \dots, r(i),$$

where $s_1, s_2, \dots, s_k$ are distinct eigenvalues of the matrix $\mathbf{A}$ with multiplicities n_i, $i = 1, 2, \dots, k$; $\mathbf{J}_i$, $i = 1, 2, \dots, k$ are $(n_i \times n_i)$-dimensional matrices containing all the Jordan blocks associated with the eigenvalues s_i; $\mathbf{J}_{ij}$, $i = 1, 2, \dots, k, j = 1, 2, \dots, r(i)$ are $(n_{ij} \times n_{ij})$-dimensional Jordan blocks in $\mathbf{J}_i$; $r(i)$ is the number of Jordan blocks in the submatrix $\mathbf{J}_i$, $i = 1, 2, \dots, k$; $\mathbf{G}_i$, $i = 1, 2, \dots, k$ are $(n_i \times m)$-dimensional submatrices of the matrix $\mathbf{G}$ corresponding to submatrices $\mathbf{J}_i$; $\mathbf{G}_{ij}$, $i = 1, 2, \ldots, k$, $j = 1, 2, \ldots, r(i)$ are $(n_{ij} \times m)$-dimensional submatrices of the matrix $\mathbf{G}_i$ corresponding to the Jordan blocks $\mathbf{J}_{ij}$; $\mathbf{g}_{ijn_{ij}}$, $i = 1, 2, \ldots, k$, $j = 1, 2, \ldots, r(i)$ are the rows of the submatrix $\mathbf{G}_{ij}$ corresponding to the rows of the Jordan blocks $\mathbf{J}_{ij}$.

The following equalities hold:

$$n = \sum_{i=1}^{i=k} n_i = \sum_{i=1}^{i=k} \sum_{j=1}^{j=r(i)} n_{ij}. \tag{1.5.7}$$

LEMMA 1.5.2 (Chen (1970, Ch. 5.5)). *The rows of the matrix* $(s\mathbf{I}-\mathbf{J})^{-1}\mathbf{G}$ *are linearly independent over the field of complex numbers if and only if, for every* $i = 1, 2, \ldots, k$, *the rows of the matrices* $(s\mathbf{I}-\mathbf{J}_i)^{-1}\mathbf{G}_i$ *are linearly independent over the field of complex numbers.*

Proof:

By (1.5.4), (1.5.5) and (1.5.6) we have

$$(s\mathbf{I}-\mathbf{J})^{-1}\mathbf{G} = \begin{bmatrix} (s\mathbf{I}-\mathbf{J}_1)^{-1}\mathbf{G}_1 \\ (s\mathbf{I}-\mathbf{J}_2)^{-1}\mathbf{G}_2 \\ \vdots \\ (s\mathbf{I}-\mathbf{J}_k)^{-1}\mathbf{G}_k \end{bmatrix},$$

$$(s\mathbf{I}-\mathbf{J}_i)^{-1}\mathbf{G}_i = \begin{bmatrix} (s\mathbf{I}-\mathbf{J}_{i1})^{-1}\mathbf{G}_{i1} \\ (s\mathbf{I}-\mathbf{J}_{i2})^{-1}\mathbf{G}_{i2} \\ \vdots \\ (s\mathbf{I}-\mathbf{J}_{ir(i)})^{-1}\mathbf{G}_{ir(i)} \end{bmatrix}, \quad i = 1, 2, \ldots, k,$$

$$(s\mathbf{I}-\mathbf{J}_{ij})^{-1}\mathbf{G}_{ij} = \begin{bmatrix} (s-s_i)^{-1} & (s-s_i)^{-2} & \ldots & (s-s_i)^{-n_{ij}} \\ 0 & (s-s_i)^{-1} & \ldots & (s-s_i)^{-n_{ij}+1} \\ \ldots & \ldots & \ldots & \ldots \\ 0 & 0 & \ldots & (s-s_i)^{-1} \end{bmatrix} \begin{bmatrix} \mathbf{g}_{ij1} \\ \mathbf{g}_{ij2} \\ \vdots \\ \mathbf{g}_{ijn_{ij}} \end{bmatrix},$$

$$i = 1, 2, \ldots, k, j = 1, 2, \ldots, r(i).$$

Therefore, the rows of the matrices $(s\mathbf{I}-\mathbf{J}_i)^{-1}\mathbf{G}_i$, $i = 1, 2, \ldots, k$, are linear combinations of the component $(s-s_i)^{-1}$. Hence from the assumption that the eigenvalues $s_1, s_2, \ldots, s_k$ are distinct we see that for each i all the rows of $(s\mathbf{I}-\mathbf{J}_i)^{-1}\mathbf{G}_i$ are linearly independent over the field of complex numbers if and only if all the rows of $(s\mathbf{I}-\mathbf{J})^{-1}\mathbf{G}$ are linearly independent over the fields of complex numbers. ■

Using Lemma 1.5.2, we formulate the necessary and sufficient condition for the controllability of the dynamical system (1.2.8) basing oneselves on the Jordan canonical form.

THEOREM 1.5.1 (Chen (1970, Ch. 5.5)). *The dynamical system* (1.2.8) *is controllable if and only if for each* $i = 1, 2, \ldots, k$ *the rows* $\mathbf{g}_{i1n_{i1}}, \mathbf{g}_{i2n_{i2}}, \ldots, \mathbf{g}_{ir(i)n_{ir(i)}}$ *of the matrix* $\mathbf{G}$ *are linearly independent over the field of complex numbers.*

Proof:

Necessity. Lemma 1.5.2 directly implies that in the matrix $(s\mathbf{I}-\mathbf{J}_i)^{-1}\mathbf{G}_i$ there are $r(i)$ rows of the form

$$(s-s_i)^{-1}\mathbf{g}_{i1n_{i1}}, \quad (s-s_i)^{-1}\mathbf{g}_{i2n_{i2}}, \quad \ldots, \quad (s-s_i)^{-1}\mathbf{g}_{ir(i)n_{ir(i)}}.$$

Hence, if the rows $\mathbf{g}_{i1n_{i1}}, \mathbf{g}_{i2n_{i2}}, \ldots, \mathbf{g}_{ir(i)n_{ir(i)}}$ are not linearly independent, then the rows of $(s\mathbf{I}-\mathbf{J}_i)^{-1}\mathbf{G}_i$ are not linearly independent either. Consequently, if the rows of $(s\mathbf{I}-\mathbf{J}_i)^{-1}\mathbf{G}_i$ are not linearly independent for some i, then by Lemma 1.5.2 the dynamical system (1.2.8) is not controllable. ■

Sufficiency. From Lemma 1.5.2 it follows that the rows of the matrix $(s\mathbf{I}-\mathbf{J}_{ij})^{-1}\cdot \mathbf{G}_{ij}$ contain the components depending on $\mathbf{g}_{ijn_{ij}}$. Moreover, in the l-th row we have the term of the form $(s-s_i)^{-n_{ij}+l-1}\mathbf{g}_{ijn_{ij}}$. Hence all the rows of $(s\mathbf{I}-\mathbf{J}_{ij})^{-1}\mathbf{G}_{ij}$ are linearly independent. If $\mathbf{g}_{i1n_{i1}}, \mathbf{g}_{i2n_{i2}}, \ldots, \mathbf{g}_{ir(i)n_{ir(i)}}$ are linearly independent, it can be shown that all the rows of $(s\mathbf{I}-\mathbf{J}_i)^{-1}\mathbf{G}_i$ are linearly independent. Consequently, by Lemmas 1.5.1 and 1.5.2, if the rows of $(s\mathbf{I}-\mathbf{J}_i)^{-1}\mathbf{G}_i$ are linearly independent for each i, then the dynamical system (1.2.8) is controllable. ■

Theorem 1.5.1 may also be stated as follows: the dynamical system (1.2.8) is controllable if and only if for each $i = 1, 2, \ldots, k$ the rows of the matrix $\mathbf{G}$ corresponding to the last rows of the Jordan blocks $\mathbf{J}_{ij}$, $j = 1, 2, \ldots, r(i)$ are linearly independent over the field of complex numbers.

COROLLARY 1.5.1. *If the dynamical system* (1.2.8) *is controllable then for each* $i = 1, 2, \ldots, k$ *we have*

$$r(i) \leqslant m. \tag{1.5.8}$$

Proof:

Observe that in order for the rows $\mathbf{g}_{i1n_{i1}}, \mathbf{g}_{i2n_{i2}}, \ldots, \mathbf{g}_{ir(i)n_{ir(i)}}$ to be linearly independent over the field of complex numbers, it is necessary that $r(i) \leqslant m$ for each $i = 1, 2, \ldots, k$. Hence our Corollary 1.5.1 follows. ■

COROLLARY 1.5.2 (Chen (1970, Ch. 5.5)). *If* $m = 1$, *then the dynamical system* (1.2.8) *is controllable if and only if all the eigenvalues are pairwise distinct and all the components of the column vector* $\mathbf{G}$ *that correspond to the last row of each Jordan block are different from zero.*

Proof:

Observe that, for the case $m = 1$, by Corollary 1.5.1 and Theorem 1.5.1 the necessary and sufficient condition for the controllability of the dynamical system

(1.2.8) is that there should be at most one Jordan block associated with one eigenvalue, i.e., $\mathbf{J}_i = \mathbf{J}_{i1}$. Hence of course $k = n$ and $r(i) = 1$ for all $i = 1, 2, \ldots, n$. ∎

Example 1.5.1

Let us consider the dynamical system (1.2.8) with matrices **A** and **B** of the following form:

$$\mathbf{A} = \begin{bmatrix} 1 & -3 & -2 \\ -2 & -1 & 1 \\ 8 & -7 & -7 \end{bmatrix}, \quad \mathbf{B} = \begin{bmatrix} 3 & 4 & 5 \\ 0 & 0 & 0 \\ 5 & 5 & 8 \end{bmatrix}. \tag{1.5.9}$$

Hence we have the characteristic polynomial of the matrix **A**, $\varphi(s) = (s+2)^2(s+3)$, and the minimal polynomial of the matrix **A**, $\psi(s) = (s+2)^2(s+3)$. Since $\varphi(s) = \psi(s)$, every submatrix $\mathbf{J}_i$ contains only one Jordan block $\mathbf{J}_{i1}$ (Chen (1970, Ch. 2.6)). In this example we have $s_1 = -3$, $s_2 = -2$, $r(1) = 1$, $r(2) = 1$, $n_{11} = 1$, $n_{21} = 2$, $n = 3$, $m = 3$. The matrices **J** and **G** are of the following form:

$$\mathbf{J} = \begin{bmatrix} -3 & 0 & 0 \\ 0 & -2 & 1 \\ 0 & 0 & -2 \end{bmatrix}, \quad \mathbf{G} = \begin{bmatrix} 2 & 1 & 3 \\ 0 & 0 & 0 \\ 1 & 3 & 2 \end{bmatrix} = \begin{bmatrix} \mathbf{g}_{111} \\ \mathbf{g}_{211} \\ \mathbf{g}_{212} \end{bmatrix}.$$

Since $\mathbf{g}_{111} \neq 0$ and $\mathbf{g}_{212} \neq 0$, then by Theorem 1.5.1 the dynamical system (1.5.9) is controllable in spite of the fact that matrix **B** and matrix **G** contain the zero row, i.e., $\mathbf{g}_{211} = 0$.

Example 1.5.2

Let us consider the dynamical system (1.2.8) with the matrices **A** and **B** of the following form:

$$\mathbf{A} = \begin{bmatrix} -9 & 1 & 5 \\ -6 & -1 & 4 \\ -8 & 1 & 4 \end{bmatrix}, \quad \mathbf{B} = \begin{bmatrix} 1 & 0 & 3 \\ 3 & 3 & 0 \\ 2 & -2 & 4 \end{bmatrix}. \tag{1.5.10}$$

Hence we have the characteristic polynomial of the matrix **A**, $\varphi(s) = (s+1)(s+2)(s+3)$, and the minimal polynomial of the matrix **A**, $\psi(s) = \varphi(s)$. Thus, every submatrix $\mathbf{J}_i$ contains only one Jordan block $\mathbf{J}_{i1}$ (Chen (1970, Ch. 2.6)). In this example we have $s_1 = -1$, $s_2 = -2$, $s_3 = -3$, $r(1) = 1$, $r(2) = 1$, $r(3) = 1$, $n_{11} = 1$, $n_{21} = 1$, $n_{31} = 1$, $n = 3$, $m = 3$. Moreover, the matrices **J** and **G** are of the following form:

$$\mathbf{J} = \begin{bmatrix} -1 & 0 & 0 \\ 0 & -2 & 0 \\ 0 & 0 & -3 \end{bmatrix}, \quad \mathbf{G} = \begin{bmatrix} 1 & -2 & 1 \\ 3 & 1 & -2 \\ -4 & 3 & 3 \end{bmatrix} = \begin{bmatrix} \mathbf{g}_{111} \\ \mathbf{g}_{211} \\ \mathbf{g}_{311} \end{bmatrix}.$$

Since $\mathbf{g}_{111} \neq 0$, $\mathbf{g}_{211} \neq 0$ and $\mathbf{g}_{311} \neq 0$, Theorem 1.5.1 immediately implies that the dynamical system (1.5.10) is controllable.

Controllability of the dynamical systems (1.5.9) and (1.5.10) may be of course considered by using the results given in Section 1.4 (Theorem 1.4.4). It should be pointed out that Example 1.5.2 has been considered in Section 1.4 (Example 1.4.1) without transformation into the Jordan canonical form. The results are of course the same in both cases.

1.6 Perturbations of controllable dynamical systems

In practice the fundamental problem is the question which bounded perturbations of the dynamical system (1.2.1) preserve controllability. In this section we shall consider this problem with respect to the controllability in $[t_0, t_1]$ of the dynamical system (1.2.1). The generalizations to other kinds of controllability are obvious and will be omitted here.

For brevity let us introduce the following notation: For the matrix $\mathbf{F} \in L^p([t_0, t_1], R^{n\times m})$, $||\mathbf{F}||_p$, $1 \leqslant p \leqslant \infty$, denotes the norm of the matrix $\mathbf{F}(t)$ in the space $L^p([t_0, t_1], R^{n\times m})$, Moreover, for the $(n\times m)$-dimensional constant matrix $\mathbf{G}$, the term $||\mathbf{G}||$ denotes the ordinary norm of this matrix.

THEOREM 1.6.1 (Dauer (1971)). *Suppose that* $\mathbf{A} \in L^1([t_0, t_1], R^{n\times m})$ *and* $\mathbf{B} \in L^2([t_0, t_1], R^{n\times m})$, *and, moreover, the dynamical system* (1.2.1) *is controllable in* $[t_0, t_1]$.*Then there exists* $\varepsilon > 0$ *such that if*

$$||\mathbf{A}-\mathbf{F}||_1+||\mathbf{B}-\mathbf{C}||_2 < \varepsilon \tag{1.6.1}$$

then the dynamical system of the form

$$\dot{\mathbf{y}}(t) = \mathbf{F}(t)\mathbf{y}(t)+\mathbf{G}(t)\mathbf{u}(t), \quad \mathbf{y}(t) \in R^n, \quad t \in [t_0, t_1] \tag{1.6.2}$$

is also controllable in $[t_0, t_1]$.

Proof:

The complete proof of Theorem 1.6.1 is based on the theory of measure and Lebesgue's integral. This proof is presented in the paper of Dauer (1971). ∎

COROLLARY 1.6.1 (Dauer (1971), Lee and Markus (1968, Ch. 2.3, Th. 11)). *Suppose the dynamical system* (1.2.8) *is controllable. Then there exists an* $\varepsilon > 0$ *such that if*

$$||\mathbf{A}-\mathbf{F}||+||\mathbf{B}-\mathbf{G}|| < \varepsilon \tag{1.6.3}$$

then the time-invariant dynamical system of the form

$$\dot{\mathbf{y}}(t) = \mathbf{F}\mathbf{y}(t)+\mathbf{G}\mathbf{u}(t), \quad \mathbf{y}(t) \in R^n, \quad t \geqslant 0 \tag{1.6.4}$$

is also controllable.

Proof:

Since the matrices $\mathbf{A}$ and $\mathbf{B}$ are constant matrices, the assumptions of Theorem 1.6.1 are satisfied. Hence we can choose the matrices $\mathbf{F}(t)$ and $\mathbf{G}(t)$ in such a way that inequality (1.6.1) holds. Particularly, we can choose $\mathbf{F}(t) = \mathbf{F}$ and $\mathbf{G}(t) = \mathbf{G}$ in such a way that inequality (1.6.3) is satisfied. The controllability of the dynamical system (1.6.4) follows from Theorem 1.6.1. ■

In the book of Lee and Markus (1968, Ch. 2.3, Th. 11) Corollary 1.6.1 is proved without using Theorem 1.6.1. The proof is entirely based on the properties of non-singular matrices.

THEOREM 1.6.2 (Dauer (1971)). *Suppose that the dynamical system* (1.2.1) *is given. Then for every real number* $\varepsilon > 0$ *there exists a matrix* $\mathbf{G} \in L^2([t_0, t_1], R^{n\times m})$ *such that*

$$||\mathbf{B}-\mathbf{G}||_2 < \varepsilon \tag{1.6.5}$$

and the dynamical system of the form

$$\dot{\mathbf{y}}(t) = \mathbf{A}(t)\mathbf{y}(t)+\mathbf{G}(t)\mathbf{u}(t), \quad \mathbf{y}(t) \in R^n, t \in [t_0, t_1] \tag{1.6.6}$$

is controllable in $[t_0, t_1]$.

Proof:

The proof of Theorem 1.6.2 is based on the theory of measure and Lebesgue's integral and is presented in the paper by Dauer (1971). ■

COROLLARY 1.6.2 (Chu (1966)). *Suppose that the dynamical system* (1.2.8) *is given. Then for every real number* $\varepsilon > 0$ *there exists a constant matrix* $\mathbf{G}$ *such that*

$$||\mathbf{B}-\mathbf{G}|| < \varepsilon \tag{1.6.7}$$

and the time-invariant dynamical system of the form

$$\dot{\mathbf{y}}(t) = \mathbf{A}\mathbf{y}(t)+\mathbf{G}\mathbf{u}(t), \quad \mathbf{y}(t) \in R^n, t \in [t_0, t_1] \tag{1.6.8}$$

is controllable.

Proof:

Since the dynamical system (1.2.8) is time-invariant ($\mathbf{A}(t) = \mathbf{A}$ and $\mathbf{B}(t) = \mathbf{B}$), Corollary 1.6.2 immediately follows from Theorem 1.6.2. ■

Corollary 1.6.2 is also presented in the book by Lee and Markus (1968, Ch. 2.3) and proved by using the fundamental properties of nonsingular matrices. However, it should be stressed that the proofs given in the book by Lee and Markus (1968, Ch. 2.3) cannot be extended to cover the case of time-varying dynamical systems (1.2.1).

Theorems 1.6.1 and 1.6.2 are used in the investigations of the properties of the set of dynamical systems controllable in $[t_0, t_1]$. In order to do so we shall use the topological notions, namely the density and the openness of the sets (Dunford and Schwartz (1958, Ch. 1), Warga (1972, Ch. 1.2)). Let us denote by $S\left([t_0, t_1], \mathbf{A}(t), \mathbf{B}(t)\right)$

the set of all dynamical systems of the form (1.2.1) defined in the time interval $[t_0, t_1]$. Moreover, let $S_1(\mathbf{A}_1(t), \mathbf{B}_1(t))$ and $S_2(\mathbf{A}_2(t), \mathbf{B}_2(t))$ denote any two elements of the set $S([t_0, t_1], \mathbf{A}(t), \mathbf{B}(t))$. In the set $S([t_0, t_1], \mathbf{A}(t), \mathbf{B}(t))$ we introduce the metric $\varphi(\cdot, \cdot)$ in the following way:

$$\varphi(S_1(\mathbf{A}_1(t), \mathbf{B}_1(t)), \ S_2(\mathbf{A}_2(t), \mathbf{B}_2(t))) = \|\mathbf{A}_1 - \mathbf{A}_2\|_1 + \|\mathbf{B}_1 - \mathbf{B}_2\|_2 . \tag{1.6.9}$$

It is obvious that the set $S([t_0, t_1], \mathbf{A}(t), \mathbf{B}(t))$ together with the metric φ is a topological metric space (Rolewicz (1974, Ch. 1.1), Warga (1972, Ch. 1.3)).

THEOREM 1.6.3 (Dauer (1971)). *The set of dynamical systems* (1.2.1) *which are controllable in* $[t_0, t_1]$ *is open and dense in the topological space* $S([t_0, t_1], \mathbf{A}(t), \mathbf{B}(t))$.

Proof:

The openness of the set of dynamical systems (1.2.1) which are controllable in $[t_0, t_1]$ immediately follows from Theorem 1.6.1, and the definition of the metric φ and the definition of the open set in the metric space. The density of this set is a consequence of Theorem 1.6.2, the definition of the metric φ and the definition of the dense set in the metric space. ∎

COROLLARY 1.6.3 (Chen and Desoer (1967), Dauer (1971), Lee and Markus (1968), Ch. 2.3). *The set of time-invariant dynamical systems which are controllable, is open and dense in the set of all dynamical systems of the form* (1.2.8).

Proof:

Corollary 1.6.3 immediately follows from the definition of the metric φ and Corollaries 1.6.1 and 1.6.2. ∎

Theorem 1.6.3 is of great practical importance. It states that almost all dynamical systems of the form (1.2.1) are controllable in $[t_0, t_1]$. The same is of course true for time-invariant dynamical systems of the form (1.2.8). Hence the Lebesgue measure of the set of noncontrollable dynamical systems is equal to zero.

In other words, in the terminology given in the papers by Hosoe (1980), Olbrot (1980), controllability is the so-called "generic property" of the dynamical systems (1.2.1) and (1.2.8). Intuitively this means that almost all dynamical systems are controllable and, moreover, that for almost all dynamical systems there exist open neighbourhoods containing entirely only controllable dynamical systems.

Theorem 1.6.3 enables us to define the so-called controllability margin (Klamka, (1974b)). The *controllability margin* for dynamical system (1.2.1) is defined as the distance between the given dynamical system of the form (1.2.1) and the nearest dynamical system (1.2.1) which is noncontrollable in $[t_0, t_1]$. The controllability margin may be defined only under the condition that the set of controllable dynamical systems is open and dense. It is obvious that a dynamical system (1.2.1) which is noncontrollable in $[t_0, t_1]$ has the controllability margin equal to zero. In practice the precise values of the controllability margin can be computed only for very special kinds of the dynamical systems (1.2.1) or (1.2.8) (Klamka (1974b), Olbrot (1980)). The idea of the controllability margin is developed in the papers of Klamka (1974b) and Olbrot (1980) in a more rigorous way.

Similar considerations concerning time-invariant dynamical systems were presented in the paper of Eising (1979a) by using the theory of polynomials with many variables.

1.7 Minimum energy control

In the case where a linear dynamical system (1.2.1) is controllable in the time interval $[t_0, t_1]$, there exist generally many different controls $\mathbf{u}(t)$, $t \in [t_0, t_1]$, that can steer the initial state $\mathbf{x}(t_0)$ to the desired final state $\mathbf{x}_1 \in \boldsymbol{R}^n$ at time t_1. We may ask which of these possible controls that achieve the same result is the optimal one according to some criterion given a priori.

Let us consider the minimum energy cost function (performance index)

$$J(\mathbf{u}) = \int_{t_0}^{t_1} ||\mathbf{u}(t)||^2_{\mathbf{R}(t)} dt, \tag{1.7.1}$$

where $\mathbf{R}(t)$ is an $(m \times m)$-dimensional, continuous, symmetric weighting matrix which is positive-definite for all $t \in [t_0, t_1]$. Moreover,

$$||\mathbf{u}(t)||^2_{\mathbf{R}(t)} = \langle \mathbf{u}(t), \mathbf{R}(t)\mathbf{u}(t) \rangle_{\mathbf{R}^m} = \mathbf{u}^{\mathrm{T}}(t)\mathbf{R}(t)\mathbf{u}(t), \quad t \in [t_0, t_1]. \tag{1.7.2}$$

The performance index $J(\mathbf{u})$ defines the control energy in the time inretval $[t_0, t_1]$ with the weight determined by the weighting matrix $\mathbf{R}(t)$, $t \in [t_0, t_1]$. The control $\mathbf{u}$ which minimizes the performance index $J(\mathbf{u})$ is called the *minimum energy control* (Chen (1970, pp. 401–403)). The performance index $J(\mathbf{u})$ given by the formula (1.7.1) is the special case of the general quadratic performance index, and hence the existence of a minimizing control function is assured (Athans and Falb (1966, Ch. 9), Lee and Markus (1968, Ch. 3.2)).

In this section, using the controllability matrix $\mathbf{W}(t_0, t_1)$ defined by (1.3.8), we shall present an analytic formula for minimum energy control and, moreover, we shall compute effectively the minimal value of the performance index $J(\mathbf{u})$ corresponding to optimal control.

To abbreviate the notation, let us denote:

$$\mathbf{W}_{\mathbf{R}}(t_0, t_1) = \int_{t_0}^{t_1} \mathbf{F}(t_1, t)\mathbf{B}(t)\mathbf{R}^{-1}(t)\mathbf{B}^{\mathrm{T}}(t)\mathbf{F}^{\mathrm{T}}(t_1, t)\,dt, \tag{1.7.3}$$

where $\mathbf{W}_{\mathbf{R}}(t_0, t_1)$ is an $(n \times n)$-dimensional symmetric matrix.

If we assume that the dynamical system (1.2.1) is controllable in $[t_0, t_1]$, then the controllability matrix $\mathbf{W}(t_0, t_1)$ is nonsingular and there exists an inverse matrix $\mathbf{W}^{-1}(t_0, t_1)$. Taking into account formula (1.7.3), we can easily conclude, that in this case the inverse matrix $\mathbf{W}_{\mathbf{R}}^{-1}(t_0, t_1)$ also exists. Therefore we may define the control $\mathbf{u}^0(t)$, $t \in [t_0, t_1]$ in the following manner:

$$\mathbf{u}^0(t) = \mathbf{R}^{-1}(t)\mathbf{B}^{\mathrm{T}}(t)\mathbf{F}^{\mathrm{T}}(t_1, t)\mathbf{W}_{\mathbf{R}}^{-1}(t_0, t_1)(\mathbf{x}_1 - \mathbf{F}(t_1, t_0)\mathbf{x}(t_0)),$$
$$t \in [t_0, t_1]. \tag{1.7.4}$$

Since the matrix $\mathbf{R}(t)$ is continuous in $[t_0, t_1]$, also the inverse matrix $\mathbf{R}^{-1}(t)$ is continuous in $[t_0, t_1]$. Hence by (1.7.4) we have $\mathbf{u}^0 \in L^2([t_0, t_1], \boldsymbol{R}^m)$.

THEOREM 1.7.1. *Suppose that the dynamical system* (1.2.1) *is controllable in* $[t_0, t_1]$ *and let* $\mathbf{u}' \in L^2([t_0, t_1], \boldsymbol{R}^m)$ *be any control which steers the dynamical system* (1.2.1) *from the initial state* $\mathbf{x}(t_0) \in \boldsymbol{R}^n$ *to the desired final state* $\mathbf{x}_1 \in \boldsymbol{R}^n$ *at time* t_1. *Then the control* $\mathbf{u}^0 \in L^2([t_0, t_1], \boldsymbol{R}^m)$ *given by* (1.7.4) *accomplishes the same transfer and*

$$\int_{t_0}^{t_1} \|\mathbf{u}'(t)\|^2_{\mathbf{R}(t)} dt \geqslant \int_{t_0}^{t_1} \|\mathbf{u}^0(t)\|^2_{\mathbf{R}(t)} dt. \tag{1.7.5}$$

Moreover, the minimal value of the performance index $J(\mathbf{u}^0)$ *is given by the formula*

$$J(\mathbf{u}^0) = \int_{t_0}^{t_1} \|\mathbf{u}^0(t)\|^2_{\mathbf{R}(t)} dt = \|\mathbf{x}_1 - \mathbf{F}(t_1, t_0)\mathbf{x}(t_0)\|^2_{\mathbf{W}_{\mathbf{R}}^{-1}(t_0, t_1)}. \tag{1.7.6}$$

Proof:

In the first step of the proof we shall show, that the control $\mathbf{u}^0(t)$ $t \in [t_0, t_1]$ given by formula (1.7.4) steers the dynamical system (1.2.1) from the initial state $\mathbf{x}(t_0) \in \boldsymbol{R}^n$ to the desired final state $\mathbf{x}_1 \in \boldsymbol{R}^n$ at time t_1. From the controllability assumptiom it follows that at least one such control exists and, moreover, there exists on inverse matrix $\mathbf{W}_{\mathbf{R}}^{-1}(t_0, t_1)$. Therefore the control $\mathbf{u}^0(t)$ is well-defined. Substituting (1.7.4) into (1.2.2), we obtain for $t = t_1$ the equality

$$\begin{aligned}
&\mathbf{x}(t_1, \mathbf{x}(t_0), \mathbf{u}^0)\\
&= \mathbf{F}(t_1, t_0)\mathbf{x}(t_0) + \int_{t_0}^{t_1} \mathbf{F}(t_1, t)\mathbf{B}(t)\mathbf{R}^{-1}(t)\mathbf{B}^{\mathrm{T}}(t)\mathbf{F}^{\mathrm{T}}(t_1, t)\, dt\, \mathbf{W}_{\mathbf{R}}^{-1}(t_0, t_1)(\mathbf{x}_1 - \\
&\quad - \mathbf{F}(t_1, t_0)\mathbf{x}(t_0))\\
&= \mathbf{F}(t_1, t_0)\mathbf{x}(t_0) + \mathbf{W}_{\mathbf{R}}(t_0, t_1)\mathbf{W}_{\mathbf{R}}^{-1}(t_0, t_1)(\mathbf{x}_1 - \mathbf{F}(t_1, t_0)\mathbf{x}(t_0)) = \mathbf{x}_1.
\end{aligned} \tag{1.7.7}$$

From (1.7.7) it follows that the control $\mathbf{u}^0 \in L^2([t_0, t_1], \boldsymbol{R}^m)$ steers the dynamical system (1.2.1) from the initial state $\mathbf{x}(t_0)$ to the final state $\mathbf{x}_1$ at time t_1. Hence the first part of the proof is proved.

In the next step we shall show the validity of inequality (1.7.5). In order to do that we shall use the properties of the norm and the scalar product in the spaces R^n and $L^2([t_0, t_1], R^m)$. Suppose the control $\mathbf{u}' \in L^2([t_0, t_1], R^m)$ is any control which steers the dynamical system (1.2.1) from the initial state $\mathbf{x}(t_0)$ to the fiinal state $\mathbf{x}_1$ at time t_1. Hence we have

$$\int_{t_0}^{t_1} \mathbf{F}(t_1, t)\mathbf{B}(t)\mathbf{u}'(t)\,dt = \int_{t_0}^{t_1} \mathbf{F}(t_1, t)\mathbf{B}(t)\mathbf{u}^0(t)\,dt. \tag{1.7.8}$$

Subtracting from both sides, we obtain

$$\int_{t_0}^{t_1} \mathbf{F}(t_1, t)\mathbf{B}(t)\big(\mathbf{u}'(t)-\mathbf{u}^0(t)\big)dt = 0 \tag{1.7.9}$$

which implies that

$$\Big\langle \int_{t_0}^{t_1} \mathbf{F}(t_1, t)\mathbf{B}(t)\big(\mathbf{u}'(t)-\mathbf{u}^0(t)\big)dt,\ \mathbf{W}_{\mathbf{R}}^{-1}(t_0, t_1)\big(\mathbf{x}_1-\mathbf{F}(t_1, t_0)\mathbf{x}(t_0)\big)\Big\rangle_{R^n} = 0. \tag{1.7.10}$$

Using the properties of scalar product in the space R^n, by (1.7.10), we have

$$\int_{t_0}^{t_1} \langle \big(\mathbf{u}'(t)-\mathbf{u}^0(t)\big), \mathbf{B}^{\mathrm{T}}(t)\mathbf{F}^{\mathrm{T}}(t_1, t)\mathbf{W}_{\mathbf{R}}^{-1}(t_0, t_1)\big(\mathbf{x}_1-\mathbf{F}(t_1, t_0)\mathbf{x}(t_0)\big)\rangle_{R^n} dt$$

$$= \int_{t_0}^{t_1} \langle \big(\mathbf{u}'(t)-\mathbf{u}^0(t)\big), \mathbf{R}(t)\mathbf{R}^{-1}(t)\mathbf{B}^{\mathrm{T}}(t)\mathbf{F}^{\mathrm{T}}(t_1, t)\mathbf{W}_{\mathbf{R}}^{-1}(t_0, t_1)\big(\mathbf{x}_1 -$$

$$-\mathbf{F}(t_1, t_0)\mathbf{x}(t_0)\big)\rangle_{R^n} dt = 0. \tag{1.7.11}$$

Using (1.7.4) and (1.7.11), we obtain the following equality:

$$\int_{t_0}^{t_1} \langle \big(\mathbf{u}'(t)-\mathbf{u}^0(t)\big), \mathbf{R}(t)\mathbf{u}^0(t)\rangle_{R^n} dt = 0. \tag{1.7.12}$$

By the properties of the scalar product and taking into account formula (1.7.12) we obtain after some easy computations the following relations:

$$\int_{t_0}^{t_1} \|\mathbf{u}'(t)\|^2_{\mathbf{R}(t)}\,dt = \int_{t_0}^{t_1} \langle \mathbf{u}'(t), \mathbf{R}(t)\mathbf{u}'(t)\rangle_{R^n} dt$$

$$= \int_{t_0}^{t_1} \langle \big(\mathbf{u}'(t)-\mathbf{u}^0(t)+\mathbf{u}^0(t)\big), \mathbf{R}(t)\big(\mathbf{u}'(t)-\mathbf{u}^0(t)+\mathbf{u}^0(t)\big)\rangle_{R^n} dt$$

$$= \int_{t_0}^{t_1} \langle \big(\mathbf{u}'(t)-\mathbf{u}^0(t)\big), \mathbf{R}(t)\big(\mathbf{u}'(t)-\mathbf{u}^0(t)\big)\rangle_{R^n} dt + \int_{t_0}^{t_1} \langle \mathbf{u}'(t), \mathbf{R}(t)\big(\mathbf{u}'(t)-$$

$$-\mathbf{u}^0(t)\big)\rangle_{R^n} dt + \int_{t_0}^{t_1} \langle \big(\mathbf{u}'(t)-\mathbf{u}^0(t)\big), \mathbf{R}(t)\mathbf{u}^0(t)\rangle_{R^n} dt +$$

$$+\int_{t_0}^{t_1}\langle \mathbf{u}^0(t), \mathbf{R}(t)\mathbf{u}^0(t)\rangle_{R^n}dt$$

$$=\int_{t_0}^{t_1}\langle (\mathbf{u}'(t)-\mathbf{u}^0(t)), \mathbf{R}(t)(\mathbf{u}'(t)-\mathbf{u}^0(t))\rangle_{R^n}dt+\int_{t_0}^{t_1}\langle \mathbf{u}^0(t), \mathbf{R}(t)\mathbf{u}^0(t)\rangle_{R^n}dt$$

$$=\int_{t_0}^{t_1}||\mathbf{u}'(t)-\mathbf{u}^0(t)||^2_{\mathbf{R}(t)}dt+\int_{t_0}^{t_1}||\mathbf{u}^0(t)||^2_{\mathbf{R}(t)}dt. \tag{1.7.13}$$

Since $\int_{t_0}^{t_1}||\mathbf{u}'(t)-\mathbf{u}^0(t)||^2_{\mathbf{R}(t)}dt \geqslant 0$, we conclude by (1.7.13) that

$$\int_{t_0}^{t_1}||\mathbf{u}'(t)||^2_{\mathbf{R}(t)}dt \geqslant \int_{t_0}^{t_1}||\mathbf{u}^0(t)||^2_{\mathbf{R}(t)}dt. \tag{1.7.14}$$

Hence we have proved inequality (1.7.5), which states that the control $\mathbf{u}^0(t)$, $t \in [t_0, t_1]$, is the optimal one according to the performance index (1.7.1). Hence follows the second step of our proof.

In the last part of the proof we shall demonstrate that the equality (1.7.6) is valid. Let us recall that formula (1.7.6) defines the minimal value of the performance index $J(\mathbf{u}^0)$ corresponding to the optimal control $\mathbf{u}^0 \in L^2([t_0, t_1], \boldsymbol{R}^m)$, given by the formula (1.7.4). Substituting (1.7.4) in (1.7.1), we have

$$J(\mathbf{u}^0) = \int_{t_0}^{t_1}||\mathbf{u}^0(t)||^2_{\mathbf{R}(t)}$$

$$=\int_{t_0}^{t_1}||\mathbf{R}^{-1}(t)\mathbf{B}^{\mathrm{T}}(t)\mathbf{F}^{\mathrm{T}}(t_1, t_1)\mathbf{W}_{\mathbf{R}}^{-1}(t_0, t_1)(\mathbf{x}_1-\mathbf{F}(t_1, t_0)\mathbf{x}(t_0))||^2_{\mathbf{R}(t)}dt$$

$$=\int_{t_0}^{t_1}\langle \mathbf{R}^{-1}(t)\mathbf{B}^{\mathrm{T}}(t)\mathbf{F}^{\mathrm{T}}(t_1, t)\mathbf{W}_{\mathbf{R}}^{-1}(t_0, t_1)(\mathbf{x}_1-\mathbf{F}(t_1, t_0)\mathbf{x}(t_0)),$$

$$\mathbf{B}^{\mathrm{T}}(t)\mathbf{F}^{\mathrm{T}}(t_1, t)\mathbf{W}_{\mathbf{R}}^{-1}(t_0, t_1)(\mathbf{x}_1-\mathbf{F}(t_1, t_0)\mathbf{x}(t_0))\rangle_{R^n}dt. \tag{1.7.15}$$

Since the matrix $\mathbf{W}_{\mathbf{R}}(t_0, t_1)$ is symmetric, by (1.7.3) and (1.7.15) the performance index $J(\mathbf{u}^0)$ can be expressed as follows:

$$J(\mathbf{u}^0) = \int_{t_0}^{t_1}\langle (\mathbf{x}_1-\mathbf{F}(t_1, t_0)\mathbf{x}(t_0)),$$

$$\mathbf{W}_{\mathbf{R}}^{-1}(t_0, t_1)\mathbf{F}(t_1, t)\mathbf{B}(t)\mathbf{R}^{-1}(t)\mathbf{B}^{\mathrm{T}}(t)\mathbf{F}^{\mathrm{T}}(t_1, t)\times$$

$$\times\mathbf{W}_{\mathbf{R}}^{-1}(t_0, t_1)(\mathbf{x}_1-\mathbf{F}(t_1, t_0)\mathbf{x}(t_0))\rangle_{R^n}dt$$

$$=\langle (\mathbf{x}_1-\mathbf{F}(t_1, t_0)\mathbf{x}(t_0)),$$

$$\mathbf{W}_{\mathbf{R}}^{-1}(t_0, t_1)\int_{t_0}^{t_1}\mathbf{F}(t_1, t)\mathbf{B}(t)\mathbf{R}^{-1}(t)\mathbf{B}^{\mathrm{T}}(t)\mathbf{F}^{\mathrm{T}}(t_1, t)dt\times$$

$$\times\mathbf{W}_{\mathbf{R}}^{-1}(t_0, t_1)(\mathbf{x}_1-\mathbf{F}(t_1, t_0)\mathbf{x}(t_0))\rangle_{R^n}$$

$$= \langle (\mathbf{x}_1 - \mathbf{F}(t_1, t_0)\mathbf{x}(t_0)), \mathbf{W}_\mathbf{R}^{-1}(t_0, t_1)\mathbf{W}_\mathbf{R}(t_0, t_1)\mathbf{W}_\mathbf{R}^{-1}(t_0, t_1)(\mathbf{x}_1 - \mathbf{F}(t_1, t_0)\mathbf{x}(t_0))\rangle_{R^n}$$
$$= \langle (\mathbf{x}_1 - \mathbf{F}(t_1, t_0)\mathbf{x}(t_0)), \mathbf{W}_\mathbf{R}^{-1}(t_0, t_1)(\mathbf{x}_1 - \mathbf{F}(t_1, t_0)\mathbf{x}(t_0))\rangle_{R^n}$$
$$= \|(\mathbf{x}_1 - \mathbf{F}(t_1, t_0)\mathbf{x}(t_0))\|^2_{\mathbf{W}_\mathbf{R}^{-1}(t_0, t_1)} \tag{1.7.16}$$

Thus the equality (1.7.6) has been proved and our theorem follows. ∎

Theorem 1.7.1 was proved under the following assumptions:
(1) the dynamical system is linear,
(2) there are no constraints in control,
(3) the dynamical system is controllable,
(4) the performance index does not contain the state variable $\mathbf{x}(t)$,
(5) the performance index is quadratic with respect to control $\mathbf{u}(t)$.

It should be pointed out that Theorem 1.7.1 was proved by using only the fundamental properties of the norms and the scalar product in the Hilbert spaces R^n and $L^2([t_0, t_1], R^m)$, without applying complicated results from the mathematical optimal control theory. This was possible since all the five assumptions mentioned before were satisfied.

The performance index (1.7.1) can also be used in determining the so-called *controllability measures*, which characterize qualitatively the dynamical system (1.2.1). Using the definitions of various kinds of controllability given in Section 1.2, we can divide the set af all dynamical systems into two disjoint parts: the set of controllable dynamical systems and the set of noncontrollable dynamical systems. In the set of controllable dynamical systems we define the controllability measures, which are equal to zero on the set of noncontrollable dynamical systems. There are many different measures of controllability (Müller and Weber (1972)). In this section we concentrate on the controllability measures which are strongly related to minimum energy control and the performance index (1.7.1) (Müller and Weber (1972)).

For controllable dynamical systems (1.2.1) we can obtain by Theorem 1.7.1 the optimal control $\mathbf{u}^0(t)$, which steers the system from the initial state $\mathbf{x}(t_0)$ to a given final state $\mathbf{x}_1$ at time t_1. This optimal control is given by the formula (1.7.4) and is called the *minimum energy control.* Let us fix the times t_0 and t_1 and, moreover, let us assume that the final state $\mathbf{x}_1$ is equal to 0. Hence by (1.3.8) and (1.7.6) we obtain the following equality:

$$J(t_0, t_1; \mathbf{x}(t_0)) = \mathbf{x}(t_0)^\mathrm{T}\mathbf{W}_\mathbf{R}^{-1}(t_0, t_1)\mathbf{x}(t_0), \tag{1.7.17}$$

where $J(t_0, t_1; \mathbf{x}(t_0))$ may be interpreted as the value of the energy of the control $\mathbf{u}(t)$ which transforms the initial state $\mathbf{x}(t_0)$ to zero at time t_1.

The first controllability measure μ_1 of the dynamical system (1.2.1) is defined as the reciprocal of the maximal value of the index (1.7.17) for all initial states taken from the unit ball, i.e., $\{\mathbf{x}(t_0) \in R^n: \|\mathbf{x}(t_0)\| = 1\}$ (Müller and Weber (1972)); hence

$$1/\mu_1 = \max_{\|\mathbf{x}(t_0)\|=1} J(t_0, t_1; \mathbf{x}(t_0)) = s_{\max}(\mathbf{W}_{\mathbf{R}}^{-1}(t_0, t_1))$$
$$= 1/s_{\min}(\mathbf{W}_{\mathbf{R}}(t_0, t_1)), \tag{1.7.18}$$

where $s_{\max}(\mathbf{W}_{\mathbf{R}}^{-1}(t_0, t_1))$ and $s_{\min}(\mathbf{W}_{\mathbf{R}}(t_0, t_1))$ denote, respectively, the maximal and the minimal eigenvalue of a given matrix. Since by assumption the dynamical system (1.2.1) is controllable in the time interval $[t_0, t_1]$, the matrix $\mathbf{W}(t_0, t_1)$ is positive definite and therefore has only positive and real eigenvalues. Thus the measure $\mu_1 = s_{\min}(\mathbf{W}_{\mathbf{R}}(t_0, t_1))$ is well defined and satisfies all the measure axioms (Müller and Weber (1972)). The measure μ_1 may also be interpreted as the minimal distance between the quadratic form represented by the matrix $\mathbf{W}_{\mathbf{R}}(t_0, t_1)$ and zero. The eigenvector of the matrix $\mathbf{W}_{\mathbf{R}}(t_0, t_1)$ corresponding to $s_{\min}(\mathbf{W}\ (t_0, t_1))$ represents the direction which is the worst controllable. This means that the control energy needed in steering to zero the initial state $\mathbf{x}(t_0)$, which is proportional to this direction, is relatively the maximal one. So, the measure μ_1 has a good physical interpretation. However, it should be stressed that the measure μ_1 has also a disadvantage. Namely, it requires computing the eigenvalues of the matrix $\mathbf{W}_{\mathbf{R}}(t_0, t_1)$, which is generally rather tedious and difficult.

We can also define another measure of controllability for the dynamical system (1.2.1). Let us define the measuer μ_2 as follows (Müller and Weber (1972)):

$$\mu_2 = \frac{n}{\operatorname{tr}(\mathbf{W}_{\mathbf{R}}^{-1}(t_0, t_1))}, \tag{1.7.19}$$

where the symbol $\operatorname{tr}(\cdot)$ denotes the trace of the square matrix.

It should be pointed out that the measure μ_2 does not require computing the eigenvalues of the matrix $\mathbf{W}_{\mathbf{R}}^{-1}(t_0, t_1)$. This is an important advantage of this measure. The controllability measure μ_2 can be interpreted as the minimal value of the performance index (1.7.1) for the initial states given on the unit ball, i.e., $\{\mathbf{x}(t_0) \in \mathbf{R}^n: \|\mathbf{x}(t_0)\| = 1\}$ (Müller and Weber (1972)).

The next controllability measure μ_3 for the dynamical system (1.2.1) is defined as follows (Müller and Weber (1972)):

$$\mu_3 = \det(\mathbf{W}_{\mathbf{R}}(t_0, t_1)), \tag{1.7.20}$$

where the symbol $\det(\cdot)$ denotes the determinant of a square matrix.

The measure μ_3 does not need the computation of the eigenvalues. Physically the measure μ_3 characterizes the set of the initial states $\mathbf{x}(t_0)$ which can be steered to zero at time t_1 by using the controls corresponding to a given value of the performance index (1.7.1).

For noncontrollable dynamical systems we have: $\mu_1 = \mu_2 = \mu_3 = 0$. In the case of stationary dynamical systems (1.2.8) it can be proved (Müller and Weber (1972)) that instead of the matrix $\mathbf{W}_{\mathbf{R}}(t_0, t_1)$ in the formulas (1.7.18), (1.7.19) and (1.7.20) we can consider the matrix of the following form:

$$[\mathbf{B} \vdots \mathbf{AB} \vdots \mathbf{A}^2\mathbf{B} \vdots \dots \vdots \mathbf{A}^{n-1}\mathbf{B}][\mathbf{B} \vdots \mathbf{AB} \vdots \mathbf{A}^2\mathbf{B} \vdots \dots \vdots \mathbf{A}^{n-1}\mathbf{B}]^{\mathrm{T}}$$
$$= \mathbf{BB}^{\mathrm{T}} + \mathbf{ABB}^{\mathrm{T}}\mathbf{A}^{\mathrm{T}} + \dots + \mathbf{A}^{n-1}\mathbf{BB}^{\mathrm{T}}(\mathbf{A}^{\mathrm{T}})^{n-1}.$$

It should be stressed that the above matrix is easily computable.

1.8 Controllability with constrained controls

In practice admissible controls are required to satisfy the magnitude constraints. In this section the necessary and sufficient conditions for the various kinds of constrained controllability for the dynamical systems (1.2.1) will be presented. In order to do so, let us introduce some notations and definitions.

Let $U \subset R^m$ be an arbitrary set and let $M(U)$ denote the set of functions measurable in the time interval (t_0, ∞) with values in the set U. Then any control $\mathbf{u} \in M(U)$ is said to be *admissible*. Moreover, let $S \subset R^n$ be an arbitrary set. We shall now define various notions of constrained controllability or, more precisely, U-controllability.

DEFINITION 1.8.1. The dynamical system (1.2.1) is said to be *U-controllable to the set S from $\mathbf{x}_0 \in R^n$* if, for any initial state $\mathbf{x}(t_0) = \mathbf{x}_0$, there exists an admissible control $\mathbf{u} \in M(U)$ such that the corresponding trajectory $\mathbf{x}(t, \mathbf{x}(t_0), \mathbf{u})$ of the dynamical system (1.2.1) satisfies for some $t_1 \in [t_0, \infty)$ the condition

$$\mathbf{x}(t_1, \mathbf{x}(t_0), \mathbf{u}) \in S.$$

DEFINITION 1.8.2. The dynamical system (1.2.1) is said to be *locally U-controllable to the set S* if there exists a nonempty open set $V \subset R^n$ such that the dynamical system (1.2.1) is U-controllable to the set S for each initial state $\mathbf{x}_0 \in V$ and the following inclusion is satisfied:

$$S \subset V \subset R^n.$$

DEFINITION 1.8.3. The dynamical system (1..21) is said to be *globally U-controllable to the set S* if it is U-controllable to the set S from each initial state $\mathbf{x}_0 \in R^n$.

In the case where $S = R^n$ we speak shortly about *global U-controllability.* In the special case where $S = \{0\}$, we speak about *null U-controllability, from $\mathbf{x}_0$, local null U-controllability* and *global null U-controllability*, respectively.

Very often the set S is a linear variety of the following form:

$$S = \{\mathbf{x} \in R^n \colon \mathbf{Lx} = \mathbf{c}\}, \tag{1.8.1}$$

where $\mathbf{L}$ is a given $(p \times n)$-dimensional matrix of rank p, and $\mathbf{c} \in R^p$ is a given vector. In the special case where $\mathbf{L} = \mathbf{I}_n$ ($\mathbf{I}_n$ is the $(n \times n)$-dimensional identity matrix) and $\mathbf{c} = 0$ we have $S = \{0\}$.

Let $K_U([t_0, t], \mathbf{x}_0)$ denote the *attainable set at time $t > t_0$* from the initial state $\mathbf{x}_0 = \mathbf{x}(t_0)$ for the dynamical system (1.2.1), i.e.,

$$K_U([t_0, t], \mathbf{x}_0) = \left\{\mathbf{x} \in R^n:\ \mathbf{x} = \mathbf{F}(t, t_0)\mathbf{x}_0 + \int_{t_0}^{t} \mathbf{F}(t, \tau)\mathbf{B}(\tau)\mathbf{u}(\tau)d\tau:\ \mathbf{u} \in M(U)\right\}. \quad (1.8.2)$$

DEFINITION 1.8.4. The dynamical system (1.2.1) is said to be *locally U-controllable from* $\mathbf{x}_0$ if there exists a $t \in (t_0, \infty)$ such that the attainable set $K_U([t_0, t], \mathbf{x}_0)$ contains the neighbourhood of the point $\mathbf{x}_0$.

DEFINITION 1.8.5. The dynamical system (1.2.1) is said to be *globally U-controllable from* $\mathbf{x}_0$ if the following condition holds:

$$K_U(\mathbf{x}_0) = \bigcup_{t \geq t_0} K_U([t_0, t], \mathbf{x}_0) = R^n. \quad (1.8.3)$$

In the special case where $\mathbf{x}_0 = 0$ we speak about *local U-controllability from zero*, and *global U-controllability from zero*, respectively.

Now we shall formulate the criteria for investigating various kinds of constrained controllability. It is assumed that the target set S is of the form (1.8.1). In order to do so let us introduce the following notation: for an arbitrary vector $\mathbf{a} \in R^p$ we define the support hyperplane $H_U(\mathbf{a})$ for the set U as follows:

$$H_U(\mathbf{a}) = \sup\{\mathbf{w}^T\mathbf{L}^T\mathbf{a}:\ \mathbf{w} \in U\}.$$

Our criteria for U-controllability will be given in terms of the scalar function J: $R^n \times R \times R^p \to R$ defined by

$$J(\mathbf{x}_0, t, \mathbf{a}) = \mathbf{x}_0^T\mathbf{F}^T(t, t_0)\mathbf{L}^T\mathbf{a} + \int_{t_0}^{t} H_U\left(\mathbf{B}^T(\tau)\mathbf{F}^T(t, \tau)\mathbf{L}^T\mathbf{a}\right)d\tau - \mathbf{c}^T\mathbf{a}. \quad (1.8.4)$$

We note that the compactness assumption on U and continuity of the matrix function $\mathbf{A}(\cdot)$ and $\mathbf{B}(\cdot)$ guarantee that the integrand in (1.8.4) is a continuous function of τ. Hence the integral in formula (1.8.4) is well defined. The scalar function $J(\mathbf{x}_0, t, \mathbf{a})$ may be treated as the so-called *support function* for the attainable set $K_U(t_0, t, \mathbf{x}_0, \mathbf{u})$.

THEOREM 1.8.1 (Schmitendorf and Barmish (1980, 1981)). *Suppose U is a compact set and* $E \subset R^n$ *is an arbitrary set which contains zero as an interior point. Then the dynamical system* (1.2.1) *is U-controllable to the set S given by* (1.8.1) *from the initial state* $\mathbf{x}_0 \in R$ *if and only if for a certain* $t_1 \in [t_0, \infty)$

$$\min\{J(\mathbf{x}_0, t_1, \mathbf{a}):\ \mathbf{a} \in \mathbf{E}\} = 0 \quad (1.8.5)$$

or equivalently if and only if

$$J(\mathbf{x}_0, t_1, \mathbf{a}) \geq 0 \quad \textit{for all } \mathbf{a} \in E. \quad (1.8.6)$$

Proof:

Since the set $K_U([t_0, t_1], \mathbf{x}_0)$ is convex and compact then also the set $\tilde{K}_U([t_0, t_1], \mathbf{x}_0)$ given by the formula

$$\tilde{K}_U([t_0, t_1], \mathbf{x}_0) = \{\mathbf{y} \in \boldsymbol{R}^p\colon \mathbf{y} = \mathbf{L}\mathbf{x}, \mathbf{x} \in K_U([t_0, t_1], \mathbf{x}_0)\} \tag{1.8.7}$$

is convex and compact. An initial state $\mathbf{x}_0$ can be steered to the set S at time t_1 if and only if the vector $\mathbf{c}$ and the set $\tilde{K}_U([t_0, t_1], \mathbf{x}_0)$ are not exactly separable by hyperplane or, equivalently, if and only if for every vector $\mathbf{a} \in \boldsymbol{R}^p$ we have

$$\mathbf{a}^{\mathrm{T}}\mathbf{c} \leqslant \sup\{\mathbf{a}^{\mathrm{T}}\mathbf{b}\colon \mathbf{b} \in \tilde{K}_U([t_0, t_1], \mathbf{x}_0)\}. \tag{1.8.8}$$

By (1.8.7) inequality (1.8.8) can be rewritten as follows:

$$\mathbf{a}^{\mathrm{T}}\mathbf{L}\mathbf{F}(t_1, t_0)\mathbf{x}_0 + \sup\left\{\int_{t_0}^{t_1} \mathbf{a}^{\mathrm{T}}\mathbf{L}\mathbf{F}(t_1, \tau)\mathbf{B}(\tau)\mathbf{u}(\tau)d\tau\colon \mathbf{u} \in M(U)\right\} - \mathbf{a}^{\mathrm{T}}\mathbf{c} \geqslant 0. \tag{1.8.9}$$

As a consequence of the measurable selection theory (Warga (1972, Ch. 1.7)), we can commute the supremum and the integral operations in (1.8.9). Thus $\mathbf{c} \in \tilde{K}_U([t_0, t_1], \mathbf{x}_0)$ if and only if $J(\mathbf{x}_0, t_1, \mathbf{a}) \geqslant 0$ for all vectors $\mathbf{a} \in \boldsymbol{R}^p$. By (1.8.4) it is easy to verify that

$$kJ(\mathbf{x}_0, t_1, \mathbf{a}) = J(\mathbf{x}_0, t_1, k\mathbf{a}) \quad \text{for every real number } k \geqslant 0.$$

Hence by $\mathbf{a} \in E$ our theorem follows. ∎

COROLLARY 18.1 (Schmitendorf and Barmish (1980)). *Suppose U is a compact set and $E \subset \boldsymbol{R}^n$ is an arbitrary set containing zero as an interior point. Then the dynamical system* (1.2.1) *is U-controllable to zero from the initial state $\mathbf{x}_0$ if and only if for a certain $t_1 \in [t_0, \infty)$*

$$\min\{J(\mathbf{x}_0, t_1, \mathbf{a})\colon \mathbf{a} \in E\} = 0 \tag{1.8.10}$$

or, equivalently, if and only if

$$J(\mathbf{x}_0, t_1, \mathbf{a}) \geqslant 0 \quad \textit{for each } \mathbf{a} \in E. \tag{1.8.11}$$

Proof:

Since $S = \{0\}$, we have $\mathbf{L} = \mathbf{I}_n$ and $\mathbf{c} = 0$. Since the set E satisfies the hypothesis of Theorem 1.8.1, our Corollary follows by that theorem. ∎

COROLLARY 1.8.2 (Schmitendorf and Barmish (1980)). *Suppose that the assumptions of Corollary* 1.8.1 *are satisfied. Then the dynamical system* (1.2.1) *is U-controllable to zero from the initial state $\mathbf{x}_0$ if and only if for a certain $t_1 \in [t_0, \infty)$ the equality*

$$\mathbf{x}_0^{\mathrm{T}}\mathbf{z}(t_0) + \int_{t_0}^{t_1} H_U(\mathbf{B}^{\mathrm{T}}(\tau)\mathbf{z}(\tau))d\tau \geqslant 0 \tag{1.8.12}$$

holds for all solutions $\mathbf{z}(\cdot)$ *of the adjoint system*

$$\dot{\mathbf{z}}(t) = -\mathbf{A}^{\mathrm{T}}\mathbf{z}(t), \qquad t \in [t_0, \infty). \tag{1.8.13}$$

Proof:

Since $\mathbf{z}(t) = \mathbf{F}^{\mathrm{T}}(t_0, t)\mathbf{z}(t_0)$, we obtain the desired result by substituting $\mathbf{z}(t) = \mathbf{F}^{\mathrm{T}}(t_1, t)\mathbf{a}$ and using Corollary 1.6.1. ∎

COROLLARY 1.8.3 (Schmitendorf and Barmish (1980)). *Suppose that the assumptions of Corollary* 1.8.1 *are satisfied. Then the dynamical system* (1.2.1) *is globally U-controllable to zero if and only if for every* $\mathbf{x}_0 \in \mathbf{R}^n$ *there exists a time* $t_{1,\mathbf{x}_0} \in [t_0, \infty)$ *such that*

$$\min\{J(\mathbf{x}_0, t_{1,\mathbf{x}_0}, \mathbf{a}) \colon \mathbf{a} \in E \subset \mathbf{R}^n\} = 0. \tag{1.8.14}$$

Proof:

Corollary 1.8.3 immediately follows by Definition 1.8.3 and Corollary 1.8.1. ∎

Now we shall formulate the criteria of global U-controllability. In order to do so let us introduce two additional functions of time. For $\mathbf{g} \in \mathbf{R}^n$ we define the function $V\colon \mathbf{R}^n \times [t_0, \infty) \to \mathbf{R}$ as follows:

$$V(\mathbf{g}, t) = \int_{t_0}^{t} \max\{\mathbf{w}^{\mathrm{T}}\mathbf{B}^{\mathrm{T}}(\tau)\mathbf{F}^{\mathrm{T}}(t_0, \tau)\mathbf{g} \colon \mathbf{g} \in U\}d\tau - $$
$$-\inf\{\mathbf{x}^{\mathrm{T}}\mathbf{F}^{\mathrm{T}}(t_0, t)\mathbf{g} \colon \mathbf{x} \in S\}. \tag{1.8.15}$$

Moreover, let us define the function $W\colon [t_0, \infty) \to \mathbf{R}$ as follows:

$$W(t) = \min\{V(\mathbf{F}^{\mathrm{T}}(t, t_0)\mathbf{L}^{\mathrm{T}}\mathbf{a}, t) \colon \mathbf{a} \in \mathbf{R}^p,\ \|\mathbf{F}^{\mathrm{T}}(t, t_0)\mathbf{L}^{\mathrm{T}}\mathbf{a}\| = 1\}. \tag{1.8.16}$$

THEOREM 1.8.2 (Schmitendorf and Barmish (1981)). *The following condition is necessary for the dynamical system* (1.2.1) *to be globally U-controllable to the set S* (*given by* (1.8.1)):

$$\sup_{t \geq t_0} V(\mathbf{L}^{\mathrm{T}}\mathbf{a}, t) = +\infty \qquad \textit{for all nonzero vectors } \mathbf{a} \in \mathbf{R}^p. \tag{1.8.17}$$

Moreover, the following condition is sufficient for the dynamical system (1.2.1) *to be globally U-controllable to the set S* (*given by* (1.8.1)):

$$\sup_{t \geq t_0} W(t) = +\infty. \tag{1.8.18}$$

Proof:

Necessity. By contradiction. Let us assume that the dynamical system (1.2.1) is globally U-controllable to the set S but condition (1.8.17) is not satisfied. Then there exist a constant k, $0 < k < \infty$, and a nonzero vector $\tilde{\mathbf{a}} \in \mathbf{R}^p$ such that for all $t \geq t_0$ we have

$$V(\mathbf{L}^{\mathrm{T}}\tilde{\mathbf{a}}, t) < k.$$

Thus $\inf\{\mathbf{x}^T\mathbf{F}^T(t_0, t)\mathbf{L}^T\tilde{\mathbf{a}}: \mathbf{x} \in S\} > -\infty$ for all $t \geqslant t_0$ and for every $t \in [t_0, \infty)$ there exists a vector $\mathbf{a}_t \in \boldsymbol{R}^p$ such that $\mathbf{F}^T(t_0, t)\mathbf{L}^T\tilde{\mathbf{a}} = \mathbf{L}^T\mathbf{a}_t$. Let

$$\tilde{\mathbf{x}}_0 = \frac{-2k\mathbf{L}^T\tilde{\mathbf{a}}}{\mathbf{a}^T\mathbf{L}\mathbf{L}^T\tilde{\mathbf{a}}}.$$

Since $\operatorname{rank}\mathbf{L} = p$ and $\tilde{\mathbf{a}} \neq 0$, we have $\mathbf{L}^T\tilde{\mathbf{a}} \neq 0$ and $\tilde{\mathbf{x}}_0 \neq 0$. For the initial state $\tilde{\mathbf{x}}_0$ we obtain the equalities

$$\begin{aligned} J(\tilde{\mathbf{x}}_0, t, \mathbf{a}_t) &= \tilde{\mathbf{x}}_0^T\mathbf{F}^T(t, t_0)\mathbf{L}^T\mathbf{a}_t - \mathbf{a}_t^T\mathbf{c} + \int_{t_0}^{t} H_U\big(\mathbf{B}^T(\tau)\mathbf{F}^T(t, \tau)\mathbf{L}^T\mathbf{a}_t\big)d\tau \\ &= -2k - \inf\{\mathbf{a}_t^T\mathbf{L}\mathbf{x}: \mathbf{x} \in S\} + \\ &\quad + \int_{t_0}^{t} H_U\big(\mathbf{B}^T(\tau)\mathbf{F}^T(t_0, \tau)\mathbf{L}^T\tilde{\mathbf{a}}\big)d\tau \\ &= -2k + V(\mathbf{L}^T\tilde{\mathbf{a}}, t) < -k. \end{aligned}$$

Thus for all $t \geqslant t_0$ we have $J(\tilde{\mathbf{x}}_0, t, \mathbf{a}_t) < 0$. The last inequality implies that the dynamical system (1.2.1) is not U-controllable to the set S from the initial state $\tilde{\mathbf{x}}_0$. This contradicts the assumption of global U-controllability to the set S. Thus our necessary condition follows.

Sufficiency. By contradiction. Let us assume that condition (1.8.18) is satisfied, but the dynamical system (1.2.1) is not globally U-controllable to the set S. Then there exists an initial state $\tilde{\mathbf{x}}_0$ which cannot be steered to the set S. Thus for an arbitrary $t \in [t_0, \infty)$ there exists a nonzero vector $\mathbf{a}_t$ such that

$$\tilde{\mathbf{x}}_0^T\mathbf{F}^T(t, t_0)\mathbf{L}^T\mathbf{a}_t - \mathbf{a}_t^T\mathbf{c} + \int_{t_0}^{t} H_U\big(\mathbf{B}^T(\tau)\mathbf{F}^T(t, \tau)\mathbf{L}^T\mathbf{a}_t\big)d\tau < 0. \tag{1.8.19}$$

By the Schwartz inequality (Dunford and Schwartz (1958, p. 248)) and (1.8.19) for all $t \in [t_0, \infty)$ we obtain the inequality

$$\begin{aligned} &\int_{t_0}^{t} H_U\big(\mathbf{B}^T(\tau)\mathbf{F}^T(t, \tau)\mathbf{L}^T\mathbf{a}_t\big)d\tau - \inf\{\mathbf{a}_t^T\mathbf{L}\mathbf{x};\ \mathbf{x} \in S\} \\ &\leqslant \|\tilde{\mathbf{x}}_0\|\,\|\mathbf{F}^T(t, t_0)\mathbf{L}^T\mathbf{a}_t\|. \end{aligned} \tag{1.8.20}$$

For every $t \in (t_0, \infty)$ let us define a vector $\tilde{\mathbf{a}}_t \in \boldsymbol{R}^p$ as follows:

$$\tilde{\mathbf{a}}_t = \frac{\mathbf{a}_t}{\|\mathbf{F}^T(t, t_0)\mathbf{L}^T\mathbf{a}_t\|}. \tag{1.8.21}$$

Thus $\|\mathbf{F}^T(t, t_0)\mathbf{L}^T\tilde{\mathbf{a}}_t\| = 1$. Dividing inequality (1.8.20) by $\|\mathbf{F}^T(t, t_0)\mathbf{L}^T\mathbf{a}_t\|$ and noting that $\mathbf{F}^T(t, \tau) = \mathbf{F}^T(t_0, \tau)\mathbf{F}^T(t, t_0)$, we obtain for all $t \geqslant t_0$ the inequality

$$\int_{t_0}^{t} H_U\big(\mathbf{B}^T(\tau)\mathbf{F}^T(t_0, \tau)\mathbf{F}^T(t, t_0)\mathbf{L}^T\tilde{\mathbf{a}}_t\big)d\tau - \inf\{\mathbf{x}^T\mathbf{F}^T(t_0, t)\mathbf{F}^T(t, t_0)\mathbf{L}^T\tilde{\mathbf{a}}_t:\ \mathbf{x} \in S\} \leqslant \|\tilde{\mathbf{x}}_0\|.$$

By (1.8.15) the above inequality is equivalent to

$$V(\mathbf{F}^{\mathrm{T}}(t, t_0)\mathbf{L}^{\mathrm{T}}\tilde{\mathbf{a}}_t, t) \leqslant \|\tilde{\mathbf{x}}_0\| \quad \text{for all } t \geqslant t_0.$$

Thus by (1.8.16) $W(t) \leqslant \|\tilde{\mathbf{x}}_0\|$ for all $t \geqslant t_0$. This contradicts condition (1.8.18). Hence our theorem follows. ∎

It is the difference between the conditions given in (1.8.17) and (1.8.18) that gives rise to the so-called "controllability gap". In the paper of Bermish and Schmitendorf (1980) it was shown that for the special case $S = \{0\}$ the gap between the necessary condition (1.8.17) and the sufficient condition (1.8.18) disappears. Moreover, in the paper of Petersen and Barmish (1983a) the controllability gap was investigated without this restriction on the target set S. The fact that a separate necessary condition and sufficient condition was obtained leads one to investigate the possibility of a controllability gap. In other words, it is of interest to know if there exist systems which satisfy the sufficient condition but not the necessary condition. For a large class of systems the sufficient condition is also necessary.

COROLLARY 1.8.4 (Schmitendorf and Barmish (1980)). *Suppose the set U is compact and contains the point zero (not necessarily as an interior point). Then the dynamical system* (1.2.1) *is globally U-controllable to zero if and only if*

$$\int_{t_0}^{\infty} H_U(\mathbf{B}^{\mathrm{T}}(\tau)\mathbf{z}(\tau))d\tau = +\infty \tag{1.8.22}$$

for all nonzero solutions $\mathbf{z}(\cdot)$ of the adjoint system (1.8.13) *or, equivalently, if and only if*

$$\int_{t_0}^{\infty} \sup\{\mathbf{w}^{\mathrm{T}}\mathbf{B}^{\mathrm{T}}(\tau)\mathbf{F}^{\mathrm{T}}(t_0, \tau)\mathbf{a}\colon \mathbf{w} \in U\}d\tau = +\infty \tag{1.8.23}$$

for all nonzero vectors $\mathbf{a} \in R^n$.

Proof:

Necessity. The proof of the necessary condition is analogous to the proof of the necessary condition in Theorem 1.8.2. It should be pointed out that we use equality $\mathbf{L} = \mathbf{I}_n$ and the substitution given in the proof of Corollary 1.8.2.

Sufficiency. The proof of the sufficiency condition is similar to the proof of the sufficient condition in Theorem 1.8.2. However, it should be stressed that the positive homogeneity of the function $H_U(\cdot)$ and the equality $\mathbf{L} = \mathbf{I}_n$ are used. ∎

COROLLARY 1.8.5 (Schmitendorf and Barmish (1980)). *Suppose there exists a compact set U such that*

(1) *zero is an interior point of U,*

(2) *the dynamical system* (1.2.1) *is globally U-controllable to zero.*

Then the dynamical system (1.2.1) *is also globally U-controllable to zero for any other set U′* (*not necessarily compact*) *which contains zero as an interior point.*

Proof:

Suppose U and U' satisfy the hypotheses of Corollary 1.8.5. We are going to show that the dynamical system (1.2.1) is globally U'-controllable to zero. To prove this it is sufficient to find a subset $U'_r \subset U'$ such that the dynamical system (1.2.1) is globally U'_r-controllable to zero. Choose a real number $r > 0$ such that

$$U'_r = \{\mathbf{w} \in \boldsymbol{R}^m \colon \|\mathbf{w}\| \leqslant r\} \subset U'.$$

This can be accomplished because zero is interior to U'. To prove that U'_r has the desired property, we choose $R > 0$ such that

$$U_R = \{\mathbf{w} \in \boldsymbol{R}^m \colon \|\mathbf{w}\| \leqslant R\} \supset U.$$

This can also be done since U is compact and hence bounded. Let $\mathbf{z}(\cdot)$ be any non-zero solution of the adjoint system (1.8.13). Then since the dynamical system (1.2.1) is globally U-controllable to zero (U_R-controllability to zero follows from U-controllability to zero in conjunction with the fact that $U_R \supset U$), we have

$$\begin{aligned}\int_{t_0}^{\infty} H_{U'_r}(\mathbf{B}^{\mathrm{T}}(\tau)\mathbf{z}(\tau))d\tau &= \int_{t_0}^{\infty} \sup\{\mathbf{w}^{\mathrm{T}}\mathbf{B}^{\mathrm{T}}(\tau)\mathbf{z}(\tau) \colon \|\mathbf{w}\| \leqslant r\}d\tau \\ &= r\int_{t_0}^{\infty} \|\mathbf{B}^{\mathrm{T}}(\tau)\mathbf{z}(\tau)\|\,d\tau = \frac{r}{R}\int_{t_0}^{\infty} R\|\mathbf{B}^{\mathrm{T}}(\tau)\mathbf{z}(\tau)\|d\tau \\ &= \frac{r}{R}\int_{t_0}^{\infty} \sup\{\mathbf{w}^{\mathrm{T}}\mathbf{B}^{\mathrm{T}}(\tau)\mathbf{z}(\tau) \colon \|\mathbf{w}\| \leqslant R\}d\tau \\ &= \frac{r}{R}\int_{t_0}^{\infty} H_{U_R}(\mathbf{B}^{\mathrm{T}}(\tau)\mathbf{z}(\tau))d\tau = +\infty.\end{aligned}$$

Thus by Theorem 1.8.2 and Corollary 1.8.4 we conclude that the dynamical system (1.2.1) must be globally U'_r-controllable to zero and hence globally U'-controllable to zero. ∎

Corollary 1.8.5 is a special case of Corollary 1.8.4 where the point zero is an interior point of the set U. From Corollary 1.8.5 it follows that the structure of the set U (besides that it contains zero as an interior point) is not important for the global U-controllability to zero of the dynamical system (1.2.1).

An important special case of Corollary 1.8.4 occurs when U is a closed unit ball in $\boldsymbol{R}^m$, i.e.,

$$U = \{\mathbf{w} \in \boldsymbol{R}^m \colon \|\mathbf{w}\| \leqslant 1\}, \tag{1.8.24}$$

where $||\cdot||$ is a prescribed norm on $\boldsymbol{R}^m$. In this situation we have

$$H_U(\mathbf{B}^m(\tau)\mathbf{z}(\tau)) = \sup\{\mathbf{w}^T\mathbf{B}^T(\tau)\mathbf{z}(\tau):\ ||\mathbf{w}|| \leqslant 1\} = ||\mathbf{B}^T(\tau)\mathbf{z}(\tau)||. \tag{1.8.25}$$

COROLLARY 1.8.6 (Pandolfi (1976a), Schmitendorf and Barmish (1980)). *Suppose U is a closed unit ball of the form* (1.8.24). *Then the dynamical system* (1.2.1) *is globally U-controllable to zero if and only if*

$$\int_{t_0}^{\infty} ||\mathbf{B}^T(\tau)\mathbf{z}(\tau)||d\tau = +\infty \tag{1.8.26}$$

for all nonzero solutions $\mathbf{z}(\cdot)$ of the adjoint system (1.8.13).

Proof:

By (1.8.24) the set U is compact (Rolewicz (1974, Ch. 4.5), Warga (1972, Ch. 1.3)) and contains the point zero in its interior. Thus the assumptions of Corollary 1.8.4 are satisfied and by (1.8.22) and (1.8.25) we obtain the desired result. ■

COROLLARY 1.8.7 (Schmitendorf and Barmish (1980)). *Suppose the set U is an arbitrary set which contains zero as an interior point. Then the dynamical system* (1.2.1) *is globally U-controllable to zero if and only if equality* (1.8.26) *is satisfied.*

Proof:

Corollary 1.8.7 immediately follows from Corollary 1.8.5 and 1.8.6, since it is always possible to find a closed ball centred at zero which is entirely contained in a given set U. ■

Theorem 1.8.2 and the resulting corollaries give the conditions for global U-controllability to the target set S of the form (1.8.1) (i.e., S being a linear manifold in the space $\boldsymbol{R}^n$). An insignificant modification in the proof of Theorem 1.8.2 leads to the condition of global U-controllability to the target set which is an arbitrary compact and convex set in $\boldsymbol{R}^n$.

THEOREM 1.8.3 (Barmish and Schmitendorf (1980)). *Suppose the set U is compact and the set S compact and convex. Then the following condition is necessary for the dynamical system* (1.2.1) *to be globally U-controllable to the target set S*:

$$\sup_{t \geqslant t_0} V(\mathbf{z}(t_0), t) = +\infty \tag{1.8.27}$$

for all nonzero initial condition $\mathbf{z}(t_0) \in \boldsymbol{R}^n$ of the adjoint system (1.8.13), *where the scalar function V: $\boldsymbol{R}^n \times [t_0, \infty) \rightarrow \boldsymbol{R}$ is defined by the formula*

$$V(\mathbf{z}(t_0), t) = \int_{t_0}^{t} \max\{\mathbf{w}^T\mathbf{B}^T(\tau)\mathbf{z}(\tau):\ \mathbf{w} \in U\}d\tau - \{\inf \mathbf{x}^T\mathbf{z}(t):\ \mathbf{x} \in S\}. \tag{1.8.28}$$

Moreover, the following condition is sufficient for the dynamical system (1.2.1) *to be globally U-controllable to the target set S*:

$$\sup_{t \geqslant t_0} W(t) = +\infty, \tag{1.8.29}$$

where the scalar function W: $[t_0, \infty) \to \mathbf{R}$ *is defined as*

$$W(t) = \min\{V(\mathbf{z}(t_0), t) \cdot \; ||\mathbf{z}(t_0)|| = 1\}. \tag{1.8.30}$$

Proof:

Substituting $\mathbf{F}^{\mathrm{T}}(t, t_0)\mathbf{L}^{\mathrm{T}}\mathbf{a} = \mathbf{z}(t)$ (hence $\mathbf{z}(t_0) = \mathbf{L}^{\mathrm{T}}\mathbf{a}$) and using the proof of Theorem 1.8.2, we obtain the desired result. ■

COROLLARY 1.8.8 (Barmish and Schmitendorf (1980)). *Suppose the set U is compact and contains the point zero and, moreover, the transition matrix* $\mathbf{F}(t, t_0)$ *is uniformly bounded. Then the dynamical system* (1.2.1) *is globally U-controllable if and only if*

$$\int_{t_0}^{\infty} \max\{\mathbf{w}^{\mathrm{T}}\mathbf{B}^{\mathrm{T}}(\tau)\mathbf{z}(\tau) \colon \; \mathbf{w} \in U\} d\tau = +\infty \tag{1.8.31}$$

for all nonzero trajectories $\mathbf{z}(\cdot)$ *of the adjoint system* (1.8.13).

Proof:

Necessity. Since global U-controllability implies global U-controllability to zero, the necessary condition immediately follows from Corollary 1.8.4.

Sufficiency. In order to prove the sufficient condition we shall use the sufficient condition (1.8.29) from Theorem 1.8.3 with the modified target set $S = \mathbf{a} \in \boldsymbol{R}^n$. The proof will proceed by contradiction. Suppose condition (1.8.29) is not satisfied, i.e., $\sup_{t \geqslant t_0} W(t) \leqslant k$ for $k \in (0, \infty)$. Thus by (1.8.30) there exist an increasing sequence $t_i \to \infty$ as $i \to \infty$ and a sequence of the initial states $\{\mathbf{z}_i(t_0)\}$, $i = 1, 2, \ldots$ such that $||\mathbf{z}_i(t_0)|| = 1$ and $V(\mathbf{z}_i(t_0), t_i) \leqslant k$. Let $M = \sup_{t \geqslant t_0} ||\mathbf{F}(t, t_0)||$. Since the matrix $\mathbf{F}(t, t_0)$ is uniformly bounded by assumption, we have $M < \infty$. Hence by (1.8.28)

$$\begin{aligned} V(\mathbf{z}_i(t_0), t_i) + \inf_i\{\mathbf{a}^{\mathrm{T}}\mathbf{z}_i(t_i)\} &= \int_{t_0}^{t_i} \max\{\mathbf{w}^{\mathrm{T}}\mathbf{B}^{\mathrm{T}}(\tau)\mathbf{z}_i(\tau) \colon \; \mathbf{w} \in U\} d\tau \\ &\leqslant k + \mathbf{a}^{\mathrm{T}}\mathbf{z}_i(t_i) \leqslant k + ||\mathbf{F}(t_i, t_0)|| \; ||\mathbf{z}_i(t_0)|| \; ||a|| = k + M||\mathbf{a}||. \end{aligned} \tag{1.8.32}$$

Since $\mathbf{z}_i(t_0)$, $i = 1, 2, \ldots$ is an infinite sequence in the compact set, there must be a subsequence which converges to a certain $\mathbf{z}(t_0) \in \boldsymbol{R}^n$ such that for a certain $\bar{t} > t_0$ we have

$$\int_{t_0}^{\bar{t}} \max\{\mathbf{w}^{\mathrm{T}}\mathbf{B}^{\mathrm{T}}(\tau)\mathbf{z}(\tau) \colon \; \mathbf{w} \in U\} d\tau \leqslant k + M||\mathbf{a}||. \tag{1.8.33}$$

This clearly contradicts condition (1.8.31). ■

COROLLARY 1.8.9 (Barmish and Schmitendorf (1980)). *Suppose that the hypotheses of Corollary* 1.8.8 *are satisfied. Then the dynamical system* (1.2.1) *is globally U-controllable to zero if and only if it is globally U-controllable.*

Proof:

Corollary 1.8.9 follows immediately from Corollaries 1.8.4 and 1.8.8. ∎

COROLLARY 1.8.10 (Barmish and Schmitendorf (1980)). *Suppose that the hypotheses of Corollary* 1.8.8 *are satisfied and the target set S is convex, closed, symmetric with respect to zero and positively invariant with respect to the transition matrix* $\mathbf{F}(t, t_0)$ *for* $t \geqslant t_0$. *Then the necessary condition* (1.8.27) *is equivalent to the sufficient condition* (1.8.29). *Hence the dynamical system* (1.2.1) *is globally U-controllable to the set S if and only if*

$$\sup_{t \geqslant t_0} V(\mathbf{z}(t_0), t) = +\infty \tag{1.8.34}$$

for all nonzero initial conditions $\mathbf{z}(t_0) \in \mathbf{R}^n$ *of the adjoint system* (1.8.13).

Proof:

By Corollary 1.8.8 it follows that the necessary condition (1.8.29) implies the necessary condition (1.8.27). The proof of the converse implication is similar to that of Corollary 1.8.8, i.e., by contradiction. Preserving the notation from the proof of Corollary 1.8.8, we have

$$\int_{t_0}^{t_i} \max\{\mathbf{w}^{\mathrm{T}}\mathbf{B}^{\mathrm{T}}(\tau)\mathbf{z}_i(\tau):\ \mathbf{w} \in U\}d\tau - \inf\{\mathbf{x}^{\mathrm{T}}\mathbf{z}_i(t_i):\ \mathbf{x} \in S\} \leqslant k, \tag{1.8.35}$$

$$i = 1, 2, \ldots$$

Next, choosing suitably the convergent subsequence of initial conditions $\mathbf{z}_i(t_0) \to \mathbf{z}(t_0)$, we can prove that for arbitrary but fixed $\bar{t} \in [t_0, \infty)$ we have

$$V(\mathbf{z}(t_0), \bar{t}) \leqslant 2k \tag{1.8.36}$$

which contradics condition (1.8.34). Hence our corollary follows. ∎

Example 1.8.1 (Schmitendorf and Barmish (1980)).

The following example illustrates the application of Corollary 1.8.4 to the investigation of the global U-controllability to zero of a nonautonomous linear dynamical system. We consider the time-varying two-dimensional dynamical system described by

$$\begin{aligned} \dot{x}_1(t) &= u(t)\sin t, \\ \dot{x}_2(t) &= \frac{1}{(t+1)^2}\, x_1(t) + u(t)\, t \sin t, \end{aligned} \qquad t \in [0, \infty). \tag{1.8.37}$$

The control constraint set is taken to be $U = [0, 1]$. By straightforward computation, the state transition matrix for the adjoint system (1.8.13) is found to be

$$\mathbf{F}(t, t_0) = \begin{bmatrix} 1 & \dfrac{t-t_0}{(t+1)(t_0+1)} \\ 0 & 1 \end{bmatrix}.$$

Hence, in accordanec with Corollary 1.8.4, the dynamical system (1.8.37) is globally U-controllable to zero at $t_0 = 0$ if and only if

$$\int_0^{\infty} \left(\sup_{w \in [0,1]} w[\sin\tau, \ \tau\sin\tau] \begin{bmatrix} 1 & \dfrac{\tau}{\tau+1} \\ 0 & 1 \end{bmatrix} \begin{bmatrix} z_{01} \\ z_{02} \end{bmatrix} \right) d\tau = +\infty \tag{1.8.38}$$

for all nonzero initial conditions $\mathbf{z}(0) = \mathbf{z}_0 = [z_{01} \ z_{02}]^{\mathrm{T}}$. This implies the following requirement:

$$\int_0^{\infty} I(\tau)\,d\tau = \int_0^{\infty} \max\left\{0, \ z_{01}\sin\tau + z_{02}\,\tau\sin\tau\left(1 + \frac{1}{\tau+1}\right)\right\} d\tau = +\infty \tag{1.8.39}$$

for all nonzero initial conditions $\mathbf{z}(0) = \mathbf{z}_0 \neq 0$. Hence the necesary and sufficient condition for the global U-controllability to zero of the dynamical system (1.8.37) is that the formula (1.8.39) should be satisfied for all nonzero initial conditions $\mathbf{z}_0 \neq 0$. We shall show that this condition is indeed satisfied.

Case 1. $z_{01} \neq 0$ and $z_{02} = 0$. For this case, we have

$$\int_0^{\infty} I(\tau)\,d\tau = \int_0^{\infty} \max\{0, \ z_{01}\sin\tau\}\,d\tau = \int_{I_1} z_{01}\sin\tau\,d\tau,$$

where $I_1 = \{t \geqslant 0\colon \ z_{01}\sin\tau > 0\}$. Since the range set I_1 of integration is the union of infinitely many intervals of length π, it follows that

$$\int_0^{\infty} I(\tau)\,d\tau = +\infty.$$

Case 2. z_{01} = anything, $z_{02} \neq 0$. Let $\bar{t} = (|z_{01}|+1)/|z_{02}|$. Then to verify (1.8.39) it sufficies to show that

$$\int_{I_2} I(\tau)\,d\tau = +\infty,$$

where $I_2 = \{\tau \geqslant \bar{t}\colon \ z_{02}\sin\tau > 0\}$. (Recall that the integrand is nonnegative.) Now, for $\tau \in I_2$, we notice that the integrand $I(\tau)$ can be bounded from below as follows:

$$\begin{aligned} z_{01}\sin\tau + z_{02}\,\tau\sin\tau\left(1+\frac{1}{\tau+1}\right) &\geqslant |z_{02}||\sin\tau|\tau\left(1+\frac{1}{\tau+1}\right) - |z_{01}||\sin\tau| \\ &\geqslant (|z_{02}|\tau - |z_{01}|)|\sin\tau| \\ &\geqslant (|z_{02}|\bar{t} - |z_{01}|)|\sin\tau| = |\sin\tau|. \end{aligned}$$

Hence

$$\int_{I_2} I(\tau)d\tau \geqslant \int_{I_2} |\sin\tau|d\tau = +\infty$$

because the range of integration is once again the union of infinitely many intervals of length π. We conclude that the dynamical system (1.8.37) is globally U-controllable to zero.

Example 1.8.2 (Barmish and Schmitendorf (1980), Schmitendorf and Barmish (1981)). Let $x(t)$ and $u(t)$ be scalars and suppose that the dynamical system (1.2.1) is the standard double integrator described by

$$\begin{aligned} \dot{x}_1(t) &= x_2(t), \\ \dot{x}_2(t) &= u(t), \end{aligned} \qquad t \in [0, \infty) \tag{1.8.40}$$

with an unbouded target set $S = \{(x_1, x_2) \in \boldsymbol{R}^2\colon x_1 + x_2 = 1\}$ and admissible control set $U = [-1, +1]$. Hence the target set S is of the form (1.8.1) with $c = 1$ and $L = [-1+1]$. Even for such a simple example, neither the uniform boundedness condition nor the positive invariance condition holds. Consequently, the theory presented in Corollaries 1.8.8, 1.8.9, and 1.8.10 does not apply. Hence, in order to investigate the global U-controllability to the target set S of the dynamical system (1.8.40), we shall use Theorem 1.8.2.

In accordance with Theorem 1.8.2, we must compute the function $V(\mathbf{L}^{\mathrm{T}}\mathbf{a}, t)$. By (1.8.17) we have

$$\begin{aligned} V(\mathbf{L}^{\mathrm{T}}\mathbf{a}, t) &= \int_0^t \max\{w(\tau+1)a\colon w \in [-1, +1]\}d\tau - \inf\{(-x_1 + x_2 t + \\ &\quad + x_2)a\colon -x_1 + x_2 = 1\} \\ &= a\left(\frac{t^2}{2} + \right) - \inf\{(x_2 t + 1)a\colon x_2 \in \boldsymbol{R}\}. \end{aligned}$$

Since $\inf\{(x_2 t+1)a\colon x_2 \in \boldsymbol{R}\} = +\infty$ for all real numbers $a \neq 0$, we have $V(\mathbf{L}^{\mathrm{T}}a, t) = +\infty$ for all $a \neq 0$. Thus we conclude that the necessary condition (1.8.17) in Theorem 1.8.2 is satisfied and the dynamical system (1.8.40) is globally U-controllable to the set S.

In order to test the sufficient condition (1.8.18) in Theorem 1.8.2, we first compute the function $V(\mathbf{F}^{\mathrm{T}}(t, 0)\mathbf{L}^{\mathrm{T}}\mathbf{a}, t)$

$$\begin{aligned} V(\mathbf{F}^{\mathrm{T}}(t, 0)\mathbf{B}^{\mathrm{T}}\mathbf{a}, t) &= |a| \int_0^t |\tau + 1 - t| d\tau - a \\ &= \begin{cases} |a|(t - t^2/2) - a & \text{for } 0 \leqslant t \leqslant 1, \\ |a|(t^2/2 + 1 - t) - a & \text{for } t \geqslant 1. \end{cases} \end{aligned}$$

Thus for $t \in [0, 1]$ the function $W(t)$ is of the form

$$\begin{aligned} W(t) &= \min\{|a|(t - t^2/2) - a\colon a^2(1 + (1-t)^2) = 1\} \\ &= (1 + (1-t)^2)^{-0.5}(t - t^2/2 - 1). \end{aligned}$$

However, for $t \geqslant 1$ we have

$$W(t) = \min\{|a|(t^2/2+1-t)-a\colon\ a^2(1+(1-t)^2) = 1\}$$
$$= (1+(1-t)^2)^{-0.5}(t^2/2-t),$$

Since $\lim_{t\to\infty} W(t) = +\infty$, we conclude by (1.8.18) that the dynamical system (1.8.40) is globally U-controllable to the target set S.

The dynamical system (1.8.40) has also been studied in the paper of Barmish and Schmitendorf (1980) with a slight different type of the target set S (rectangular on the plane) and another set of admissible controls U (a sector without the point zero). All these special cases can be analysed by using Theorem 1.8.2 and subsequent corollaries.

By Definitions 1.8.4 and 1.8.5 it immediately follows that controllability with constrained controls is strongly related to the notion of the so-called attainable sets of the dynamical system (1.2.1). Next we shall present the main qualitative properties of attainable sets, such as algebraic structure, topological structure and changeability in time.

THEOREM 1.8.4 (Lee and Markus (1968), Warga (1972)). *Suppose the set U is convex. Then the attainable set $K_U([t_0, t_1], \mathbf{x}_0)$ is also convex for all $t_1 \geqslant t_0$ and each $\mathbf{x}_0 \in \mathbf{R}^n$.*

Proof:

Let $\mathbf{y}_1, \mathbf{y}_2 \in K_U([t_0, t_1], \mathbf{x}_0)$ and $r \in [0, 1]$. By Definition 1.8.4 and relation (1.8.2) there exist admissible controls $\mathbf{u}_1, \mathbf{u}_2 \in M(U)$ such that

$$\mathbf{y}_1 = \mathbf{F}(t_1, t_0)\mathbf{x}_0 + \int_{t_0}^{t_1} \mathbf{F}(t_1, t)\mathbf{B}(t)\mathbf{u}_1(t)\,dt, \tag{1.8.41}$$

$$\mathbf{y}_2 = \mathbf{F}(t_1, t_0)\mathbf{x}_0 + \int_{t_0}^{t_1} \mathbf{F}(t_1, t)\mathbf{B}(t)\mathbf{u}_2(t)\,dt. \tag{1.8.42}$$

Since the set U is convex, the set $M(U)$ is also convex in the linear space of measurable functions defined in $[t_0, t_1]$. Thus by (1.8.2), (1.8.41) and (1.8.42) we have

$$r\mathbf{y}_1 + (1-r)\mathbf{y}_2 = \mathbf{F}(t_1, t_0)\mathbf{x}_0 + \int_{t_0}^{t_1} \mathbf{F}(t_1, t)\mathbf{B}(t)\big(r\mathbf{u}_1(t) +$$
$$+ (1+r)\mathbf{u}_2(t)\big)dt \in K_U([t_0, t_1], \mathbf{x}_0).$$

Hence the attainable set $K_U([t_0, t_1], \mathbf{x}_0)$ is convex for each $\mathbf{x}_0 \in \mathbf{R}$ and all $t_1 \geqslant t_0$ ($\mathbf{x}_0$ and t_1 are arbitrary in the above considerations). ∎

COROLLARY 1.8.11 (Lee and Markus (1968), Warga (1972)). *Suppose the set U is absolutely convex. Then the attainable set $K_U([t_0, t_1], 0)$ is also absolutely convex.*

Moreover, the attainable set $K_U([t_0, t_1], \mathbf{x}_0)$ is symmetric with respect to the point $\mathbf{F}(t_1, t_0)\mathbf{x}_0$ for each $\mathbf{x}_0 \in \mathbf{R}^n$ and all $t_1 \geqslant t_0$.

Proof:

By (1.8.2) changing the sign of the control, we immediately conclude that for $\mathbf{x}_0 = 0$ the inclusion $\mathbf{y} \in K_U([t_0, t_1], 0)$ implies $-\mathbf{y} \in K_U([t_0, t_1], 0)$. For $\mathbf{x}_0 \neq 0$ we conclude that by shifting the set $K_U([t_0, t_1], 0)$ along the vector $\mathbf{F}(t_1, t_0)\mathbf{x}_0$ we do not change its symmetry and, moreover, such shift by (1.8.2) gives the set $K_U([t_0, t_1], \mathbf{x}_0)$. Since the above considerations do not depend on the length of the time interval, our corollary follows. ∎

Subsequently we shall show that the assumption about the convexity of the set U is not needed for the convexity of the attainable set $K_U([t_0, t_1], \mathbf{x}_0)$.

THEOREM 1.8.5 (Lee and Markus (1968), Warga (1972)). *Suppose the set U is convex and the matrix function $\mathbf{B}(t)$ is bounded in $[t_0, t_1]$. Then the attainable set $K_U([t_0, t_1], \mathbf{x}_0)$ is bounded for each $\mathbf{x}_0 \in \mathbf{R}^n$ and all $t_1 \geqslant t_0$.*

Proof:

Let $\mathbf{y} \in K_U([t_0, t_1], \mathbf{x}_0)$ be an arbitrary point of the attainable set. By (1.8.2) and the definition of the attainable set, there exists a control $\mathbf{u} \in M(U)$ such that

$$\mathbf{y} = \mathbf{F}(t_1, t_0)\mathbf{x}_0 + \int_{t_0}^{t_1} \mathbf{F}(t_1, t)\mathbf{B}(t)\mathbf{u}(t)\,dt.$$

Since the set U is bounded, we have $\|\mathbf{u}\| \leqslant K$ for any $\mathbf{u} \in M(U)$. Since the matrix function $\mathbf{B}(t)$ is bounded in $[t_0, t_1]$ and the transition matrix $\mathbf{F}(t_1, t)$ is continuous for all $t \in \mathbf{R}$, the integral in (1.8.2) is bounded and hence the set $K_U([t_0, t_1], \mathbf{x}_0)$ is also bounded. ∎

THEOREM 1.8.6 (Lee and Markus (1968), Warga (1972)). *If the set U is compact, then the attainable set $K_U([t_0, t_1], \mathbf{x}_0)$ is also compact.*

Proof:

Let the sequence $\{\mathbf{y}_n\} \in K_U([t_0, t_1], \mathbf{x}_0)$ be convergent to $\bar{\mathbf{y}} \in \mathbf{R}^n$. Then by (1.8.2) for every natural number n there exists an admissible control $\mathbf{u}_n \in M(U)$ such that

$$\mathbf{y}_n = \mathbf{F}(t_1, t_0)\mathbf{x}_0 + \int_{t_0}^{t_1} \mathbf{F}(t_1, t)\mathbf{B}(t)\mathbf{u}_n(t)\,dt. \tag{1.8.43}$$

Since the set U is compact (hence closed and bounded), $M(U)$ is bounded in the space of measurable functions. Hence, by the Weierstrass theorem, from the sequence $\{\mathbf{u}_n\}$ we can choose a convergent subsequence which converges to the function $\bar{\mathbf{u}} \in M(\mathbf{u})$. Hence by (1.8.43) we obtain

$$\bar{\mathbf{y}} = \mathbf{F}(t_1, t_0)\mathbf{x}_0 + \int_{t_0}^{t_1} \mathbf{F}(t_1, t)\mathbf{B}(t)\bar{\mathbf{u}}(t)dt \in K_U([t_0, t_1], \mathbf{x}_0).$$

Thus the set $K_U([t_0, t_1], \mathbf{x}_0)$ is closed. Since the set U is bounded, by Theorem 1.8.4 $K_U([t_0, t_1], \mathbf{x}_0)$ is also bounded and thus compact. ∎

COROLLARY 1.8.12 (Lee and Markus (1968, Ch. 2.2, Th. 1)). *Suppose the assumptions of Theorem* 1.8.6 *are satisfied. Then the attainable set* $K_U([t_0, t_1], \mathbf{x}_0)$ *is continuously dependent on* t_1 *for all* $t_1 \geqslant t_0$.

Proof:

The detailed proof of this corollary is based on the topological properties of the sets in R^n and is given in the book of Lee and Markus (1968, Ch. 2.2, Th. 1). ∎

THEOREM 1.8.7 (Lee and Markus (1968)). *Suppose the set U is compact and* $U_0 \subset U$ *is its compact subset with a convex hull* $\mathrm{CH}(U_0) = U$. *Then the attainable set* $K_U([t_0, t_1], \mathbf{x}_0)$ *is compact and convex and moreover, we have*

$$K_U([t_0, t_1], \mathbf{x}_0) = K_{U_0}([t_0, t_1], \mathbf{x}_0). \tag{1.8.44}$$

Proof:

The compacteness of the set $K_U([t_0, t_1], \mathbf{x}_0)$ immediately follows from Theorem 1.8.6, whose assumptions are all satisfied. The convexity of $K_U([t_0, t_1], \mathbf{x}_0)$ and equality (1.8.44) are proved in detail in Theorem 1A and Corollary 1A in the book of Lee and Markus (1968, Ch. 2). ∎

COROLLARY 1.8.13 (Lee and Markus (1968)). *Suppose the assumptions of Theorem* 1.8.7 *are satisfied. Then the set* $K_{U_0}([t_0, t_1], \mathbf{x}_0)$ *depends continuously on* t_1 *for all* $t_1 \geqslant t_0$.

Proof:

Corollary 1.8.13 follows from equality (1.8.44) and Corollary 1.8.12. ∎

THEOREM 1.8.8. *The multifunction* $t \to K_U([t_0, t], \mathbf{x}_0)$ *has the semigroup property, i.e., for every* $t_2 > t_1 > t_0$ *we have*

$$K_U([t_0, t_2], \mathbf{x}_0) = \bigcup_{\mathbf{x} \in K_U([t_0, t_1], \mathbf{x}_0)} K_U([t_1, t_2], \mathbf{x}). \tag{1.8.45}$$

Proof:

The inclusion $K_U([t_0, t_2], \mathbf{x}_0) \subset \bigcup_{\mathbf{x} \in K_U([t_0, t_1], \mathbf{x}_0)} K_U([t_1, t_2], \mathbf{x})$ is obvious. The converse inclusion will be proved by contradiction. Let us assume that the converse inclusion does not hold. Then there exists an $\bar{\mathbf{x}} \in K_U([t_0, t_1], \mathbf{x}_0)$ and an admissible control $\bar{\bar{\mathbf{u}}}(t)$ defined in $[t_1, t_2]$ such that the trajectory of the dynamical system (1.2.1) with the initial state $\bar{\bar{\mathbf{x}}}(t_1) = \bar{\mathbf{x}}$ and corresponding to the control $\bar{\bar{\mathbf{u}}}(t)$ has the property $\bar{\bar{\mathbf{x}}}(t_2, \bar{\mathbf{x}}, \bar{\bar{\mathbf{u}}}) \notin K_U([t_0, t_2], \mathbf{x}_0)$. Since $\bar{\mathbf{x}} \in K_U([t_0, t_1], \mathbf{x}_0)$, then there exists a control $\bar{\mathbf{u}} \in M(U)$ such that $\bar{\mathbf{x}}(t_1, \mathbf{x}_0, \bar{\mathbf{u}}) = \bar{\mathbf{x}}$. Moreover, the control $\mathbf{u}(t)$ given by

$$\mathbf{u}(t) = \begin{cases} \bar{\mathbf{u}}(t) & \text{for } t \in [t_0, t_1], \\ \bar{\bar{\mathbf{u}}}(t) & \text{for } t \in (t_1, t_2] \end{cases}$$

is an admissible control in $[t_0, t_2]$ and the trajectory corresponding to this control satisfies the condition $\mathbf{x}(t_2, \mathbf{x}_0, \mathbf{u}) = \bar{\bar{\mathbf{x}}}(t_2, \bar{\mathbf{x}}, \bar{\bar{\mathbf{u}}}) \in K_U([t_0, t_2], \mathbf{x}_0)$. This is a contradiction, and hence we conclude that equality (1.8.45) is true. ∎

COROLLARY 1.8.14 (Warga (1972)). *If for every $t_1 \geqslant t_0$ and each $\mathbf{x} \in R^n$ there exists a control $\mathbf{u} \in M(U)$ such that $\mathbf{A}(t)\mathbf{x}+\mathbf{B}(t)\mathbf{u}(t) = 0$ for $t \in [t_0, t_1]$, then the multifunction $t \to K_U([t_0, t], \mathbf{x}_0)$ is increasing.*

Proof:

Let $\bar{\mathbf{x}} \in K_U([t_0, t_1], \mathbf{x}_0)$. Let us choose a control $\bar{\mathbf{u}}(t)$, $t \in [t_1, t_2]$, such that $\mathbf{A}(t)\bar{\mathbf{x}}+\mathbf{B}(t)\bar{\mathbf{u}}(t) = 0$ for $t \in [t_1, t_2]$. Then we have $\mathbf{x}(t, \bar{\mathbf{x}}, \bar{\mathbf{u}}) = \bar{\mathbf{x}}$ for $t \in [t_1, t_2]$. Thus $\bar{\mathbf{x}} \in K_U([t_0, t_2], \mathbf{x}_0)$ and

$$K_U([t_0, t_1], \mathbf{x}_0) \subset K_U([t_0, t_2], \mathbf{x}_0) \quad \text{for } t_2 \geqslant t_1. \tag{1.8.46}$$

Then we conclude that the multifunction $t \to K_U([t_0, t], \mathbf{x}_0)$ is increasing.

THEOREM 1.8.9 (Warga (1972)). *Suppose that $\mathbf{x}_0 = 0$ and $Q \subset R^m$ is a compact and convex set which contains zero as an interior point. If for every such set $Q \subset U \subset R^m$ and all $t > \tau$ the following inclusion is satisfied:*

$$\mathbf{F}(t, \tau)Q \supset Q, \tag{1.8.47}$$

then the multifunction $t \to K_U([t_0, t], 0)$ is increasing, i.e., for every $t_1 \leqslant t_2$ we have

$$K_U([t_0, t_1], 0) \subset K_U([t_0, t_2], 0). \tag{1.8.48}$$

Proof:

Theorem 1.8.9 immediately follows from the relations

$$\begin{aligned} K_U([t_0, t_2], 0) &= \{\mathbf{x} \in R^n \colon \mathbf{x} = \int_{t_0}^{t_2} \mathbf{F}(t_2, t)\mathbf{B}(t)\mathbf{u}(t)dt \colon \mathbf{u} \in M(U)\} \\ &\supset \left\{\mathbf{x} \in R^n \colon \mathbf{x} = \mathbf{F}(t_2, t_1) \int_{t_0}^{t_1} \mathbf{F}(t_1, t)\mathbf{B}(t)\mathbf{u}(t)dt \colon \mathbf{u} \in M(U)\right\} \\ &= \mathbf{F}(t_2, t_1)\left(K_U([t_0, t_1], 0)\right) \supset K_U([t_1, t_0], 0). \ \blacksquare \end{aligned}$$

THEOREM 1.8.19 (Lee and Markus (1968)). *Suppose the set U has a nonempty interior and the rows of the matrix* $\mathbf{F}(t_1, t)\mathbf{B}(t)$ *are linearly independent over the time interval* $[t_0, t_1]$. *Then the attainable set* $K_U([t_0, t_1], \mathbf{x}_0)$ *has a nonempty interior.*

Proof:

If $U = R^m$, then by Theorem 1.3.1 the linear independence of the rows of the matrix $\mathbf{F}(t_1, t)\mathbf{B}(t)$ over $[t_0, t_1]$ is equivalent to the controllability of the dynamical system (1.2.1) in $[t_0, t_1]$. Thus the attainable set $K_U([t_0, t_1], \mathbf{x}_0)$ is equal to R^n and hence of course has a nonempty interior. In the case where $U \subsetneqq R^m$ the attainable set $K_U([t_0, t_1], \mathbf{x}_0)$ also has a nonempty interior. This is a consequence of the facts that the set U has a nonempty interior and the multifunction $t \to K_U([t_0, t_1], \mathbf{x}_0)$ is conntiuous. ■

It is of practical interest to investigate the so-called *boundary controls*. This leads to the bang-bang principle considered in detail in, e.g., Artstein (1980), Lee and Markus (1968), Warga (1972).

The bang-bang principle in linear control systems is stronly related to the problems of time-optimal controls.

1.9 Controllability of time-inavriant dynamical systems with constrained controls

All the results obtained in the Section 1.8 for time-varying dynamical systems can also be applied to time-invariant dynamical systems of the form (1.2.8). However, it should be pointed out that for dynamical systems (1.2.8) some special results concerning various kinds of controllability can be derived by using different methods. These results will be presented in Section 1.9.

THEOREM 1.9.1 (Brammer (1972), Heymann and Stern (1975), Schmitendorf and Barmisch (1980)). *The dynamical system* (1.2.8) *is globally U-controllable to zero if and only if all the following conditions are satisfied simultaneously*:

(1) *there exists a* $\mathbf{w} \in U$ *such that* $\mathbf{B}\mathbf{w} = 0$,

(2) *the convex hull* $\mathrm{CH}(U)$ *has a nonempty interior in the space* R^m,

(3) $\mathrm{rank}[\mathbf{B} \vdots \mathbf{A}\mathbf{B} \vdots \mathbf{A}^2\mathbf{B} \vdots \ldots \vdots \mathbf{A}^{n-1}\mathbf{B}] = n$,

(4) *there is no real eigenvector* $\mathbf{v} \in R^n$ *of* $\mathbf{A}^T$ *satisfying* $\mathbf{v}^T\mathbf{B}\mathbf{w} \leqslant 0$ *for all* $\mathbf{w} \in U$,

(5) *no eigenvalue of* $\mathbf{A}$ *has a positive real part.*

Proof:

There exist three versions of the proof of Theorem 1.9.1 given in the papers mentioned above. The proof presented in the paper of Brammer (1972) is entirely based on the theory of almost periodic functions and on the investigation of the scalar product of the form $\mathbf{v}^T\exp(\mathbf{A}t)\mathbf{B}\mathbf{w}$. The proof given in the paper of Heymann and Stern (1975) is based on the geometrical properties of the attainable sets and on some theorems from topology. Finally, the proof presented in the paper of Schmitendorf and Barmish (1980) utilizes some results given in Section 1.8, particularly Corollary 1.8.3. All these three versions of the proof of Theorem 1.9.1 are rather long, tedious and laborious, and so they will be omitted here. ■

COROLLARY 1.9.1 (Brammer (1972), Heymann and Stern (1975), Schmitendorf and Barmish (1980)). *The dynamical system* (1.2.8) *is locally U-controllable to zero if and only if the conditions* (1), (2), (3) *and* (4) *of Theorem* 1.9.1 *are simultaneously satisfied.*

Proof:

Condition (5) in Theorem 1.9.1 is used only for steering the dynamical system (1.2.8) into the neighbourhood of the point zero. The remaining conditions guarantee the existence of a neighbourhood of zero from which we can reach exactly the point zero. ■

COROLLARY 1.9.2 (Brammer (1972)). *Suppose the set U contains zero as its interior point. Then the dynamical system* (1.2.8) *is locally U-controllable to zero if and only if*

$$\operatorname{rank}[\mathbf{B} \vdots \mathbf{AB} \vdots \mathbf{A}^2\mathbf{B} \vdots \dots \vdots \mathbf{A}^{n-1}\mathbf{B}] = n. \tag{1.9.1}$$

Proof:

Since the set U contains zero as its interior point, the assumptions (1) and (2) of Theorem 1.9.1 are satisfied. Moreover, there exists a symmetric neighbourhood $V \subset U$ of the point zero. Hence, if $\mathbf{v}^T\mathbf{Bw} \leqslant 0$ for all $\mathbf{w} \in U$, then $\mathbf{v}^T\mathbf{Bw} = 0$ for each $\mathbf{w} \in V$. Thus, if $\mathbf{v}$ is the eigenvector of $\mathbf{A}^T$, then $\exp(st)\mathbf{v}^T\mathbf{Bw} = \exp(\mathbf{A}^T\mathbf{v})^T\mathbf{Bw} = \mathbf{v}^T\exp(\mathbf{A}^T)\mathbf{Bw} = 0$ for every $\mathbf{w} \in V$. Since the neighbourhood V is an open set in the space $\mathbf{R}^m$, we have $\operatorname{rank}[\mathbf{B} \vdots \mathbf{AB} \vdots \mathbf{A}^2\mathbf{B} \vdots \dots \vdots \mathbf{A}^{n-1}\mathbf{B}] < n$. Thus, if the point zero is an interior point in the set U and if the condition (1.9.1) is satisfied, then the condition (4) of Theorem 1.9.1 is also satisfied. Hence by Corollary 1.9.1 the dynamical system (1.2.8) is locally U-controllable to zero. ■

COROLLARY 1.9.3 (Brammer (1972)). *Suppose the $m = 1$ and $U = [0, 1]$. Then the dynamical system* (1.2.8) *is globally U-controllable to zero if and only if condition* (1.9.1) *is satisfied and matrix* $\mathbf{A}$ *has only complex eigenvalues.*

Proof:

By contradiction. Suppose $m = 1$, $U = [0, 1]$ and the matrix $\mathbf{A}^T$ has a real eigenvector $\mathbf{v}$. Then $\mathbf{v}^T\mathbf{B}w = \mathbf{v}^T\mathbf{b}w$ (since $\mathbf{B} = \mathbf{b}$). Moreover, either $\mathbf{v}^T\mathbf{b} \leqslant 0$ or $-\mathbf{v}^T\mathbf{b} \leqslant 0$. Thus either $\mathbf{v}^T\mathbf{B}w \leqslant 0$ or $-\mathbf{v}^T\mathbf{Bw} \leqslant 0$ for each $w \in U = [0, 1]$. Thus, if the matrix $\mathbf{A}$ has a real eigenvalue s corresponding to the real eigenvector $\mathbf{v}$, then condition (4) of Theorem 1.9.1 is not satisfied and hence the dynamical system

(1.2.8) is not locally U-controllable to zero. Thus the dynamical system (1.2.8) is locally U-controllable to zero if and only if the matrix $\mathbf{A}$ has only complex eigenvalues. ∎

COROLLARY 1.9.4 (Brammer (1972)). *Suppose the set U is a cone with vertex at zero and a nonempty interior in the space R^m. Then the attainable set $K_U(0)$ is equal to R^n if and only if the conditions* (3) *and* (4) *of Theorem* 1.9.1 *are satisfied.*

Proof:

If conditions (3) and (4) of Theorem 1.9.1 are not satisfied, then it is easy to show that no neighbourhood of zero is reachable. We shall show that if conditions (3) and (4) of Theorem 1.9.1 are satisfied, then $K_U(0) = R^n$. Since U is a cone with vertex at zero, for an admissible control $\mathbf{u}(\cdot)$ the control $k\mathbf{u}(\cdot)$ is also admissible for every $k \geqslant 0$. The attainable set $K_U(0)$ is a convex set containing zero as an interior point and it is a cone with vertex at zero (since it is linear with respect to the controls $\mathbf{u}(\cdot)$). Thus the attainable set must be the whole space R^n. ∎

COROLLARY 1.9.5 (Saperstone (1973), Saperstone and Yorke (1971)). *Suppose $m = 1$ and $U = [0, \infty)$. Then the attainable set $K_U(0)$ equals R^n if and only if the condition* (1.9.1) *is satisfied and the matrix $\mathbf{A}$ has only complex eigenvalues.*

Proof:

By Corollary 1.9.3 the condition (4) of Theorem 1.9.1 is equivalent to the statement that matrix $\mathbf{A}$ has only complex eigenvalues. Since $U = [0, \infty)$ is a cone in the space R^m with a nonempty interior and the vertex at zero, by Corollary 1.9.4 we obtain Corollary 1.9.5. ∎

Now we shall show that the minimal time of reaching the neighbourhood of zero for the case of $U = [0, c]$, $c > 0$, is the same as the minimal time needed for reaching all points in the space R^n in the case where $U = [0, \infty)$.

COROLLARY 1.9.5 (Saperstone (1973), Saperstone and Yorke (1971)). *If*

$$T_c = \inf\{t \in R^+ : 0 \in \operatorname{int} K_{[0,c]}([0, t], 0)\},$$
$$T_\infty = \inf\{t \in R^+ : K_{[0,\infty)}([0, t], 0) = R^n\},$$

then $T_c = T_\infty$ for each $c > 0$.

Proof:

By the linearity of relation (1.8.2) we deduce that $cK_{[0,1]}([0,t],0) = K_{[0,c]}([0,t],0)$. Thus $T_\infty \leqslant T_c$. Suppose that $T_\infty < T_c$. Then there exists a $t \in (T_\infty, T_c)$ such that $0 \in \partial K_{[0,c]}([0, t], 0)$, which implies, in turn that $0 \in \partial K_{[0,\infty]}([0, t], 0)$ (the symbol ∂ denotes the boundary of the set). Thus $K_{[0,\infty)}([0, t], 0) \neq R^n$, whence $t \leqslant T_\infty$, which is a contradiction. Thus $T_c = T_\infty$. ∎

It is rather difficult to compute the time T_c (Saperstone (1973), Saperstone and Yorke (1971)). However, in some special cases this can be done relatively easily. For example let us consider a dynamical system which is a model of the mathematical pendulum. The dynamical equation of such a model is as follows (Saperstone and Yorke (1971)):

$$\ddot{x}(t) = -k^2x(t)+u(t),$$

where k is constant and $u \in U = [0, c]$. Let us assume that the initial state $(x, \dot{x})$ equals $(0, 0)$ and we want to steer it to the final state with a negative value of x and a positive value of $\dot{x}$. Intuitively in order to do that, we first shift the pendulum away from the stability point and then wait until it reaches the desired state. In this case $T_c = \pi/k$.

COROLLARY 1.9.6 (Bianchini (1983a, 1983b)). *The dynamical system* (1.2.8) *is locally U-controllable to zero if and only if it is locally* CH(U)*-controllable to zero.*

Proof:

For each $t > 0$ we have $\overline{K_U([0, t], 0)} = \overline{K_{CH(U)}([0, t], 0)}$. Thus we have obtained the desired result. ■

Now we shall present some useful lemmas concerning U-controllability. We shall study the situation where the signs of the coefficients in the state equations (1.2.8) change.

LEMMA 1.9.1 (Bianchini (1983a), Brammer (1972)). *The dynamical system* (1.2.8) *is locally U-controllable to zero if and only if the dynamical system*

$$\dot{\mathbf{x}}(t) = -\mathbf{A}\mathbf{x}(t)-\mathbf{B}\mathbf{u}(t) \tag{1.9.2}$$

is locally U-controllable to zero.

Proof:

Lemma 1.9.1 immediately follows from the equality

$$\begin{aligned} &\operatorname{rank}[\mathbf{B} \vdots \mathbf{AB} \vdots \mathbf{A}^2\mathbf{B} \vdots \dots \vdots \mathbf{A}^{n-1}\mathbf{B}] \\ &= \operatorname{rank}[-\mathbf{B} \vdots -\mathbf{A}(-\mathbf{B}) \vdots (-\mathbf{A})^2(-\mathbf{B}) \vdots \dots \vdots (-\mathbf{A})^{n-1}(-\mathbf{B})]. \ ■ \end{aligned} \tag{1.9.3}$$

LEMMA 1.9.2 (Bianchini (1983a), Brammer (1972)). $0 \in \operatorname{int} K_U(0)$ *if and only if the dynamical system* (1.2.8) *is locally U-controllable to zero.*

Proof:

By Lemma 1.9.1 the local U-controllability to zero of the dynamical system (1.2.8) is equivalent to the local U-controllability to zero of the dynamical system (1.9.2). Thus the control $\mathbf{u}(t)$ steers the dynamical system (1.9.2) from the initial state $\mathbf{x}_0$ to zero at time t_1 if and only if the control $\mathbf{u}(t_1-t)$ steers the dynamical system (1.2.8) from zero to $\mathbf{x}_0$ at time t_1. Thus the attainable set $K_U(0)$ for the dynamical system (1.2.8) contains zero as its interior point if and only if the system is locally U-controllable to zero. ■

LEMMA 1.9.3 (Bianchini (1983a), Brammer (1972)). *The dynamical system* (1.2.8) *is locally U-controllable to zero if and only if the dynamical system*

$$\dot{\mathbf{x}}(t) = -\mathbf{A}\mathbf{x}(t)+\mathbf{B}\mathbf{u}(t) \tag{1.9.4}$$

is locally $(-U)$*-controllable to zero.*

Proof:

Lemma 1.9.3 follows immediately from Lemmas 1.9.1 and 1.9.2. ■

THEOREM 1.9.2 (Brammer (1972), Heymann and Stern (1975), Schmitendorf and Barmish (1980)). *The dynamical system* (1.2.8) *is globally U-controllable if and only if the conditions* (1), (2), (3), (4) *of Theorem* 1.9.1 *are satisfied simultaneously and, moreover, the matrix* **A** *has only the eigenvalues with a zero real part and a nonzero imaginary part.*

Proof:

As in Theorem 1.9.1, there exist three versions of the proof of Theorem 1.9.2. Since all these proofs are very long and tedious, they will be omitted here. However, it should be pointed out that the matrix **A** has in this case dimensions which are even numbers (since all eigenvalues are complex numbers and hence the characteristic polynomial has an even degree). ■

In investigating global U-controllability of the dynamical system (1.2.8) to the sets we can directly use Theorem 1.8.3 and its corollaries. However, it should be stressed that this method requires the verification of many embarrassing assumptions concerning the set S, namely, the symmetry with respect to zero and the positive invariance condition with respect to the transition matrix. For the time-invariant dynamical systems (1.2.8) these assumptions are equivalent to the following condition:

$$\exp(-\mathbf{A}^{\mathrm{T}}\tau)S \subset \exp(-\mathbf{A}^{\mathrm{T}}t)S \quad \text{for all } t \geqslant \tau.$$

However, by Theorem 1.8.3 we can formulate a relatively simple and useful necessary and sufficient condition for the global U-controllability of the dynamical system (1.2.8) to the sets.

THEOREM 1.9.3 (Petersen and Barmish (1983a)). *Assume that*

(1) *S is convex and compact,*

(2) *U is compact with nonempty interior in R^m,*

(3) $\operatorname{rank}[\mathbf{B} \vdots \mathbf{A}\mathbf{B} \vdots \mathbf{A}^2\mathbf{B} \vdots \ldots \vdots \mathbf{A}^{n-1}\mathbf{B}] = n.$

Then the following condition is necessary and sufficient for the global U-controllability of the dynamical system (1.2.8) *to S*:

$$\sup_{t\geq 0} W(t) = +\infty, \tag{1.9.5}$$

where

$$W(t) = \min\{V(\mathbf{z}_0, t): \|\mathbf{z}_0\| = 1\}$$

$$= \min\left\{\int_0^t \max \mathbf{w}^{\mathrm{T}}\mathbf{B}^{\mathrm{T}}\exp(-\mathbf{A}^{\mathrm{T}}\tau)\mathbf{z}_0: \mathbf{w} \in U\right\}d\tau -$$

$$-\inf\{\mathbf{x}^{\mathrm{T}}\exp(-\mathbf{A}^{\mathrm{T}}t)\mathbf{z}_0: \mathbf{x} \in S, \|\mathbf{z}_0\| = 1\}, \tag{1.9.6}$$

and the terms $\exp(-\mathbf{A}^{\mathrm{T}}\tau)\mathbf{z}_0$ *and* $\exp(-\mathbf{A}^{\mathrm{T}}t)\mathbf{z}_0$ *are free solutions of the adjoint dynamical system*

$$\dot{\mathbf{z}}(t) = -\mathbf{A}^{\mathrm{T}}\mathbf{z}(t), \quad t \in [0, \infty) \tag{1.9.7}$$

with initial condition $\mathbf{z}(0) = \mathbf{z}_0 \in \boldsymbol{R}^n$.

Proof:

Sufficiency. Suppose that condition (1.9.5) is satisfied. Then by Theorem 1.8.3 the dynamical system (1.2.8) is globally U-controllable to the set S.

Necessity. By contradiction. Let us assume that condition (1.9.5) is not satisfied, i.e., that $\sup_{t\geq 0} W(t) \leq k < +\infty$. Thus there exists an initial state $\tilde{\mathbf{x}}_0 \in \boldsymbol{R}^n$, which cannot be steered to the target set S at times t_i, $i = 1, 2, \ldots$, $t_i \to \infty$ as $i \to \infty$. Since the dynamical system (1.2.8) is by our assumption globally U-controllable to zero, there exists a point $(\mathbf{x}', \mathbf{u}') \in \boldsymbol{R}^n \times \boldsymbol{R}^m$ such that $\mathbf{A}\mathbf{x}' + \mathbf{B}\mathbf{u}' = 0$ and $\mathbf{u}' \in \operatorname{int}\mathrm{CH}(U)$. Substituting $\mathbf{y} = \mathbf{x} - \mathbf{x}'$ and $\mathbf{v} = \mathbf{u} - \mathbf{u}'$, we obtain the so-called *shifted dynamical system* of the following form:

$$\dot{\mathbf{y}}(t) = \mathbf{A}\mathbf{y}(t) + \mathbf{B}\mathbf{v}(t), \quad t \in [0, \infty). \tag{1.9.8}$$

For the dynamical system (1.9.8) the set of control values is given by the formula $V = \{\mathbf{v} \in \boldsymbol{R}^m: \mathbf{v} + \mathbf{u}' \in U\}$ and the target set is of the form $Y = \{\mathbf{y} \in \boldsymbol{R}^n: \mathbf{y} + \mathbf{x}' \in S\}$. Since the set S is compact and convex, also the set Y is compact and convex. Similarly, since the set U is compact, also the set V is compact. Moreover, since $\mathbf{u}' \in \operatorname{int}\mathrm{CH}(U)$, we have $0 \in \operatorname{int}\mathrm{CH}(V)$. The dynamical system (1.9.8) has the same matrices $\mathbf{A}$ and $\mathbf{B}$ as the dynamical system (1.2.8). Thus the global V-controllability to zero of the dynamical system (1.9.8) implies the global U-controllability to the point $\mathbf{x}'$ of the dynamical system (1.2.8). By Corollaries 1.9.2 and 1.9.6 the dynamical system (1.9.8) is locally V-controllable to zero. Since the set Y is compact and the dynamical system (1.9.8) is globally V-controllable to the set Y, it is easy to show that this system is also globally V-controllable to zero. Thus the dynamical system (1.2.1) is globally U-controllable to the point $\mathbf{x}'$, whence there exists a control $\tilde{\mathbf{u}} \in M(U)$ steering the initial state $\tilde{\mathbf{x}}_0$ to the point $\mathbf{x}'$ in finite time $\tilde{t} > 0$. Since the dynamical system (1.2.8) is globally U-controllable to the set S, there exists a control $\bar{\mathbf{u}} \in M(U)$ steering the state $\mathbf{x}'$ to the set S in finite time $\bar{t} > 0$. Thus we can choose an index $\tilde{i}$ such that $t_{\tilde{i}} > \tilde{t}+\bar{t}$. Now we shall prove that the state $\mathbf{x}_0$ can be steered to the set S in time $t_{\tilde{i}}$. It will be done in three steps, as follows:

(1) using control $\tilde{\mathbf{u}}(t)$, $t \in [0, \tilde{t}]$ we obtain that $\mathbf{x}(\tilde{t}, \tilde{\mathbf{x}}_0, \mathbf{u}) = \mathbf{x}'$,

(2) using control $\mathbf{u}(t) = \mathbf{u}' = \text{const} \in \text{CH}(U)$ for $t \in (\tilde{t}, t_{\tilde{i}}, -\bar{t})$, we obtain $\mathbf{x}(t_{\tilde{i}}, -t, \mathbf{x}', \mathbf{u}') = \bar{\mathbf{x}}$,

(3) using control $\bar{\mathbf{u}}(t) = \mathbf{u}(t-(t_{\tilde{i}}, -\bar{t}))$ for $t \in [t_i, -\bar{t}, t_i,]$, we obtain $\mathbf{x}(t_i, \bar{\mathbf{x}}, \bar{\mathbf{u}}) \in S$.

This contradicts the assumption that there exists an initial state $\tilde{\mathbf{x}}_0 \in \boldsymbol{R}^n$, which cannot be steered to the set S. Thus condition (1.9.5) is a necessary condition for the global U-controllability of the dynamical system (1.2.8) to the set S. ∎

The criteria for global and local U-controllability to the target set S or to zero which are given in this section depend substantially on the properties of the sets U and S. In the remainder of this section we shall present some computable examples concerning various types of global and local controllability with constrained controls.

Example 1.9.1 (Schmitendorf and Barmish (1980))
Let us consider a dynamical system (1.2.8) of the form

$$\dot{x}(t) = x(t)+u(t), \qquad t \in [0, \infty) \tag{1.9.9}$$

and the control contrained set $U = [0, 1]$. Let us consider the global U-controllability of the dynamical system (1.9.9) to zero. In order to do that, let us construct an adjoint system (1.8.13) of the following form:

$$\dot{z}(t) = -z(t), \qquad t \in [0, \infty). \tag{1.9.10}$$

The solution $z(t)$ of the dynamical system (1.9.10) with initial condition $z(0) = z_0 \in \boldsymbol{R}$ is given by

$$z(t) = \exp(-t)z_0,$$

Thus by the definition of the function $H_U(\cdot)$ given in Section 1.8 we can obtain

$$H_U(z(t)) = \sup\{wz(t)\colon w \in [0, 1]\} = \sup\{w\exp(-t)z_0\colon w \in [0, 1]\} = 0$$

for $z_0 < 0$. Hence

$$H_U(\mathbf{B}^{\mathrm{T}}(t)z(t))dt = H_U(z(t))dt < +\infty,$$

and by Corollary 1.8.4 the dynamical system (1.9.9) is not globally U-controllable to zero since condition (1.8.22) is not satisfied. The same conclusion follows from Corollary 1.9.3 since the only eigenvalue of the dynamical system (1.9.9) is $s_1 = 1$.

Now we shall prove that the dynamical system (1.9.9) is also not locally U-controllable to zero either. In order to do that we shall use Corollary 1.9.1 and Theorem 1.9.1. Since condition (4) of Theorem 1.9.1 is not satisfied, by Corollary 1.9.1 the dynamical system (1.9.9) is not locally U-controllable to zero. It should be pointed out, however, that there exist initial states $x_0 \in \boldsymbol{R}$, which can be steered to zero in finite time. In order to calculate these states we shall use Theorem 1.8.1. By (1.8.4), (1.8.1) and (1.9.9) we have

$$J(x_0, t, a) = x_0 \exp(t)a + \int_0^t \sup\{w \exp(t-\tau)a : w \in [0, 1]\} d\tau, \quad a \in E \subset \boldsymbol{R}. \tag{1.9.11}$$

In the case where the set $E = [-1, +1]$ the relation (1.9.11) has the form:

$$J(x_0, t, a) = \begin{cases} x_0 \exp(t) a & \text{for } a \in [-1, 0], \\ x_0 \exp(t) a + a(\exp(t) - 1) & \text{for } a \in (0, 1]. \end{cases}$$

Thus the formula

$$\min\{J(x_0, t, a) : a \in [-1, +1]\} = 0$$

is satisfied for a certain $t \in [0, \infty)$ if and only if $x_0 \leqslant 0$ and $x_0 \geqslant (\exp(-t) - 1)$, i.e., if and only if $-1 < x_0 \leqslant 0$. Hence, although the dynamical system (1.9.9) is not locally U-controllable to zero, it is by Theorem 1.8.1 U-controllable to zero from initial states $x_0 \in \boldsymbol{R}$ such that $-1 < x_0 \leqslant 0$.

Now let us consider the case where the set $U = [1, 2]$. Since the set U does not contain the point zero, we cannot use Theorems 1.9.1 and 1.9.2 and Corollaries 1.9.1, 1.9.2 and 1.9.3. However, we can use Corollary 1.8.1 in order to obtain the initial states which are U-controllable to zero. By (1.8.1) we have $L = 1$ and $c = 0$ and thus

$$\begin{aligned} H_U(\mathbf{B}^{\mathrm{T}}(t)\mathbf{F}^{\mathrm{T}}(t, \tau)\mathbf{L}^{\mathrm{T}}\mathbf{a}) &= H_U(\exp(t-\tau)a) \\ &= \sup\{w \exp(t-\tau)a : w \in [1, 2]\} \\ &= \begin{cases} 2a \exp(t-\tau) & \text{for } a > 0, \\ a \exp(t-\tau) & \text{for } a \leqslant 0. \end{cases} \end{aligned} \tag{1.9.12}$$

Thus by (1.8.4) and (1.9.12) we find that the function $J(x_0, t, a)$ has the form

$$J(x_0, t, a) = \begin{cases} x_0 \exp(t) a + 2a(\exp(t) - 1) & \text{for } a > 0, \\ x_0 \exp(t) a + a(\exp(t) - 1) & \text{for } a \leqslant 0. \end{cases} \tag{1.9.13}$$

Hence the relation

$$\min\{J(x_0, t, a) : a \in [-1, +1]\} = 0$$

is satisfied for a certain $t \in [0, \infty)$ if and only if $2(\exp(-1) - 1) \leqslant x_0 \leqslant (\exp(-t) - 1)$. Thus the dynamical system (1.9.9) is, by Theorem 1.8.1, U-controllable to zero from the initial states $x_0 \in \boldsymbol{R}$ whenever $-2 < x_0 \leqslant 0$.

As a final variation of this problem, suppose $U = [-d, +d]$. It is easy to verify that conditions (1), (2), (3) and (4) of Theorem 1.9.1 are satisfied. However, condition (5) is not satisfied, because there exists an eigenvalue $s_1 = 1$ with a positive real part. Thus by Theorem 1.9.1 the dynamical system (1.9.9) is not globally U-controllable to zero, but by Corollary 1.9.1 it is locally U-controllable to zero. Theorem 1.8.1 tells us not only that the dynamical system (1.9.9) is locally U-controllable to zero but also that the initial states x_0 which can be steered to the origin are obtained from the following calculations. By (1.8.1) and (1.9.12) we have

$$H_U(\exp(t-\tau)a) = \sup\{w\exp(t-\tau)a\colon\ w \in [-d, +d]\}$$
$$= \begin{cases} d\exp(t-\tau)a & \text{for } a > 0, \\ -d\exp(t-\tau)a & \text{for } a \leqslant 0. \end{cases}$$

Using (1.8.4), we calculate the function $J(x_0, t, a)$ as follows:

$$J(x_0, t, a) = \begin{cases} x_0\exp(t)a + da(\exp(t)-1) & \text{for } a > 0, \\ x_0\exp(t)a + da(1-\exp(t)) & \text{for } a \leqslant 0. \end{cases}$$

Thus relation

$$\min\{J(x_0, t, a)\colon\ a \in [-d, +d]\} = 0$$

is satisfied for a certain $t \in [0, \infty)$ if and only if $d(\exp(-t)-1) \leqslant x_0 - \leqslant d(\exp(-t)-$ $-1)$. Then the dynamical system (1.9.9) is locally U-controlable to zero from the initial states $x_0 \in \boldsymbol{R}$ satisfying $-d < x_0 < +d$.

Example 1.9.2 (Petersen and Barmish (1983a))

Let us consider the dynamical system (1.2.8) described by

$$\begin{aligned} \dot{x}_1(t) &= x_1(t) + x_2(t), \\ \dot{x}_2(t) &= -x_1(t) + x_2(t) + u(t), \end{aligned} \qquad t \in [0, \infty) \tag{1.9.14}$$

with the set $U = [-1, +1]$ and an unbounded target set $S = \{[x_1, x_2]^{\mathrm{T}} \in \boldsymbol{R}^2\colon$ $x_1 \geqslant 0, x_2 = 0\}$. Let us consider the global U-controllability of the dynamical system (1.9.14) to the set S. Note that for this system the matrices $\mathbf{A}$ and $\mathbf{B}$ are of the form

$$\mathbf{A} = \begin{bmatrix} 1 & 1 \\ -1 & 1 \end{bmatrix}, \quad \mathbf{B} = \begin{bmatrix} 0 \\ 1 \end{bmatrix},$$

and thus the controllability matrix

$$[\mathbf{B} \vdots \mathbf{AB}] = \begin{bmatrix} 0 & 1 \\ 1 & 1 \end{bmatrix} \tag{1.9.15}$$

has full rank equal to $2 = n$. Thus condition (3) of Theorem 1.9.1 is satisfied.

Since the set S is not compact, we cannot use Theorem 1.9.3. On the other hand, since the set S is not of the form (1.8.1), we cannot also use Theorem 1.8.2 and the subsequent corollaries. Thus we shall investigate the global U-controllability to the set S of the dynamical system (1.9.14) by using direct calculations.

For this system the state transition matrix exp(At) is given by the formula

$$\exp(\mathbf{A}t) = \exp(t)\begin{bmatrix} \cos t & \sin t \\ -\sin t & \cos t \end{bmatrix}.$$

The solution $\mathbf{x}(t, \mathbf{x}_0, u)$ of equations (1.9.14) for $x_0 = [x_{01}, x_{02}]^T \in \boldsymbol{R}^2$ and $u(t) = 0$, $t \geqslant 0$ has the following form:

$$\begin{bmatrix} x_1(t) \\ x_2(t) \end{bmatrix} = \exp(t)\begin{bmatrix} \cos t & \sin t \\ -\sin t & \cos t \end{bmatrix}\begin{bmatrix} x_{01} \\ x_{02} \end{bmatrix}.$$

Thus for an arbitrary initial state $\mathbf{x}_0 \in \boldsymbol{R}^2$ we can choose time $t_1 > 0$ such that $x_{01}\sin t_1 = x_{02}\cos t_1$ and $x_{01}\cos t_1 + x_{02}\sin t_1 \geqslant 0$. Hence for $t = t_1$ we have

$$\begin{bmatrix} x_1(t) \\ x_2(t) \end{bmatrix} = \begin{bmatrix} \exp(t)x_{01}\cos t_1 + \exp(t_1)x_{02}\sin t_1 \\ 0 \end{bmatrix}. \tag{1.9.16}$$

Thus $\mathbf{x}(t_1) = [x_1(t_1), x_2(t_1)]^T \in S$ for any initial conditions $\mathbf{x}_0 = [x_{01}, x_{02}]x^T \in \boldsymbol{R}^2$. Hence by Definition 1.8.3 the dynamical system (1.9.14) is globally U-controllable to the set S, which is unbounded and noncompact. This example shows that all the conditions of Theorem 1.9.3 are important for investigating the global U-controllability of the dynamical system. Moreover, this example illustrates some phenomenons which can occur for dynamical systems which satisfy condition (3) of Theorem 1.9.3.

1.10 Observability and duality of dynamical systems

The idea of observability is closely related to that of controllability (Kalman (1960, 1963)). Observability means that the knowledge of the matrices characterizing the system and the zero-input response over a finite time interval is sufficient to determine uniquely the initial state of the dynamical system. The notion of observability is also strongly related to the notion of state observers of dynamical systems (Chen (1970, Ch. 5.4), Kaczorek (1974, Ch. 1.9.5), Niederliński (1974, Ch. 9.2), Zadeh and Desoer (1963, Ch. 11.4)). Roughly speaking, observability denotes studying the possibility of estimating the state from the output. If a dynamical system is observable, all the modes of the dynamical equation can be observed at the output. The concept of observability is defined under the assumption that we have complete knowledge of a given dynamical equation; that is, the matrices **A**, **B**, **C** and **D** are known beforehand. In the study of the observability of a linear dynamical system it is always assumed that the state variables are not available for direct measurements. If they are available for direct measurements, the estimation of the initial state is not needed; we just measure them directly.

In this section we shall present the basic definition and the necessary and sufficient condition for the observability of the linear, time-inavriant dynamical system described by the state equation (1.2.8) and the output equation (1.2.9). Next also the duality principle will be formulated. Finally, the concept of so-called "detectability" will be considered.

DEFINITION 1.10.1. The dynamical system (1.2.8), (1.2.9) is said to be *observable* if the knowledge of the input $\mathbf{u}(t)$ and the output $\mathbf{y}(t)$ over the finite-time interval suffices to determine uniquely the initial state $\mathbf{x}(0)$.

THEOREM 1.10.1 (Chen (1970, Ch. 5.4), Kaczorek (1974, Ch. 1.9.5), Zadeh and Desoer (1963, Ch. 11.4)). *The dynamical system* (1.2.8), (1.2.9) *is observable if and only if*

$$\operatorname{rank}\begin{bmatrix}\mathbf{C}\\ \mathbf{CA}\\ \mathbf{CA}^2\\ \vdots\\ \mathbf{CA}^{n-1}\end{bmatrix} = \operatorname{rank}[\mathbf{C}^T \vdots \mathbf{A}^T\mathbf{C}^T \vdots (\mathbf{A}^2)^T\mathbf{C}^T \vdots \dots \vdots (\mathbf{A}^{n-1})^T\mathbf{C}^T] = n. \quad (1.10.1)$$

A detailed proof of Theorem 1.10.1 can be found in the book of Chen (1970, Ch. 5.4) or in the book of Zadeh and Desoer (1963, Ch. 11.4).

From Theorem 1.10.1 it immediately follows that observability does not depend on the matrices **B** and **D**. Moreover, without loss of generality we can take the input $\mathbf{u}(t) = 0$ for $t \geqslant 0$. This simplifies all our considerations. Condition (1.10.1) is similar to the necessary and sufficient condition of controllability (1.4.1). This observation immediately leads to the concept of duality and to the duality principle (Chen (1970, Ch. 5.4), Niederliński (1974, p. 112), Zadeh and Desoer (1963, p.504)).

THEOREM 1.10.2 (Chen (1970, Ch. 5.4), Niederliński (1974, p. 112), Zadeh and Desoer (1963, p. 504)). *The dynamical system* (1.2.8), (1.2.9) *is observable if and only if the dynamical system*

$$\dot{\mathbf{z}}(t) = \mathbf{A}^T\mathbf{z}(t)+\mathbf{C}^T\mathbf{v}(t), \quad \mathbf{z} \in \boldsymbol{R}^n,\ \mathbf{v} \in \boldsymbol{R}^p, \quad (1.10.2)$$

$$\mathbf{w}(t) = \mathbf{B}^T\mathbf{z}(t)+\mathbf{D}^T\mathbf{v}(t), \quad \mathbf{w} \in \boldsymbol{R}^m \quad (1.10.3)$$

is controllable. The dynamical system (1.10.2), (1.10.3) *is called the dual system.*

Proof:

By Theorem 1.4.1 the dynamical system (1.10.2), (1.10.3) is controllable if and only if

$$\operatorname{rank}[\mathbf{C}^T \vdots \mathbf{A}^T\mathbf{C}^T \vdots (\mathbf{A}^T)^2\mathbf{C}^T \vdots \dots \vdots (\mathbf{A}^T)^{n-1}\mathbf{C}^T] = n.$$

Thus, by Theorem 1.10.1 the above condition is equivalent to the observability of the dynamical system (1.2.8), (1.2.9). ∎

COROLLARY 1.10.1. *The dynamical system* (1.2.8), (1.2.9) *is controllable if and only if the dynamical system* (1.10.2), (1.10.3) *is observable.*

The duality principle formulated in Theorem 1.10.2 enables us to reduce optimal control problems to optimal observation problems and vice versa (Kalman (1963), Zadeh and Desoer (1963, Ch. 11)).

In the case where the dynamical system (1.2.8), (1.2.9) is not controllable and/or not observable, we can transform it by the similarity transformation to the so-called *Kalman canonical form* (Zadeh and Desoer (1963, p. 505)):

$$\begin{bmatrix} \dot{\mathbf{x}}_{so}(t) \\ \dot{\mathbf{x}}_{s}(t) \\ \dot{\mathbf{x}}_{o}(t) \\ \dot{\mathbf{x}}_{ns,no}(t) \end{bmatrix} = \begin{bmatrix} \mathbf{A}_{so,so} & 0 & \mathbf{A}_{so,o} & 0 \\ \mathbf{A}_{s,so} & \mathbf{A}_{ss} & \mathbf{A}_{so} & \mathbf{A}_{sn} \\ 0 & 0 & \mathbf{A}_{oo} & 0 \\ 0 & 0 & \mathbf{A}_{no} & \mathbf{A}_{nn} \end{bmatrix} \begin{bmatrix} \mathbf{x}_{so}(t) \\ \mathbf{x}_{s}(t) \\ \mathbf{x}_{o}(t) \\ \mathbf{x}_{ns,no}(t) \end{bmatrix} + \begin{bmatrix} \mathbf{B}_{so} \\ \mathbf{B}_{s} \\ 0 \\ 0 \end{bmatrix} \mathbf{u}(t), \tag{1.10.4}$$

$$\mathbf{y}(t) = [\mathbf{C}_{so} \;\; 0 \;\; \mathbf{C}_{o} \;\; 0] \begin{bmatrix} \mathbf{x}_{so}(t) \\ \mathbf{x}_{s}(t) \\ \mathbf{x}_{o}(t) \\ \mathbf{x}_{ns,no}(t) \end{bmatrix} + \mathbf{D}\mathbf{u}(t), \tag{1.10.5}$$

where $\mathbf{x}_{so} \in \mathbf{R}^{n_{so}}$ is controllable and observable state vector, $\mathbf{x}_{s} \in \mathbf{R}^{n_{s}}$ is controllable and unobservable state vector, $\mathbf{x}_{o} \in \mathbf{R}^{n_{o}}$ is uncontrollable and observable state vector, $\mathbf{x}_{ns,no} \in \mathbf{R}^{n_{ns,no}}$ is uncontrollable and unobservable state vector, $\mathbf{A}_{so,so}$, $\mathbf{A}_{ss}$, $\mathbf{A}_{oo}$ and $\mathbf{A}_{nn}$ are constant matrices of dimensions $n_{so} \times n_{so}$, $n_s \times n_s$, $n_o \times n_o$ and $n_n \times n_n$, respectively. Moreover, $n_{so}+n_s+n_o+n_n = n$.

Example 1.10.1
Let us consider the dynamical system (1.2.8), (1.2.9) with the matrices

$$\mathbf{A} = \begin{bmatrix} a_{11} & a_{12} \\ a_{21} & a_{22} \end{bmatrix}, \quad \mathbf{B} = \begin{bmatrix} 0 \\ b \end{bmatrix}, \quad \mathbf{C} = [c \;\; 0], \quad \mathbf{D} = [0], \tag{1.10.6}$$

where $a_{11}, a_{12}, a_{21}, a_{22}, b$, and c are real constant numbers.

In order to verify the controllability and observability of the dynamical system (1.10.6) we shall make use of the conditions given in Theorems 1.4.1 and 1.10.1. Since

$$\operatorname{rank}[\mathbf{B} \vdots \mathbf{AB}] = \operatorname{rank}\begin{bmatrix} 0 & a_{12}b \\ b & a_{22}b \end{bmatrix} = 2 = n \quad \text{for } a_{12} \neq 0 \text{ and } b \neq 0,$$

the dynamical system (1.10.6) is controllable if and only if $a_{12} \neq 0$ and $b \neq 0$. Since

$$\operatorname{rank}[\mathbf{C}^{\mathrm{T}} \vdots \mathbf{A}^{\mathrm{T}}\mathbf{C}^{\mathrm{T}}] = \operatorname{rank}\begin{bmatrix} c & 0 \\ ca_{11} & ca_{12} \end{bmatrix} = 2 = n \quad \text{for } a_{12} \neq 0 \text{ and } c \neq 0,$$

the dynamical system (1.10.6) is observable if and only if $a_{12} \neq 0$ and $c \neq 0$. To sum up, the dynamical system (1.10.6) is controllable and observable if and only if $a_{12} \neq 0$, $b \neq 0$, and $c \neq 0$.

Example 1.10.2

Let us consider the dynamical system (1.2.8), (1.2.9) with the matrices

$$\mathbf{A} = \begin{bmatrix} a_1 & 0 & 0 & 0 \\ 0 & a_2 & 0 & 0 \\ 0 & 0 & a_3 & 0 \\ 0 & 0 & 0 & a_4 \end{bmatrix}, \quad \mathbf{B} = \begin{bmatrix} b_1 \\ b_2 \\ 0 \\ 0 \end{bmatrix}, \quad \mathbf{C} = [c_1 \ 0 \ c_3 \ 0], \quad \mathbf{D} = [0]. \tag{1.10.7}$$

By (1.10.4) and (1.10.5) the submatrices of the Kalman canonical form are $\mathbf{A}_{so,so} = a_1$, $\mathbf{A}_{so,o} = 0$, $\mathbf{A}_{s,so} = 0$, $\mathbf{A}_{ss} = a_2$, $\mathbf{A}_{so} = 0$, $\mathbf{A}_{sn} = 0$, $\mathbf{A}_{oo} = a_3$, $\mathbf{A}_{no} = 0$, $\mathbf{A}_{nn} = a_4$. The transfer function matrix $\mathbf{K}(s) = \mathbf{C}(s\mathbf{I}-\mathbf{A})^{-1}\mathbf{B}$ of the dynamical system (1.10.7), which represents the controllable and observable part of the system (Chen (1970, p. 201)) is of the following form:

$$K(s) = \frac{y_1(s)}{u_1(s)} = \frac{c_1 b_1}{s-a_1}.$$

An essentially weaker concept than observability is the so-called "detectability" of the dynamical system (1.2.8), (1.2.9). Detectability ensures (at least) the observability of all the unstability modes of the dynamical system (1.2.8), (1.2.9), i.e., all the modes corresponding to the unstable eigenvalues of the matrix **A**. The notion of detectability is closely related to the concept of stabilizability of dynamical system and hence to the notion of controllability. In the special case where the spectrum of the matrix **A** lies completely in the right-hand side part of the complex plane detectability and observability are equivalent.

DEFINITION 1.10.2. The dynamical system (1.2.8), (1.2.9) is said to be *detectable* if the linear transformation represented by the matrix **A** is stable on the unobservable linear subspaces of the pair (**C**, **A**).

In other words, detectability of the pair (**C**, **A**) (or equivalently of the dynamical system (1.2.8), (1.2.9)) means the observability of at least all the unstable modes of the dynamical system (1.2.8), (1.2.9).

In the special case where the dynamical system (1.2.8), (1.2.9) is stable, it is of course also detectable for an arbitrary matrix **C**.

The concepts of detectability and stabilizability are dual concepts. This statement is illustrated by the following corollary:

COROLLARY 1.10.2 (Voronov (1979, Ch. 3.6)). *The pair* (**C**, **A**) *is detectable if and only if the pair* ($\mathbf{A}^T$, $\mathbf{C}^T$) *is stabilizable.*

By Corollary 1.10.2, Definition 1.4.1 and Theorem 1.4.7 we can formulate a corollary which gives the immediate relation between detectability and stabilizability of the dynamical system (1.2.8), (1.2.9).

COROLLARY 1.10.3 (Voronov (1979, Ch. 3.6)). *The pair* (**C**, **A**) *is detectable if and only if there exists an* $(n \times m)$*-dimensional matrix* $\bar{\mathbf{F}}$ *such that the spectrum of the matrix* $\mathbf{A}+\bar{\mathbf{F}}\mathbf{C}$ *lies entirely on the left-hand side of the complex plane.*

In the special case where all the eigenvalues of the matrix **A** are unstable, detectability and observability of the dynamical system (1.2.8), (1.2.9) are equivalent.

Observability and detectability of the dynamical system (1.2.8), (1.2.9) can also be investigated by using the Jordan canonical form of the dynamical system (Chen (1970, Ch. 5.5)).

1.11 Structural Controllability

In this section we shall present the concepts of "structure" and "structural controllability" for a linear, time-invariant dynamical system (1.2.8). Practically, for every pair (**A**, **B**), most of the entries of **A** and **B** are known only approximately with an accuracy up to the errors of measurement. Only some of these entries are known with infinite accuracy. Normally, this happens for some entries which are equal to zero. Thus we shall assume here that there are some entries of **A** and **B** which are precisely zero, while all the other entries are known only approximately.

In order to define structural controllability we first introduce some definitions and background information associated with structural matrices and the structure of the dynamical system (1.2.8) (Anderson and Hui (1982), Burrows and Sahinkaya (1981), Glover and Silvermann (1976), Shields and Pearson (1976), Murota (1987)).

DEFINITION 1.11.1. The matrix **M** is said to be a *structured matrix* if its entries are either fixed zeros or independent free parameters. The matrix $\bar{\mathbf{M}}$ is called *admissible* (*with respect to* **M**) if it can be obtained by fixing the free parameters of **M** at some particular values.

For example $\mathbf{M} = \begin{bmatrix} 0 & x \\ x & x \end{bmatrix}$ is a structured matrix and $\bar{\mathbf{M}} = \begin{bmatrix} 0 & 1 \\ 2 & 0 \end{bmatrix}$ is an admissible matrix with respect to the matrix **M**.

DEFINITION 1.11.2. The *generic rank of a structured matrix* **M** is defined as the maximal rank that **M** achieves as a function of its free parameters. $r_g(\mathbf{M})$ denotes the generic rank of the matrix **M**.

Therefore, a structured matrix **M** has full rank generically if and only if there exists an admissible matrix $\bar{\mathbf{M}}$ with full rank.

Now let us formulate the definition of structural controllability of the dynamical system (1.2.8).

DEFINITION 1.11.3. The dynamical system (1.2.8) is said to be *structurally controllable* if for a structured pair $(\mathbf{A}, \mathbf{B})$ there exists an admissible pair $(\bar{\mathbf{A}}, \bar{\mathbf{B}})$ which is controllable, i.e., satisfies condition (1.4.1).

In the paper of Lin (1974) the necessary and sufficient conditions for structural controllability of the dynamical sysem (1.2.8) with single input ($\mathbf{B} = \mathbf{b}$) have been derived. These conditions are based on graph theory and require a lengthy series of additional definitions. In papers of Hosoe and Matsumoto (1979), Hosoe (1980), Lin (1977), these conditions are extended to the multiinput case. Moreover, in the paper of Lin (1977) the new concept of "minimal structural controllability" relating to structural controllability is defined and studied. The paper presents also the necessary and sufficient conditions for a multiinput structured dynamical system to be minimal structurally controllable. The results are expressed in both graph-theoretic and algebraic statements.

In papers of Burrows and Sahinkaya (1981), Glover and Silvermann (1976), Shields and Pearson (1976), Tokashi and Hirokazu (1979), several necessary and sufficient conditions for structural controllability of the dynamical system (1.2.8) are derived. These conditions are obtained by purely algebraic methods which utilize Boolean operations and structured canonical forms of dynamical systems. Similar considerations are also presented in the paper of Corfmat and Morse (1976) where the concept of canonical structurally controllable dynamical systems is defined. Moreover in the paper of Davison (1977) structural controllability of composite dynamical systems are considered.

DEFINITION 1.11.4. A structured pair $(\mathbf{A}, \mathbf{B})$ is said to be in *canonical form I* if there exists a permutation matrix $\mathbf{P}$ such that

$$\mathbf{PAP}^{-1} = \begin{bmatrix} \mathbf{A}_{11} & 0 \\ \mathbf{A}_{21} & \mathbf{A}_{22} \end{bmatrix} \tag{1.11.1}$$

and

$$\mathbf{PB} = \begin{bmatrix} 0 \\ \mathbf{B}_2 \end{bmatrix}, \tag{1.11.2}$$

where $\mathbf{A}_{11}$ is a $(q \times q)$-dimensional matrix, $\mathbf{A}_{22}$ is an $(n-q) \times (n-q)$-dimensional matrix, $\mathbf{A}_{12}$ is an $(n-q) \times q$-dimensional matrix, $\mathbf{B}_2$ is an $(n-q) \times q$-dimensional matrix.

DEFINITION 1.11.5. A structured pair $(\mathbf{A}, \mathbf{B})$ is said to be in *canonical form II if*

$$r_g(\mathbf{A} \vdots \mathbf{B}) < n, \tag{1.11.3}$$

where $(\mathbf{A} \vdots \mathbf{B})$ denotes a compound matrix whose first n columns are $\mathbf{A}$ and whose last m columns are $\mathbf{B}$.

Note that a necessary and sufficient condition for a pair $(\mathbf{A}, \mathbf{B})$ to be in canonical form II is that there should exist a zero submatrix of $(\mathbf{A} \vdots \mathbf{B})$ of order $k \times i$, where $k+i \geqslant n+m+1$.

THEOREM 1.11.1 (Glover and Silvermann (1976), Lin (1974), Shields and Pearson (1976)). *The dynamical system* (1.2.8) *is structurally controllable if and only if the pair* (**A**, **B**) *is neither in canonical form I nor in canonical form II.*

Proof:

Necessity. By contradiction. Let us assume that the dynamical system (1.2.8) is structurally controllable and the structured pair (**A**, **B**) is in canonical forms I or II. If the structured pair is in canonical form I, then by Kalman's canonical form (1.10.4) the dynamical system (1.2.8) is not controllable for an arbitrary admissible pair (**A**, **B**). If the pair $(\bar{\mathbf{A}}, \bar{\mathbf{B}})$ is in canonical form II, then by definitions 1.11.2 and 1.11.5 and by condition (1.4.1) the dynamical system (1.2.8) is not controllable for an arbitrary admissible pair $(\bar{\mathbf{A}}, \bar{\mathbf{B}})$. Thus we have a contradiction and hence the dynamical system (1.2.8) is structurally controllable.

Sufficiency. The sufficiency will be proved by induction. The inductive hypothesis is that the theorem holds for dynamical system of state dimension $< n$, and with an arbitrary number of inputs, say equal to m. The result is clearly true for a dynamical system with $n = 1$. Assuming the inductive hypothesis we can prove that if the structured pair (**A**, **B**) is not in canonical forms I and II, then there exists a permutation matrix **P** such that (Glover and Silvermann (1976, Lemma 2))

$$\bar{\mathbf{A}} = \mathbf{PAP}^{-1} = \begin{bmatrix} \mathbf{A}_{11} & \mathbf{A}_{12} \\ \mathbf{A}_{21} & \mathbf{A}_{22} \end{bmatrix}, \qquad \bar{\mathbf{B}} = \mathbf{PB} = \begin{bmatrix} \mathbf{B}_1 \\ \mathbf{B}_2 \end{bmatrix}, \tag{1.11.4}$$

where $\mathbf{A}_{11}$ is a $q \times q$-dimensional matrix, $\mathbf{A}_{12}$ is a $q \times (n-q)$-dimensional matrix, $\mathbf{A}_{21}$ is an $(n-q) \times q$-dimensional matrix, $\mathbf{A}_{22}$ is an $(n-q) \times (n-q)$-dimensional matrix, $\mathbf{B}_1$ is a $q \times m$-dimensional matrix, $\mathbf{B}_2$ is an $(n-q) \times m$-dimensional matrix and $q \leqslant n-1$. In formula (1.11.4) we have two possibilities:

(1) $r_g(\mathbf{A}_{11}) = q$ and the components of $(\mathbf{A}_{22}, \mathbf{B}_2)$ can be fixed at $(\bar{\mathbf{A}}_{22}, \bar{\mathbf{B}}_2)$ such that $\operatorname{rank}[\bar{\mathbf{A}}_{22} \vdots \bar{\mathbf{B}}_2] = n-q$ and $\bar{\mathbf{A}}_{22}$ has only zero eigenvalues,

(2) the entries in **A** ans **B** can be fixed so that

$$r_g \begin{bmatrix} 0 & \bar{\mathbf{A}}_{12} & \bar{\mathbf{B}}_1 \\ 0 & \bar{\mathbf{A}}_{22} & \bar{\mathbf{B}}_2 \end{bmatrix} = n$$

and $\bar{\mathbf{A}}_{22}$ has only zero eigenvalues.

In case (1) we have

$$r_g \begin{bmatrix} \mathbf{A}_{11} & \mathbf{A}_{12} & \mathbf{B}_2 \\ 0 & \bar{\mathbf{A}}_{22} & \bar{\mathbf{B}}_2 \end{bmatrix} = n.$$

The pair $(\bar{\mathbf{A}}_{22}, \bar{\mathbf{B}}_2)$ is controllable since $\operatorname{rank}(s\mathbf{I} - \bar{\mathbf{A}}_{22} \vdots \bar{\mathbf{B}}_2) = n-q$ for all s which are eigenvalues of the matrix $\bar{\mathbf{A}}_{22}$, i.e., for $s = 0$. Since by Glover and Silvermann (1976, Lemma 1) the structured pair $(\mathbf{A}_{11}, (\mathbf{A}_{12} \vdots \mathbf{B}_1))$ is not in canonical form I the structured pair (**A**, **B**) is not in canonical form I either. Hence, by the inductive hypothesis, $q < n$ and, since he pair $(\mathbf{A}_{11}, (\mathbf{A}_{12} \,\vdots\, \mathbf{B}_1))$ is structurally controllable by Glover and Silvermann (1976, Lemma 3), we conclude that the pair

$$\left(\begin{bmatrix} \mathbf{A}_{11} & \mathbf{A}_{12} \\ 0 & \mathbf{A}_{22} \end{bmatrix}, \begin{bmatrix} \mathbf{B}_1 \\ \mathbf{B}_2 \end{bmatrix}\right)$$

is structurally controllable. Thus also the pair (**A**, **B**) is structurally controllable.

In case (2) we pick the free parameters in $(\bar{\mathbf{A}}, \bar{\mathbf{B}})$ so that

$$\operatorname{rank}\begin{bmatrix} 0 & \bar{\mathbf{A}}_{12} & \bar{\mathbf{B}}_1 \\ 0 & \bar{\mathbf{A}}_{22} & \bar{\mathbf{B}}_2 \end{bmatrix} = n. \tag{1.11.5}$$

Since matrix $\mathbf{A}_{22}$ has only zero eigenvalues, the pair $(\bar{\mathbf{A}}, \bar{\mathbf{B}})$ is controllable because we have $\operatorname{rank}(s\mathbf{I}-\bar{\mathbf{A}} \,\vdots\, \bar{\mathbf{B}}) = n$ for all the s which are eigenvalues of the matrix **A**. Thus the pair (**A**, **B**) is structurally controllable. ■

Theorem 1.11.1 states that the dynamical system (1.2.8) is not structurally controllable if and only if the structured pair (**A**, **B**) is in canonical form I or $r_g(\mathbf{A} \,\vdots\, \mathbf{B}) < n$. In the case where the structured pair (**A**, **B**) is in canonical form I in the matrix

$$\mathbf{Q} = [\mathbf{B} \,\vdots\, \mathbf{AB} \,\vdots\, \mathbf{A}^2\mathbf{B} \,\vdots\, \dots \,\vdots\, \mathbf{A}^{n-1}\mathbf{B}] \tag{1.11.6}$$

there exists at least one row which is identically zero for all parameter values. This observation leads to a simple test of structural controllability (Glover and Silvermann (1976)).

In order to present this necessary algorithm let us introduce some Boolean operations (Glover and Silvermann (1976), Hosoe (1910), Hosoe and Matsumoto (1979), Shields and Pearson (1976)). Let $\mathbf{A}'$ be the Boolean matrix obtained from the structured matrix **A** by setting $a'_{ij} = 1$ if a_{ij} is a free parameter and $a'_{ij} = 0$ otherwise. We can define the addition and multiplication of such Boolean matrices as follows:

$$a \wedge b = \begin{cases} 1 & \text{if } a = 1 \text{ and } b = 1, \\ 0 & \text{otherwise,} \end{cases}$$

$$a \oplus b = \begin{cases} 1 & \text{if } a = 1 \text{ or } b = 1, \\ 0 & \text{otherwise,} \end{cases}$$

$$\mathbf{x}^{\mathrm{T}} * \mathbf{y} = x_1 \wedge y_1 \oplus x_2 \wedge y_2 \oplus \dots \oplus x_n \wedge y_n, \qquad \mathbf{A}'^k = \mathbf{A}'^{k-1} * \mathbf{A}'.$$

Using this notation, we can immediately obtain the fixed zeros in the product of structured matrices.

COROLLARY 1.11.1 (Glover and Silvermann (1976)). *Let* $(\mathbf{A}', \mathbf{B}')$ *be Boolean matrices corresponding to the structured matrices* (**A**, **B**). *Let*

$$\mathbf{Q}'_k = [\mathbf{b}' \,\vdots\, \mathbf{A}' * \mathbf{b}' \,\vdots\, \dots \,\vdots\, \mathbf{A}'^{k-1} * \mathbf{b}'],$$

where

$$\mathbf{b}' = \mathbf{B}' * \begin{bmatrix} 1 \\ 1 \\ \vdots \\ 1 \end{bmatrix}.$$

Define p_k as the number of zero rows of the matrix $\mathbf{Q}_k$. Then, if $p_k = p_{k+1}$ for some k, then $p_{k+j} = p_k$ for all $j \geqslant 0$. Also if $r_g(\mathbf{A} \vdots \mathbf{B}) = n$, then $(\mathbf{A}, \mathbf{B})$ is structurally uncontrollable if and only if

$$p_k = p_{k+1} > 0 \quad \textit{for some } k \geqslant 1. \tag{1.11.7}$$

Proof:

The equality $p_k = p_{k+j}$ for $j \geqslant 0$ immediately follows from (1.11.6) and the Cayley–Hamilton theorem (Chen (1970, p. 53)). Thus by Theorem 1.11.1 and Definition 1.11.2 formula (1.11.7) is the necessary and sufficient condition for the structural uncontrollability of the dynamical system (1.2.8). ∎

Corollary 1.11.1 gives a computationally simple test for structural controllability of the dynamical system (1.2.8). First, check $r_g(\mathbf{A} \vdots \mathbf{B})$ by a standard algorithm given in the paper of Shields and Pearson (1976). Then, if $r_g(\mathbf{A} \vdots \mathbf{B}) = n$, use the following algorithm to check if the structured pair $(\mathbf{A}, \mathbf{B})$ is in canonical form **I**:

(1) $\mathbf{c}_0 = \mathbf{b}'$, $\mathbf{d}_0 = 0$, $k = 1$,

(2) $\mathbf{c}_k = \mathbf{A}' * \mathbf{c}_{k-1}$, $\mathbf{d}_k = \mathbf{c}_k \oplus \mathbf{d}_{k-1}$.

(3) If $\mathbf{d}_k = \mathbf{d}_{k-1}$, then stop; else $k \leftarrow k+1$. Return to (2). Then if $\mathbf{d}_k^T = [1\ 1\ \dots\ 1]$ $(\mathbf{A}, \mathbf{B})$ is structurally controllable; otherwise, it is structurally uncontrollable.

Example 1.11.1

Let us consider the dynamical system (1.2.8) with the matrices

$$\mathbf{A} = \begin{bmatrix} a_{11} & a_{12} & a_{13} \\ a_{21} & a_{22} & a_{23} \\ a_{13} & 0 & 0 \end{bmatrix}, \quad \mathbf{B} = \begin{bmatrix} b_{11} & 0 \\ 0 & b_{22} \\ 0 & 0 \end{bmatrix}, \tag{1.11.8}$$

where $\mathbf{a}_{ij}$ and $\mathbf{b}_{ij}$ are free entries. Since

$$r(\mathbf{A} \vdots \mathbf{B}) = r_g \begin{bmatrix} a_{11} & a_{12} & a_{13} & b_{11} & 0 \\ a_{21} & a_{22} & a_{23} & 0 & b_{22} \\ a_{13} & 0 & 0 & 0 & 0 \end{bmatrix} = 3 = n,$$

we can use the algorithm presented above to check the structural controllability of the dynamical system (1.11.8). By (1.11.8) we have

$$\mathbf{A}' = \begin{bmatrix} 1 & 1 & 1 \\ 1 & 1 & 1 \\ 1 & 0 & 0 \end{bmatrix}, \quad \mathbf{B}' = \begin{bmatrix} 1 & 0 \\ 0 & 1 \\ 0 & 0 \end{bmatrix}.$$

Thus

$$\mathbf{b}' = \mathbf{B}' * \begin{bmatrix} 1 \\ 1 \end{bmatrix} = \begin{bmatrix} 1 & 0 \\ 0 & 1 \\ 0 & 0 \end{bmatrix} * \begin{bmatrix} 1 \\ 1 \end{bmatrix} = \begin{bmatrix} 1 \\ 1 \\ 0 \end{bmatrix},$$

(1) $\mathbf{c}_0 = \mathbf{b}' = \begin{bmatrix} 1 \\ 1 \\ 0 \end{bmatrix}$, $\quad \mathbf{d}_0 = \begin{bmatrix} 0 \\ 0 \\ 0 \end{bmatrix}$, $k = 1$,

(2) $\mathbf{c}_1 = \mathbf{A}' * \mathbf{c}_0 = \begin{bmatrix} 1 & 1 & 1 \\ 1 & 1 & 1 \\ 1 & 0 & 0 \end{bmatrix} * \begin{bmatrix} 1 \\ 1 \\ 0 \end{bmatrix} = \begin{bmatrix} 1 \\ 1 \\ 1 \end{bmatrix}$, $\quad \mathbf{d}_1 = \mathbf{c}_1 \oplus \mathbf{d}_0 = \begin{bmatrix} 1 \\ 1 \\ 1 \end{bmatrix} \oplus \begin{bmatrix} 0 \\ 0 \\ 0 \end{bmatrix} = \begin{bmatrix} 1 \\ 1 \\ 1 \end{bmatrix}$,

(3) Since $\mathbf{d}_0 \neq \mathbf{d}_1$, we return to item (2) substituting $k = k+1 = 2$.

(2) $k = 2$, $\mathbf{c}_2 = \mathbf{A}' * \mathbf{c}_1 = \begin{bmatrix} 1 & 1 & 1 \\ 1 & 1 & 1 \\ 1 & 0 & 0 \end{bmatrix} * \begin{bmatrix} 1 \\ 1 \\ 1 \end{bmatrix} = \begin{bmatrix} 1 \\ 1 \\ 1 \end{bmatrix}$,

$$\mathbf{d}_2 = \mathbf{c}_2 \oplus \mathbf{d}_1 = \begin{bmatrix} 1 \\ 1 \\ 1 \end{bmatrix} \oplus \begin{bmatrix} 1 \\ 1 \\ 1 \end{bmatrix} = \begin{bmatrix} 1 \\ 1 \\ 1 \end{bmatrix}.$$

(3) Since $\mathbf{d}_2 = \mathbf{d}_1 = \begin{bmatrix} 1 \\ 1 \\ 1 \end{bmatrix}$, the dynamical system (1.11.8) is structurally controllable.

Example 1.11.2

Let us consider the dynamical system (1.2.8) with the matrices

$$\mathbf{A} = \begin{bmatrix} a_{11} & a_{12} & a_{13} & a_{14} \\ a_{21} & a_{22} & a_{23} & a_{24} \\ a_{31} & 0 & 0 & 0 \\ a_{41} & 0 & 0 & 0 \end{bmatrix}, \quad \mathbf{B} = \begin{bmatrix} b_{11} & 0 \\ 0 & b_{22} \\ 0 & 0 \\ 0 & 0 \end{bmatrix}. \tag{1.11.9}$$

Since $r_g(\mathbf{A} \vdots \mathbf{B}) = 3$, $n = 4$, by Theorem 1.11.1 the dynamical system (1.11.9) is not structurally controllable.

Example 1.11.3

Let us consider the dynamical system (1.2.8) with the matrices

$$\mathbf{A} = \begin{bmatrix} 0 & 0 & a_{13} \\ 0 & a_{22} & 0 \\ a_{31} & a_{32} & 0 \end{bmatrix}, \quad \mathbf{B} = \begin{bmatrix} b_{11} & 0 \\ 0 & 0 \\ 0 & b_{32} \end{bmatrix}, \tag{1.11.10}$$

where a_{ij} and b_{ij} are free parameters. Since

$$r_g(\mathbf{A} \vdots \mathbf{B}) = r_g \left[\begin{array}{ccc:cc} 0 & 0 & a_{13} & b_{11} & 0 \\ 0 & a_{22} & 0 & 0 & 0 \\ a_{31} & a_{32} & 0 & 0 & b_{32} \end{array}\right] = 3 = n$$

we can use the algorithm presented above to check the structural controllability of the dynamical system (1.11.10). By (1.11.10) we have

$$\mathbf{A}' = \begin{bmatrix} 0 & 0 & 1 \\ 0 & 1 & 0 \\ 1 & 1 & 0 \end{bmatrix}, \quad \mathbf{B}' = \begin{bmatrix} 1 & 0 \\ 0 & 0 \\ 0 & 1 \end{bmatrix}.$$

Thus

$$\mathbf{b}' = \mathbf{B}' * \begin{bmatrix} 1 \\ 1 \end{bmatrix} = \begin{bmatrix} 1 & 0 \\ 0 & 0 \\ 0 & 1 \end{bmatrix} * \begin{bmatrix} 1 \\ 1 \end{bmatrix} = \begin{bmatrix} 1 \\ 0 \\ 1 \end{bmatrix},$$

(1) $\mathbf{c}_0 = \mathbf{b}' = \begin{bmatrix} 1 \\ 0 \\ 1 \end{bmatrix}$, $\mathbf{d}_0 = \begin{bmatrix} 0 \\ 0 \\ 0 \end{bmatrix}$, $k = 1$.

(2) $\mathbf{c}_1 = \mathbf{A}' * \mathbf{c}_0 = \begin{bmatrix} 0 & 0 & 1 \\ 0 & 1 & 0 \\ 1 & 1 & 0 \end{bmatrix} * \begin{bmatrix} 1 \\ 0 \\ 1 \end{bmatrix} = \begin{bmatrix} 1 \\ 0 \\ 1 \end{bmatrix} = \begin{bmatrix} 1 \\ 0 \\ 1 \end{bmatrix}$,

$$\mathbf{d}_1 = \mathbf{c}_1 \oplus \mathbf{d}_0 = \begin{bmatrix} 1 \\ 0 \\ 1 \end{bmatrix} \oplus \begin{bmatrix} 0 \\ 0 \\ 0 \end{bmatrix} = \begin{bmatrix} 1 \\ 0 \\ 1 \end{bmatrix}.$$

(3) Since $\mathbf{d}_0 \neq \mathbf{d}_1$, we return to item (2), substituting $k = k+1 = 2$.

(2′) $\mathbf{c}_2 = \mathbf{A}' * \mathbf{c}_1 = \begin{bmatrix} 0 & 0 & 1 \\ 0 & 1 & 0 \\ 1 & 1 & 0 \end{bmatrix} * \begin{bmatrix} 1 \\ 0 \\ 1 \end{bmatrix} = \begin{bmatrix} 1 \\ 0 \\ 1 \end{bmatrix}$, $\mathbf{d}_2 = \mathbf{c}_2 \oplus \mathbf{d}_1 = \begin{bmatrix} 1 \\ 0 \\ 1 \end{bmatrix} \oplus \begin{bmatrix} 1 \\ 0 \\ 1 \end{bmatrix} = \begin{bmatrix} 1 \\ 0 \\ 1 \end{bmatrix}$.

(3′) Since $\mathbf{d}_2 = \mathbf{d}_1 = \begin{bmatrix} 1 \\ 0 \\ 1 \end{bmatrix} \neq \begin{bmatrix} 1 \\ 1 \\ 1 \end{bmatrix}$, the dynamical system (1.11.10) is not structurally controllable.

In Example 1.11.3 we can also obtain the permutation matrix **P**, transforming, according to (1.11.1) and (1.11.2), the pair (**A**, **B**) into canonical form I.

Taking $\mathbf{P} = \begin{bmatrix} 0 & 1 & 0 \\ 1 & 0 & 0 \\ 0 & 0 & 1 \end{bmatrix}$ and since $\mathbf{P}^{-1} = \mathbf{P}^{\mathrm{T}}$ we have

$$\mathbf{PAP}^{-1} = \mathbf{PAP}^{\mathrm{T}} = \begin{bmatrix} 0 & 1 & 0 \\ 1 & 0 & 0 \\ 0 & 0 & 1 \end{bmatrix} \begin{bmatrix} 0 & 0 & a_{13} \\ 0 & a_{22} & 0 \\ a_{31} & a_{32} & 0 \end{bmatrix} \begin{bmatrix} 0 & 1 & 0 \\ 1 & 0 & 0 \\ 0 & 0 & 1 \end{bmatrix}$$

$$= \left[\begin{array}{c:cc} a_{22} & 0 & 0 \\ \hdashline 0 & 0 & a_{13} \\ a_{32} & a_{31} & 0 \end{array}\right],$$

$$\mathbf{PB} = \begin{bmatrix} 0 & 1 & 0 \\ 1 & 0 & 0 \\ 0 & 0 & 1 \end{bmatrix} \begin{bmatrix} b_{11} & 0 \\ 0 & 0 \\ 0 & b_{32} \end{bmatrix} = \begin{bmatrix} 0 & 0 \\ \hdashline b_{11} & 0 \\ 0 & b_{32} \end{bmatrix}.$$

Thus we have $q = 1$, $\mathbf{A}_{11} = [a_{22}]$, $\mathbf{A}_{21} = \begin{bmatrix} 0 \\ a_{32} \end{bmatrix}$, $\mathbf{A}_{22} = \begin{bmatrix} 0 & a_{13} \\ a_{31} & 0 \end{bmatrix}$ and $\mathbf{B}_2 = \begin{bmatrix} b_{11} & 0 \\ 0 & b_{32} \end{bmatrix}$.

The concept of structural controllability may be extended in some sense. As suggested in the paper of Lin (1977), it is of interest from the physical point of view to investigate the problem of structural controllability for possible assumptions upon indeterminate parameters. For instance, with the pair $(\mathbf{A}, \mathbf{B})$ being structurally controllable, this property may or may not be true on a hypersurface in $\boldsymbol{R}^{N+M}$, where N and M are the numbers of nonfixed entries in $\mathbf{A}$ and $\mathbf{B}$, respectively. We can select the subspace of $\boldsymbol{R}^{N+M}$ by setting any one component of all the $(N+M)$-tuples of $\boldsymbol{R}^{N+M}$ equal to zero as the particular hypersurface H. Then, if structural controllability holds on H, one can fix one or possibly more approximately determined parameters to precise zeros, whereas a structurally controllable system is still attained. For the other case, it determines the "lower bound" of structural controllability so-called "minimal structural controllability" of the pair $(\mathbf{A}, \mathbf{B})$ in the sense that $(\mathbf{A}, \mathbf{B})$ is not structurally controllable on all H (Lin (1977)). The necessary and sufficent conditions for the minimal structural controllability of the dynamical system (1.2.8) are presented in the paper of Lin (1977).

1.12 Numerical algorithms related to computing controllability

There are several mathematically equivalent approaches to determining the controllability of linear, time-invariant dynamical systems (1.2.8). Unfortunately, different approaches lead to computational methods which can give markedly different results. The discrepancies are caused by the rounding errors that occur during the computation. Thus, although determining the controllability of the dynamical system (1.2.8) is based on finding the rank of the matrix $\mathbf{W}_n$ given by formula (1.4.4), and mathematically is a problem simple enough, it is not a straightforward computational problem, and it brings out some rather subtle numerical properties of algorithms.

In this section we shall present a numerically stable algorithm for computing the controllability of the dynamical system (1.2.8), which is given in the paper of Paige (1981). This algorithm is based on the so-called *singular value decomposition* (SVD) of the constant rectangular matrix. In order to present this algorithm, let us introduce some additional symbols.

Let $\mathbf{M}$ be an $(n \times m)$-dimensional constant matrix with real elements. Without loss of generality, let us assume that $n \geqslant m$. Singular value decomposition of the matrix $\mathbf{M}$ leads to the following equality (Paige (1981)):

$$\mathbf{M} = \mathbf{U}\mathbf{S}\mathbf{V}^{\mathrm{T}}, \tag{1.12.1}$$

where

$$\mathbf{U}^{\mathrm{T}}\mathbf{U} = \mathbf{V}^{\mathrm{T}}\mathbf{V} = \mathbf{V}\mathbf{V}^{\mathrm{T}} = \mathbf{I}_m,$$

$$\mathbf{S} = \mathrm{diag}(z_1, z_2, \ldots, z_m), \qquad z_1 \geqslant z_2 \geqslant \ldots \geqslant z_m \geqslant 0.$$

The real scalars z_i, $i = 1, 2, \ldots, m$, are called the *singular values of the matrix* $\mathbf{M}$ and are uniquely determined for a given matrix $\mathbf{M}$. In fact, the singular values z_i, $i = 1, 2, \ldots, m$, are the eigenvalues of the symmetric nonnegative definite $(m \times m)$-dimensional matrix $\mathbf{M}^{\mathrm{T}}\mathbf{M}$. $\mathbf{V}$ is a unitary matrix of right singular vectors. $\mathbf{V}$ is an $(m \times m)$-dimensional matrix. The left singular vectors are the columns of the $(n \times m)$-dimensional matrix $\mathbf{U}$, which is only part of a unitary matrix if $m < n$.

The important property of the singular values is that they are not very sensitive to changes in the matrix. In fact, if $\tilde{z}_1 \geqslant \tilde{z}_2 \geqslant \ldots \geqslant \tilde{z}_m$ are the singular values of the perturbed matrix $\mathbf{M}+\delta\mathbf{M}$, then (Paige (1981))

$$|z_i - \tilde{z}_i| \leqslant ||\delta\mathbf{M}||_2 \quad \text{for} \quad i = 1, 2, \ldots, m, \tag{1.12.2}$$

where

$$||\mathbf{M}||_2 = z_{\max}(\mathbf{M}) = z_1. \tag{1.12.3}$$

Thus singulares valu are "well conditioned" with respect to perturbations in the matrix. In the case, where $z_1 \geqslant z_2 \geqslant \ldots \geqslant z_r > 0$ and $z_{r+1} = z_{r+2} = \ldots = z_m = 0$, the matrix $\mathbf{M}$ has rank equal to r, and from (1.12.2), for any $j \leqslant r$, a change of $||\delta\mathbf{M}||_2 \geqslant z_j$ is required to produce a perturbed matrix $\mathbf{M}+\delta\mathbf{M}$ of rank less than j. Thus, the singular values of a matrix tell us not only its rank but also how far the matrix is from the nearest matrix of a given lower rank, and the singular value decomposition gives us such a matrix.

Computing the singular value decomposition is rather troublesome, and there is another approach to determining rank, namely to compute an orthogonal decomposition of the matrix $\mathbf{M}$, i.e.,

$$\mathbf{Q}^{\mathrm{T}}\mathbf{M}\mathbf{P} = \begin{bmatrix} \mathbf{R} & \mathbf{S} \\ 0 & 0 \end{bmatrix}, \tag{1.12.4}$$

where $\mathbf{Q}$ is an $(n \times n)$-dimensional unitary matrix, $\mathbf{R}$ is an $(r \times r)$-dimensional upper triangular and nonsingular matrix and $\mathbf{P}$ is an $(m \times m)$-dimensional permutation matrix. The dimension of the matrix $\mathbf{R}$ gives the rank r of the matrix $\mathbf{M}$, and $\mathbf{S}$ is an $r \times (m-r)$-dimensional matrix.

It should be pointed out that we cannot in general expect to find the true mathematical rank of a given matrix, as this is meaningless when we are dealing with numbers of limited precision. However, it is straightforward to compute the effective rank of the given matrix, that is the rank of a matrix which, within the precision of the data and the computation, could be the given matrix. The singular value decomposition also indicates the distance to the nearest matrix of lesser rank.

In a practical problem, the actual matrix **M** is not always be available; instead we shall often have a matrix **M** and a measure of accuracy d_{tol} such that

$$||\mathbf{M}-\mathbf{M}||_2 \leqslant d_{\text{tol}}||\mathbf{M}||_2 \leqslant d_{\text{tol}}\tilde{z}_1 . \tag{1.12.5}$$

Thus, by (1.12.2) and (1.12.5) in the case where

$$\tilde{z}_r > d_{\text{tol}}\tilde{z}_1 \quad \text{and} \quad \tilde{z}_{r+1} = d_{\text{tol}}\tilde{z}_1 \tag{1.12.6}$$

we have $z_r \neq 0$. Hence rank**M** $\geqslant r$ ($z_{r+1}, z_{r+2}, \ldots, z_m$ may be equal to zero). By (1.12.5) and (1.12.6) we conclude that only perturbations greater than $\tilde{z}_r - d_{\text{tol}}\tilde{z}_1$ may change the rank of the matrix **M**, in fact decrease the rank of the matrix **M** to the value $r-1$.

Summarizing, unless the elements of a matrix are known to infinite accuracy, the exact rank of the matrix will usually be unobtainable. However, singular value decomposition tells us how close the matrix is to one of a given rank.

In a practical problem, it is important to define a measure of the distance of a given dynamical system from the nearest uncontrollable dynamical system. This leads to the notion of the so-called *controllability margin* (Klamka (1974b)). It is well known (see e.g. Klamka (1974b), Olbrot (1980), Paige (1981)) that any uncontrollable dynamical system (1.2.8) is arbitrarily close to a controllable dynamical system. One is thus naturally led to seek some measure of how far a given dynamical system is from an uncontrollable dynamical system. The measure of controllabiliy $\mu(\mathbf{A}, \mathbf{B})$ of the dynamical system (1.2.8) is defined as follows:

$$\mu(\mathbf{A}, \mathbf{B}) = \{\min||(\delta\mathbf{A}, \delta\mathbf{B})||_2 \text{ such that the dynamical system defined by the pair of the matrices } (\mathbf{A}+\delta\mathbf{A}, \mathbf{B}+\delta\mathbf{B}) \text{ is uncontrollable}\}. \tag{1.12.7}$$

This measure with either $\delta\mathbf{A}$ or $\delta\mathbf{B}$ set to zero might also be of interest. The measure $\mu(\mathbf{A}, \mathbf{B})$ will be invariant under orthogonal transformations, but could be altered by nonunitary transformations and scaling. A numerically stable computation for finding $\mu(\mathbf{A}, \mathbf{B})$ enables to compute $\mu(\mathbf{A}, \mathbf{B})$ very close to the true value. The measure $\mu(\mathbf{A}, \mathbf{B})$ has a practical advantage. Suppose that the model matrices **A** and **B** are known to differ from the true dynamical system matrices $\bar{\mathbf{A}}$ and $\bar{\mathbf{B}}$ within

$$||\mathbf{A}-\bar{\mathbf{A}}||_2 \leqslant A_{\text{tol}}||\mathbf{A}||_2 \quad \text{and} \quad ||\mathbf{B}-\bar{\mathbf{B}}||_2 \leqslant B_{\text{tol}}||\mathbf{B}||_2 \tag{1.12.8}$$

and the computed result $\mu(\mathbf{A}, \mathbf{B})$ is exact. Then, if

$$\mu(\mathbf{A}, \mathbf{B}) > A_{\text{tol}}||\mathbf{A}||_2 + B_{\text{tol}}||\mathbf{B}||_2, \tag{1.12.9}$$

then the true dynamical system is controllable.

The measure $\mu(\mathbf{A}, \mathbf{B})$ is obtained by allowing all possible perturbations of the dynamical system coefficients. But if the bounds on the model uncertainties are dominated by the uncertainties in just a few elements of the system matrices, then criterion (1.12.9) is liable to be a pessimistic one. In fact, great changes in these few elements of the system matrices cannot even alter the controllability of the dynamical system. For this problem, good scaling appears to consist in scaling so that, as far as possible, the uncertainties in the elements of **A** and **B** are all of the same order of magnitude. In such scaling accurately known elements are ignored. The connections between scaling and coordinate transformations are presented in detail in the paper of Paige (1981), with certain examples of transformations.

1.13 Controllability of composite systems

In this section we shall study the controllability of composite systems. By a *composite system* we mean a system consisting of a collection of subsystems. Although there are many forms of composite systems, they are mainly built up through three basic connections: parallel, tandem and feedback ones. Hence we shall restrict ourselves to the studies of these three connections. All the subsystems that form a composite system will be assumed to be linear, time-invariant and finite-dimensional. Then it is easy to show that the composite system is again linear, time-invariant and finite-dimensional system. For linear composite systems, all the results given in the previous sections can be directly applied.

Consider two linear, time-invariant, finite-dimensional dynamical systems S_1 and S_2 described by the dynamical equations

$$S_1: \quad \begin{aligned} \dot{\mathbf{x}}_1(t) &= \mathbf{A}_1\mathbf{x}_1(t)+\mathbf{B}_1\mathbf{u}_1(t), \\ \mathbf{y}_1(t) &= \mathbf{C}_1\mathbf{x}_1(t)+\mathbf{D}_1\mathbf{u}_1(t) \end{aligned} \tag{1.13.1}$$

and

$$S_2: \quad \begin{aligned} \dot{\mathbf{x}}_2(t) &= \mathbf{A}_2\mathbf{x}_2(t)+\mathbf{B}_2\mathbf{u}_2(t), \\ \mathbf{y}_2(t) &= \mathbf{C}_2\mathbf{x}_2(t)+\mathbf{D}_2\mathbf{u}_2(t), \end{aligned} \tag{1.13.2}$$

where $\mathbf{x}_1 \in R^{n_1}$, $\mathbf{u}_1 \in R^{m_1}$, $\mathbf{y}_1 \in R^{p_1}$, $\mathbf{x}_2 \in R^{n_2}$, $\mathbf{u}_2 \in R^{m_2}$, $\mathbf{y}_2 \in R^{p_2}$ are, respectively, the states, the inputs and the outputs of the systems S_1 and S_2. $\mathbf{A}_1, \mathbf{B}_1, \mathbf{C}_1, \mathbf{D}_1$ and $\mathbf{A}_2, \mathbf{B}_2, \mathbf{C}_2, \mathbf{D}_2$ are real constant matrices of appropriate dimensions.

First let us consider a parallel connection of dynamical systems S_1 and S_2. Taking into account that in the case of parallel connection we have $\mathbf{u}_1(t) = \mathbf{u}_2(t) = \mathbf{u}(t)$ and $\mathbf{y}(t) = \mathbf{y}_1(t)+\mathbf{y}_2(t)$ (hence $p_1 = p_2$ and $m_1 = m_2$), we can easily show (see, e.g. Chen (1967, 1970, p. 361), Chen and Desoer (1967), Kaczorek (1974, p. 54)) that the dynamical equations of the parallel connection S_{12} of S_1 and S_2 are given in the following form:

$$S_{12}: \quad \begin{aligned} \begin{bmatrix} \dot{\mathbf{x}}_1(t) \\ \dot{\mathbf{x}}_2(t) \end{bmatrix} &= \begin{bmatrix} \mathbf{A}_1 & 0 \\ 0 & \mathbf{A}_2 \end{bmatrix} \begin{bmatrix} \mathbf{x}_1(t) \\ \mathbf{x}_2(t) \end{bmatrix} + \begin{bmatrix} \mathbf{B}_1 \\ \mathbf{B}_2 \end{bmatrix} \mathbf{u}(t), \\ \mathbf{y}(t) &= [\mathbf{C}_1 \quad \mathbf{C}_2] \begin{bmatrix} \mathbf{x}_1(t) \\ \mathbf{x}_2(t) \end{bmatrix} + (\mathbf{D}_1+\mathbf{D}_2)\mathbf{u}(t). \end{aligned} \tag{1.13.3}$$

We can easily verify that if the dynamical system S_1 or S_2 is not controllable, then the composite system S_{12} is not controllable either. This verification can be used directly by the controllability conditions given in Section 1.4 or 1.5.

Now, using the eigenvalues of the matrices $\mathbf{A}_1$ and $\mathbf{A}_2$, we shall formulate a necessary and sufficient condition for the controllability of the dynamical system S_{12}. In order to do that let us introduce the following notation: $\sigma(\mathbf{A}_1)$—spectrum of the matrix $\mathbf{A}_1$, $\sigma(\mathbf{A}_2)$—spectrum of the matrix $\mathbf{A}_2$.

THEOREM 1.13.1 (Chen (1970, p. 362), Kaczorek (1974, p. 54)). *Let us assume that the dynamical systems S_1 and S_2 are controllable. Then the parallel connection S_{12} is controllable if and only if either there are no common eigenvalues between S_1 and S_2, or in case there are common eigenvalues, i.e., $\sigma(\mathbf{A}_1)\cap\sigma(\mathbf{A}_2) = \sigma(\mathbf{A}_1\cap\mathbf{A}_2) \neq \emptyset$ for all $s \in \sigma(\mathbf{A}_1\cap\mathbf{A}_2)$ the following equality holds:*

$$\operatorname{rank}\left[\begin{array}{c:c:c} s\mathbf{I}-\mathbf{A}_1 & 0 & \mathbf{B}_1 \\ 0 & s\mathbf{I}-\mathbf{A}_2 & \mathbf{B}_2 \end{array}\right] = n_1+n_2. \tag{1.13.4}$$

Proof:

By (1.4.6) the dynamical system S_{12} is controllable if and only if the relation (1.13.4) is satisfied for all $s \in \sigma(\mathbf{A}_1)\cap\sigma(\mathbf{A}_2)$. If the sets $\sigma(\mathbf{A}_1)$ and $\sigma(\mathbf{A}_2)$ are disjoint, then the controllability of the systems S_1 and S_2 immediately implies that equality (1.13.4) is satisfied for all $s \in C$. In the case where $\sigma(\mathbf{A}_1)\cap\sigma(\mathbf{A}_2) = \sigma(\mathbf{A}_1\cap\mathbf{A}_2) \neq \emptyset$, equality (1.13.4) must be additionally satisfied for all $s \in \sigma(\mathbf{A}_1\cap\mathbf{A}_2)$. ∎

Now let us consider a tandem connection of dynamical systems S_1 and S_2, denoted by $\bar{S}_{12}$. It is assumed that system S_1 is followed by system S_2. Taking into account that in the case of a tandem connection we have $\mathbf{u}_2(t) = \mathbf{y}_1(t)$ (hence $p_1 = m_2$), we can easily show (see, e.g. (Chen (1967), Chen and Desoer (1967), Kaczorek (1974, p. 51)) that the dynamical equations of the tandem connection $\bar{S}_{12}$ of S_1 followed by S_2 are given in the form

$$\bar{S}_{12}: \quad \begin{aligned} \begin{bmatrix} \dot{\mathbf{x}}_1(t) \\ \dot{\mathbf{x}}_2(t) \end{bmatrix} &= \begin{bmatrix} \mathbf{A}_1 & 0 \\ \mathbf{B}_2\mathbf{C}_1 & \mathbf{A}_2 \end{bmatrix} \begin{bmatrix} \mathbf{x}_1(t) \\ \mathbf{x}_2(t) \end{bmatrix} + \begin{bmatrix} \mathbf{B}_1 \\ \mathbf{B}_2\mathbf{D}_1 \end{bmatrix} \mathbf{u}(t), \\ \mathbf{y}(t) &= [\mathbf{D}_2\mathbf{C}_1 \quad \mathbf{C}_2] \begin{bmatrix} \mathbf{x}_1(t) \\ \mathbf{x}_2(t) \end{bmatrix} + \mathbf{D}_2\mathbf{D}_1\mathbf{u}(t). \end{aligned} \tag{1.13.5}$$

THEOREM 1.13.2 (Davison and Wang (1975), Kaczorek (1974, p. 51)). *The dynamical system $\bar{S}_{12}$ is controllable if and only if the dynamical system S_1 is controllable and moreover*

$$\operatorname{rank}\left[\begin{array}{c:c:c} s\mathbf{I}-\mathbf{A}_1 & 0 & \mathbf{B}_1 \\ -\mathbf{B}_2\mathbf{C}_1 & s\mathbf{I}-\mathbf{A}_2 & \mathbf{B}_2\mathbf{D}_1 \end{array}\right] = n_1+n_2 \quad \textit{for all } s \in \sigma(\mathbf{A}_2). \tag{1.13.6}$$

Proof:

By (1.13.5) and condition (1.4.6) the dynamical system $\bar{S}_{12}$ is controllable if and only if relation (1.13.6) is satisfied for all $s \in \sigma(\mathbf{A}_1)\cup\sigma(\mathbf{A}_2)$. On the other hand, the controllability of the dynamical system S_1 implies that rank $[s\mathbf{I}-\mathbf{A}_1 \vdots \mathbf{B}_1] = n_1$ for all $s \in \sigma(\mathbf{A}_1)$. Hence condition (1.13.6) is satisfied if and only if the dynamical system $\bar{S}_{12}$ is controllable. ■

Next we consider the output feedback connection S'_{12} of two systems, S_1 and S_2. We assume that system S_2 is in the feedback gain and additionally has the matrix $\mathbf{D}_2 = 0$. The case $\mathbf{D}_2 \neq 0$ leads to very complicated notation and moreover requires the additional assumption of $\det(\mathbf{I}+\mathbf{D}_1\mathbf{D}_2) \neq 0$ (Chen (1970, p. 366)) and, so it will be omitted here. Taking into account that in the case of an output feedback connection we have $\mathbf{u}(t) = \mathbf{u}_1(t)+\mathbf{y}_2(t)$, $\mathbf{y}(t) = \mathbf{y}_1(t)$ and $\mathbf{u}_2(t) = \mathbf{y}_1(t)$ (hence $m_1 = p_2$ and $m_2 = p_1$), we can easily show (see, e.g. Kaczorek (1974, p. 55)) that the dynamical equations of the output feedback connection S'_{12} of S_1 and S_2 are given in the following form:

$$S'_{12}: \quad \begin{aligned} \begin{bmatrix} \dot{\mathbf{x}}_1(t) \\ \dot{\mathbf{x}}_2(t) \end{bmatrix} &= \begin{bmatrix} \mathbf{A}_1 & -\mathbf{B}_1\mathbf{C}_2 \\ \mathbf{B}_2\mathbf{C}_1 & \mathbf{A}_2-\mathbf{B}_2\mathbf{D}_1\mathbf{C}_2 \end{bmatrix} \begin{bmatrix} \mathbf{x}_1(t) \\ \mathbf{x}_2(t) \end{bmatrix} + \begin{bmatrix} \mathbf{B}_1 \\ \mathbf{B}_2\mathbf{D}_1 \end{bmatrix} \mathbf{u}(t), \\ \mathbf{y}(t) &= [\mathbf{C}_1 \quad -\mathbf{D}_1\mathbf{C}_2] \begin{bmatrix} \mathbf{x}_1(t) \\ \mathbf{x}_2(t) \end{bmatrix} + \mathbf{D}_1\mathbf{u}(t). \end{aligned} \tag{1.13.7}$$

THEOREM 1.13.3 (Davison and Wang (1975), Kaczorek (1974, p. 55)). *The dynamical system S'_{12} is controllable if and only if the dynamical system S_1 is controllable and moreover*

$$\operatorname{rank}\begin{bmatrix} s\mathbf{I}-\mathbf{A}_1 & 0 & \mathbf{D}_1 \\ -\mathbf{B}_2\mathbf{C}_1 & s\mathbf{I}-\mathbf{A}_2 & \mathbf{B}_2\mathbf{D}_1 \end{bmatrix} = n_1+n_2 \quad \textit{for all } s \in \sigma(\mathbf{A}_2) \tag{1.13.8}$$

Proof:
By (1.13.7) and Theorem 1.4.5 the dynamical system S'_{12} is controllable if and only if

$$\operatorname{rank}\begin{bmatrix} s\mathbf{I}-\mathbf{A}_1 & \mathbf{B}_1\mathbf{C}_2 & \mathbf{B}_1 \\ -\mathbf{B}_2\mathbf{C}_1 & s\mathbf{I}-\mathbf{A}_2+\mathbf{B}_2\mathbf{D}_1\mathbf{C}_2 & \mathbf{B}_2\mathbf{D}_1 \end{bmatrix} = n_1+n_2 \quad \text{for all } s \in \mathbf{C}. \tag{1.13.9}$$

Multiplying from the right by $-\mathbf{C}_2$ the last column of the above matrix and adding it to the middle column, we obtain by (1.13.9) the following formula:

$$\operatorname{rank}\begin{bmatrix} s\mathbf{I}-\mathbf{A}_1 & 0 & \mathbf{B}_1 \\ -\mathbf{B}_2\mathbf{C}_1 & s\mathbf{I}-\mathbf{A}_2 & \mathbf{B}_2\mathbf{D}_1 \end{bmatrix} = n_1+n_2 \quad \text{for all } s \in \mathbf{C}. \tag{1.13.10}$$

Controllability of system S_1 implies that $\operatorname{rank}[s\mathbf{I}-\mathbf{A}_1 \vdots \mathbf{B}_1] = n_1$ for all $s \in \mathbf{C}$. Hence condition (1.13.10) is satisfied if and only if condition (1.13.8) holds. ■

Theorem 1.13.3 immediately implies that the controllability of dynamical system S_1 is the necessary condition for the controllability of dynamical system S'_{12}.

COROLLARY 1.13.1 (Chen (1967, 1970, p. 366), Chen and Desoer (1967), Kaczorek (1974, p. 56)). *Dynamical system S'_{12} is controllable if and only if the tandem connection $\bar{S}_{12}$ of S_1 followed by S_2 is controllable.*

Proof:

Since condition (1.13.6) is the same as condition (1.13.8), by comparing Theorem 1.13.2 and Theorem 1.13.3 we obtain the desired result. ∎

Finally, we shall formulate the necessary and sufficient conditions for the controllability of composite systems consisting of arbitrarily connected controllable and observable linear, time-invariant subsystems (Sezer and Huseyin (1979)). Let us consider N controllable and observable dynamical subsystems S_i, $i = 1, 2, \dots, N$,

$$S_i:\quad \begin{aligned} \dot{\mathbf{x}}_i(t) &= \mathbf{A}_i\mathbf{x}_i(t)+\mathbf{B}_i\mathbf{u}_i(t), \\ \mathbf{y}_i(t) &= \mathbf{C}_i\mathbf{x}_i(t), \end{aligned} \qquad i = 1, 2, \dots, N, \tag{1.13.11}$$

connected by the following relations (Sezer and Huseyin (1979)):

$$\mathbf{u}_i(t) = \mathbf{K}_i\mathbf{u}_0(t)+\sum_{\substack{j=1\\ j\neq i}}^{j=N}\mathbf{E}_{ij}\mathbf{y}_j(t)+\mathbf{F}_i\mathbf{y}_i(t), \qquad i = 1, 2, \dots, N, \tag{1.13.12}$$

where $\mathbf{x}_i \in \boldsymbol{R}^{n_i}$, $\mathbf{u}_i \in \boldsymbol{R}^{m_i}$, $\mathbf{y}_i \in \boldsymbol{R}^{p_i}$, $i = 1, 2, \dots, N$, $\mathbf{u}_0 \in \boldsymbol{R}^{m_0}$. All the matrices in (1.13.11) and (1.13.12) are constant and of appropriate dimensions.

Let us introduce following notation:

$$\begin{aligned}
&[1, k] = \{1, 2, \dots, k\}, \quad k \leqslant N,\\
&P = \{i_1, i_2, \dots, i_k\}, \quad k \leqslant N,\ i_j \in [1, N] \text{ for } j = 1, 2, \dots, k,\\
&\mathbf{x}_P = [\mathbf{x}_{i_1}^{\mathrm{T}}\ \mathbf{x}_{i_2}^{\mathrm{T}}\ \dots\ \mathbf{x}_{i_k}^{\mathrm{T}}]^{\mathrm{T}},\\
&\mathbf{u}_P = [\mathbf{u}_{i_1}^{\mathrm{T}}\ \mathbf{u}_{i_2}^{\mathrm{T}}\ \dots\ \mathbf{u}_{i_k}^{\mathrm{T}}]^{\mathrm{T}},\\
&\mathbf{y}_P = [\mathbf{y}_{i_1}^{\mathrm{T}}\ \mathbf{y}_{i_2}^{\mathrm{T}}\ \dots\ \mathbf{y}_{i_k}^{\mathrm{T}}]^{\mathrm{T}},\\
&\mathbf{A}_P = \mathrm{diag}[\mathbf{A}_{i_1}\ \mathbf{A}_{i_2}\ \dots\ \mathbf{A}_{i_k}],\\
&\mathbf{B}_P = \mathrm{diag}[\mathbf{B}_{i_1}\ \mathbf{B}_{i_2}\ \dots\ \mathbf{B}_{i_k}],\\
&\mathbf{C}_P = \mathrm{diag}[\mathbf{C}_{i_1}\ \mathbf{C}_{i_2}\ \dots\ \mathbf{C}_{i_k}],\\
&\mathbf{F}_P = \mathrm{diag}[\mathbf{F}_{i_1}\ \mathbf{F}_{i_2}\ \dots\ \mathbf{F}_{i_k}],\\
&\mathbf{E}_P = [\mathbf{E}_{ij}], \quad i, j \in P,\ i \neq j,\ \mathbf{E}_{ii} = 0,\\
&\mathbf{K}_P = [\mathbf{K}_{i_1}^{\mathrm{T}}\ \mathbf{K}_{i_2}^{\mathrm{T}}\ \dots\ \mathbf{K}_{i_k}^{\mathrm{T}}]^{\mathrm{T}}.
\end{aligned}$$

Let S_P denote the set of subsystems (1.13.11) whose indexes belong to the set P, i.e.,

$$S_P:\quad \begin{aligned} \dot{\mathbf{x}}_P(t) &= \mathbf{A}_P\mathbf{x}_P(t)+\mathbf{B}_P\mathbf{u}_P(t), \\ \mathbf{y}_P(t) &= \mathbf{C}_P\mathbf{x}_P(t). \end{aligned} \tag{1.13.13}$$

Moreover, let S_P^c denote a noncompensated composite dynamical system consisting of subsystems (1.13.11) connected by the relations (1.13.12), i.e.,

$$S_P:\quad \begin{aligned} \dot{\mathbf{x}}_P(t) &= (\mathbf{A}_P+\mathbf{B}_P\mathbf{E}_P\mathbf{C}_P)\mathbf{x}_P(t)+\mathbf{B}_P\mathbf{K}_P\mathbf{u}_0(t), \\ \mathbf{y}_P(t) &= \mathbf{C}_P\mathbf{x}_P(t). \end{aligned} \tag{1.13.14}$$

Let $S_P^c(\mathbf{F}_a, \mathbf{F}_b, \ldots)$ denote a composite system S_P^c together with local output feedback gain $\mathbf{F}_a$, $\mathbf{F}_b, \ldots$ in subsystems S_a, $S_b, \ldots$ Moreover, let $\mathbf{G}_i(s)$ and $\mathbf{G}_P(s) = \mathrm{diag}[\mathbf{G}_{i_1}(s)\ \mathbf{G}_{i_2}(s)\ \ldots\ \mathbf{G}_{i_k}(s)]$ denote respectively the transfer function matrices of the dynamical systems S_i and S_P. Let

$$\mathbf{H}_P(s) = \begin{bmatrix} \mathbf{H}_P^{i_1}(s) \\ \mathbf{H}_P^{i_2}(s) \\ \vdots \\ \mathbf{H}_P^{i_k}(s) \end{bmatrix} \tag{1.13.15}$$

be the transfer function matrix of the dynamical system S_P^c. It should be stressed that the transfer function $\mathbf{H}_P^{i_j}(s)$ connects the input $\mathbf{u}_0(s)$ with the output $\mathbf{y}_{i_j}(s)$ in the system S_P^c. Similarly, let $\mathbf{H}_P(s;\ \mathbf{F}_a, \mathbf{F}_b, \ldots)$ denote the transfer function matrix of the system $S_P^c(\mathbf{F}_a, \mathbf{F}_b, \ldots)$.

THEOREM 1.13.4 (Sezer and Huseyin (1979)). *Let us assume that the subsystems are numbered so that* $\mathbf{H}_{[1,k_1]}^1(s) \neq 0$, $\mathbf{H}_{[1,k_2]}^2(s) \neq 0, \ldots, \mathbf{H}_{[1,N]}^N(s) \neq 0$. *Then the composite system* $S_{[1,N]}^c(\mathbf{F}_{[1,N]})$ *is controllable for almost all* $\mathbf{F}_{[1,N]} = \mathrm{diag}\,[\mathbf{F}_1\ \mathbf{F}_2\ \ldots\ \mathbf{F}_N]$.

Theorem 1.13.4 gives the sufficient condition for the controllability of a composite dynamical system. Controllability is achieved by a suitable selection of the local output feedback gains represented by the matrices $\mathbf{F}_i$. Controllability for almost all $\mathbf{F}_{[1,N]}$ means that the set of output feedback gains for which the composite system is not controllable is empty or is a hypersurface in the space of the parameters of the matrix $\mathbf{F}_{[1,N]}$.

THEOREM 1.13.5 (Sezer and Huseyin (1979)). *If* $\mathbf{H}_P^i(s) \neq 0$ *for* $i = 1, 2, \ldots, N$, *then the dynamical system* $S_{[1,N]}^c(\mathbf{F}_{[1,N]})$ *is controllable for almost all* $\mathbf{F}_{[1,N]}$.

The sufficient condition given by Theorem 1.13.4 is not easy to test since one might be forced to consider all possible orderings of the subsystems. The following lemma gives a sufficient condition which guarantees the existence of an ordering of the subsystems that satisfies the condition of Theorem 1.13.4.

LEMMA 1.13.1 (Sezer and Huseyin (1979)). *If* $\mathbf{H}_P^i(s) \neq 0$, $i = 1, 2, \ldots, N$, *then it is possible to order the subsystems so as to satisfy the condition of Theorem* 1.13.4.

Theorems 1.13.4 and 1.13.5 together with the Lemma 1.13.1 give sufficient conditions for a composite system, which is formed by arbitrarily interconnecting a number of linear, time-invariant multivariable subsystems, to be made controllable by using constant local output feedback. These conditions, which are fairly easy to test, simply require that there should exist a transmission from the external input to each of the subsystem outputs.

The notion of structural controllability presented in Section 1.11 can also be extended to the case of composite dynamical systems (Davison (1977)). Let us consider the general model of a composite system of the following form:

$$\begin{aligned} \dot{\mathbf{x}}(t) &= \mathbf{A}\mathbf{x}(t)+\mathbf{B}\mathbf{u}(t), \\ \mathbf{y}(t) &= \mathbf{C}\mathbf{x}(t), \end{aligned} \tag{1.13.16}$$

where $\mathbf{x} = [\mathbf{x}_1^{\mathrm{T}}\ \mathbf{x}_2^{\mathrm{T}}\ \dots\ \mathbf{x}_N^{\mathrm{T}}]^{\mathrm{T}} \in R^n$, $\mathbf{u} \in R^m$, $\mathbf{y} \in R^p$,

$$\mathbf{A} = \begin{bmatrix} \mathbf{A}_1 & \mathbf{E}_{12} & \dots & \mathbf{E}_{1N} \\ \mathbf{E}_{21} & \mathbf{A}_2 & \dots & \mathbf{E}_{2N} \\ \dots & \dots & \dots & \dots \\ \mathbf{E}_{N1} & \mathbf{E}_{N2} & \dots & \mathbf{A}_N \end{bmatrix} \in R^{n\times n}, \quad \mathbf{B} = \begin{bmatrix} \mathbf{B}_1 \\ \mathbf{B}_2 \\ \vdots \\ \mathbf{B}_N \end{bmatrix} \in R^{n\times m},$$

$$\mathbf{C} = [\mathbf{C}_1\ \mathbf{C}_2\ \dots\ \mathbf{C}_N] \in R^{p\times n}.$$

$\mathbf{A}_i$, $i = 1, 2, \dots, N$ are $(n_i \times n_i)$-dimensional constant matrices corresponding to the i-th subsystems, $\mathbf{B}_i$, $i = 1, 2, \dots, N$ are $(n_i \times n)$-dimensional constant matrices, $\mathbf{C}_i$, $i = 1, 2, \dots, N$ are $(p \times n_i)$-dimensional constant matrices, $\mathbf{E}_{ij}$, $i, j = 1, 2, \dots, N$, $i \neq j$, are $(n_i \times n_j)$-dimensional constant matrices representing interconnection gains between dynamical subsystems.

Now we shall formulate the necessary and sufficient condition for the structural controllability of the dynamical system (1.13.16). It should be stressed that the dynamical system (1.13.16) is more general than the dynamical system represented by the equations (1.13.13). Such models of general composite dynamical systems are extensively studied in the paper of Davison (1977).

THEOREM 1.13.6 (Davison (1977)). *The composite dynamical system* (1.13.16) *is structurally controllable if and only if simultaneously the following conditions holds*:

$$(1) \quad r_g(\mathbf{A} \vdots \mathbf{B}) = n, \tag{1.13.17}$$

$$(2) \quad r_g(\bar{\mathbf{A}}_k \vdots \mathbf{B}) = n \text{ for } k = 1, 2, \dots, N, \tag{1.13.18}$$

where

$$\bar{\mathbf{A}} = \begin{bmatrix} \mathbf{A}_1^* & \mathbf{E}_{12} & \dots & \mathbf{E}_{1N} \\ \mathbf{E}_{21} & \mathbf{A}_2^* & \dots & \mathbf{E}_{2N} \\ \dots & \dots & \dots & \dots \\ \mathbf{E}_{N1} & \mathbf{E}_{N2} & \dots & \mathbf{A}_N^* \end{bmatrix},$$

$\mathbf{A}_i^* = \mathbf{A}_i+\mathbf{B}_i\mathbf{K}_i\mathbf{C}_i$ *for* $i = 1, 2, \dots, N$, $\mathbf{K}_i$, $i = 1, 2, \dots, N$ *are* $(n \times p)$-*dimensional constant matrices representing the amplifier coefficients*, $\bar{\mathbf{A}}_k \in R^{n\times(n-1)}$ *is the constant matrix obtained from the matrix* $\bar{\mathbf{A}}$ *by deleting the k-th column.*

In the paper of Davison (1977) several other criteria for the structural controllability of general composite systems are presented. The majority of those criteria are formulated in terms of graphs based on the structure of the signal flows. By using the concept of duality also the observability conditions of composite systems are also given.

Controllability of composite systems may also be considered by using the Jordan canonical form (see, e.g. Chen (1970, Ch. 9.3), Klamka (1973, 1974a, 1975d)). This method is especially convenient for deriving the necessary conditions of controllability (Klamka (1973, 1974a, 1975d)). Moreover, in the book of Chen (1970) the controllability conditions for composite systems are stated in terms of irreducible realizations of transfer function matrices of dynamical subsystems. For single-input and single-output dynamical subsystems these conditions are very simple. For multivariable dynamical subsystems, the situation is of course more complicated.

Some other aspects of controllability of composite dynamical systems are presented in the following publications: Brasch, Howze and Pearson (1971), Chen (1967), Chen and Desoer (1967), Davison and Wang (1975), Desoer and Chen (1976), Panda, Chen and Desoer (1970), Wolovich and Hwang (1974), Yonemura and Ito (1972).

1.14 Brunovsky–Luenberger and Frobenius canonical forms

The notion of controllability is strongly connected with the problems of the so-called invariants of dynamical systems, invariant subspaces, canonical forms and the classification of dynamical systems (Brunovsky (1970), Kaczorek (1983a), Popov (1972), Warren and Eckberg (1975), Voronov (1979)). Canonical forms play an important role in the construction of state observers of a dynamical system, in solving problems of optimal control, in parameter identification and in investigating of the properties of dynamical systems (Chen (1970), Dickinson (1973), Ford and Johnson (1968)).

Let $(\mathbf{A}, \mathbf{B})$ be a controllable pair of matrices representing a dynamical system of the form (1.2.8). Then the following equality holds:

$$\begin{aligned}[\mathbf{B} \vdots \mathbf{AB} \vdots \mathbf{A}^2\mathbf{B} \ldots \vdots \mathbf{A}^i\mathbf{B} \vdots \ldots \vdots \mathbf{A}^{n-1}\mathbf{B}] &= [\mathbf{b}_1\ \mathbf{b}_2 \ldots \mathbf{b}_j \ldots \mathbf{b}_m\ \mathbf{Ab}_1\ \mathbf{Ab}_2 \ldots \\ &\ldots \mathbf{Ab}_j \ldots \mathbf{Ab}_m\ \mathbf{A}^2\mathbf{b}_1\ \mathbf{A}^2\mathbf{b}_2 \ldots \mathbf{A}^2\mathbf{b}_j \ldots \mathbf{A}^2\mathbf{b}_m \ldots \mathbf{A}^i\mathbf{b}_1\ \mathbf{A}^i\mathbf{b}_2 \ldots \mathbf{A}^i\mathbf{b}_j \\ &\ldots \mathbf{A}^i\mathbf{b}_m \ldots \mathbf{A}^{n-1}\mathbf{b}_1\ \mathbf{A}^{n-1}\mathbf{b}_2 \ldots \mathbf{A}^{n-1}\mathbf{b}_j \ldots \mathbf{A}^{n-1}\mathbf{b}_m]. \end{aligned} \tag{1.14.1}$$

By virtue of the controllability assumption of the dynamical system (1.2.8), the $(n \times nm)$-dimensional matrix (1.14.1) has rank n, whence it contains n linearly independent columns. These columns can be chosen in many different ways. The most frequently used method is based on the investigation of the linear independence of the columns $\mathbf{A}^i\mathbf{b}_j$ $(i = 1, 2, \ldots, n-1, j = 1, 2, \ldots, m)$ in the order of their appearance in matrix (1.14.1), starting from the column $\mathbf{b}_1$ (Brunovsky (1970), Kaczorek (1983a, Ch. 4), Popov (1972)). In the case where a given column is found to be linearly independent of the preceding columns it is included to the linearly independent columns. Otherwise, it is eliminated from further considerations. Thus chosen n linearly independent columns of the matrix (1.14.1) are used to formulate the new $(n \times n)$-dimensional transformation matrix of the following form (Kaczorek (1983a, Ch. 4)):

$$\begin{aligned}\mathbf{L} = [&\mathbf{b}_1\ \mathbf{A}\mathbf{b}_1\ \ldots\ \mathbf{A}^{d_1-1}\mathbf{b}_1\ \mathbf{b}_2\ \mathbf{A}\mathbf{b}_2\ \ldots\ \mathbf{A}^{d_2-1}\mathbf{b}_2\ \ldots\ \mathbf{b}_j\ \mathbf{A}\mathbf{b}_j\ \ldots \\ &\ldots\ \mathbf{A}^{d_j-1}\mathbf{b}_j\ \ldots\ \mathbf{b}_m\ \mathbf{A}\mathbf{b}_m\ \ldots\ \mathbf{A}^{d_m-1}\mathbf{b}_m].\end{aligned} \tag{1.14.2}$$

The natural numbers $d_j, j = 1, 2, \ldots, m$, defining the length of the chains

$$\mathbf{b}_j, \mathbf{A}\mathbf{b}_j, \mathbf{A}^2\mathbf{b}_j, \ldots, \mathbf{A}^{d_j-1}\mathbf{b}_j, \quad j = 1, 2, \ldots, m$$

are called the *controllability indices*, or *Kronecker's invariants of the controllable pair of matrices* (**A**, **B**) (Brunovsky (1970), Kaczorek (1983a, Ch. 4), Popov (1972)). The presented above method of choosing the linearly independent columns from the matrix (1.14.1) defines the controllability indices d_j, $j = 1, 2, \ldots, m$ (Kaczorek (1983a, p. 236)) and, moreover, the following relation holds:

$$\sum_{j=1}^{j=m} d_j = n.$$

THEOREM 1.14.1 (Kaczorek (1983a, Ch. 4)). *If the dynamical system* (1.2.8) *is controllable and* rank **B** = n, *then the nonsingular transformation matrix* **L** *given by formula* (1.14.2) *transforms the pair of the matrices* (**A**, **B**) *into the form*

$$\bar{\mathbf{A}} = \mathbf{L}^{-1}\mathbf{A}\mathbf{L} = \begin{bmatrix} \bar{\mathbf{A}}_{11} & \bar{\mathbf{A}}_{12} & \cdots & \bar{\mathbf{A}}_{1m} \\ \bar{\mathbf{A}}_{21} & \bar{\mathbf{A}}_{22} & \cdots & \bar{\mathbf{A}}_{2m} \\ \cdots & \cdots & \cdots & \cdots \\ \bar{\mathbf{A}}_{m1} & \bar{\mathbf{A}}_{2m} & \cdots & \bar{\mathbf{A}}_{mm} \end{bmatrix} \in R^{n\times n} \tag{1.14.3}$$

$$\bar{\mathbf{S}} = \mathbf{L}^{-1}\mathbf{B} = [\bar{\mathbf{b}}_{11}\ \bar{\mathbf{b}}_{12}\ \ldots\ \bar{\mathbf{b}}_{1m}], \tag{1.14.4}$$

where $\bar{\mathbf{A}}_{ii}, \bar{\mathbf{A}}_{ij}, \bar{\mathbf{b}}_{1j}$ *are given, respectively, by the following formulae*:

$$\bar{\mathbf{A}}_{ii} = \begin{bmatrix} 0 & 0 & 0 & \ldots & 0 & \bar{a}_{ii1} \\ 1 & 0 & 0 & \ldots & 0 & \bar{a}_{ii2} \\ 0 & 1 & 0 & \ldots & 0 & \bar{a}_{ii3} \\ \cdots & \cdots & \cdots & \cdots & \cdots & \cdots \\ 0 & 0 & 0 & \ldots & 0 & \bar{a}_{iid_i} \end{bmatrix} \in R^{d_i\times d_i}, \quad i = 1, 2, \ldots, m, \tag{1.14.5}$$

$$\bar{\mathbf{A}}_{ij} = \begin{bmatrix} 0 & 0 & 0 & \ldots & 0 & \bar{a}_{ij1} \\ 0 & 0 & 0 & \ldots & 0 & \bar{a}_{ij2} \\ \cdots & \cdots & \cdots & \cdots & \cdots & \cdots \\ 0 & 0 & 0 & \ldots & 0 & \bar{a}_{ijd_i} \end{bmatrix} \in R^{d_i\times d_j}, \quad i, j = 1, 2, \ldots, m, i \neq j, \tag{1.14.6}$$

$\bar{\mathbf{b}}_{1j}$ *is the column of the identity matrix* $\mathbf{I}_{n\times n}$ *with the number*

$$d_1 + d_2 + \ldots + d_{j-1} + 1. \tag{1.14.7}$$

DEFINITION 1.14.1 (Kaczorek (1983a)). The pair of matrices $(\bar{\mathbf{A}}, \bar{\mathbf{B}})$ given by the formulae (1.14.3) and (1.14.4) is called the *first Brunovsky–Luenberger canonical form.*

Besides the first Brunovsky–Luenberger canonical form there exists also a second Brunovsky–Luenberger canonical form. In order to obtain it we form the $(n \times n)$-dimensional nonsingular transformation matrix

$$\mathbf{S} = \begin{bmatrix} \mathbf{S}_1 \\ \mathbf{S}_2 \\ \vdots \\ \mathbf{S}_m \end{bmatrix}, \tag{1.14.8}$$

where

$$\mathbf{S}_i = \begin{bmatrix} \mathbf{s}_i \\ \mathbf{s}_i\mathbf{A} \\ \vdots \\ \mathbf{s}_i\mathbf{A}^{d_i-1} \end{bmatrix} \in \boldsymbol{R}^{d_i \times n}, \quad i = 1, 2, \dots, m, \tag{1.14.9}$$

$\mathbf{s}_i$, $i = 1, 2, \dots, m$ is the row of the matrix $\mathbf{L}^{-1}$ with the number $d_1 + d_2 + \dots + d_i$.

THEOREM 1.14.2 (Kaczorek (1983a, Ch. 4)). *If the dynamical system* (1.2.8) *is controllable and* $\operatorname{rank}\mathbf{B} = n$, *then the nonsingular transformation matrix* $\mathbf{S}$ *given by the formula* (1.14.8) *transforms the pair of the matrices* $(\mathbf{A}, \mathbf{B})$ *into the form*

$$\bar{\bar{\mathbf{A}}} = \mathbf{SAS}^{-1} = \begin{bmatrix} \bar{\bar{\mathbf{A}}}_{11} & \bar{\bar{\mathbf{A}}}_{12} & \dots & \bar{\bar{\mathbf{A}}}_{1m} \\ \bar{\bar{\mathbf{A}}}_{21} & \bar{\bar{\mathbf{A}}}_{22} & \dots & \bar{\bar{\mathbf{A}}}_{2m} \\ \dots & \dots & \dots & \dots \\ \bar{\bar{\mathbf{A}}}_{m1} & \bar{\bar{\mathbf{A}}}_{m2} & \dots & \bar{\bar{\mathbf{A}}}_{mm} \end{bmatrix}, \tag{1.14.10}$$

$$\bar{\bar{\mathbf{B}}} = \mathbf{SB} = [\bar{\bar{\mathbf{b}}}_{11}\ \bar{\bar{\mathbf{b}}}_{12}\ \dots\ \bar{\bar{\mathbf{b}}}_{1m}], \tag{1.14.11}$$

where $\bar{\bar{\mathbf{A}}}_{ii}, \bar{\bar{\mathbf{A}}}_{ij}, \bar{\bar{\mathbf{b}}}_{1j}$ *are given, respectively, by the following formulae:*

$$\bar{\bar{\mathbf{A}}}_{ii} = \begin{bmatrix} 0 & 1 & 0 & \dots & 0 \\ 0 & 0 & 1 & \dots & 0 \\ \dots & \dots & \dots & \dots & \dots \\ 0 & 0 & 0 & \dots & 1 \\ \bar{\bar{a}}_{ii1} & \bar{\bar{a}}_{ii2} & \dots & & \bar{\bar{a}}_{iid_i} \end{bmatrix} \in \boldsymbol{R}^{d_i \times d_i}, \quad i = 1, 2, \dots, m, \tag{1.14.12}$$

$$\bar{\bar{\mathbf{A}}}_{ij} = \begin{bmatrix} 0 & 0 & \dots & 0 \\ 0 & 0 & \dots & 0 \\ \dots & \dots & \dots & \dots \\ 0 & 0 & \dots & 0 \\ \bar{\bar{a}}_{ij1} & \bar{\bar{a}}_{ij2} & \dots & \bar{\bar{a}}_{ijd_j} \end{bmatrix} \in \boldsymbol{R}^{d_i \times d_j}, \quad i, j = 1, 2, \dots, m, i \neq j, \tag{1.14.13}$$

$\bar{\bar{\mathbf{b}}}_{1j}$ *is the column of the identity matrix* $\mathbf{I}_{n\times n}$ *with the number* $d_1+d_2+ \dots +d_j$, *to which the following column is added:*

$$\sum_{i=1}^{i=j-1} \mathbf{b}_{ij} \text{ (the column of the identity matrix } \mathbf{I}_{n\times n} \text{ with the number } d_1+d_2+ \dots +d_i). \tag{1.14.14}$$

DEFINITION 1.14.2 (Kaczorek (1983a)). The pair of matrices $(\bar{\bar{\mathbf{A}}}, \bar{\bar{\mathbf{B}}})$ given by the formulae (1.14.10) and (1.14.11) is called the *second Brunovsky–Luenberger canonical form.*

Theorems 1.14.1 and 1.14.2 can be appropriately formulated for the observable pair of matrices (**A**, **B**). It enables us to obtain the Brunovsky–Luenberger canonical forms for observable pairs of matrices, i.e., for observable dynamical systems. In the special case, where $m = 1$, the Brunovsky–Luenberger canonical forms are equivalent to the first and second Frobenius canonical forms, respectively (Kaczorek (1983a)). Moreover, the Frobenius canonical forms can also be obtained directly, by means of the matrix transformations utilizing the characteristic polynomial of the matrix and the Cayley–Hamilton theorem (Kaczorek (1983a), Voronov (1979)).

COROLLARY 1.14.1 (Kaczorek (1983a)). *If the pair* (**A**, **b**) *is controllable, then there exists a nonsingular* $(n\times n)$*-dimensional transformation matrix* **L**, *such that the pair* $(\bar{\mathbf{A}}, \bar{\mathbf{b}}) = (\mathbf{L}^{-1}\mathbf{A}\mathbf{L}, \mathbf{L}^{-1}\mathbf{b})$ *is in the first Frobenius canonical form, i.e.,*

$$\bar{\mathbf{A}} = \mathbf{L}^{-1}\mathbf{A}\mathbf{L} = \begin{bmatrix} 0 & 0 & \dots & 0 & -a_0 \\ 1 & 0 & \dots & 0 & -a_1 \\ 0 & 1 & \dots & 0 & -a_2 \\ \dots & \dots & \dots & \dots & \dots \\ 0 & 0 & \dots & 1 & -a_{-1} \end{bmatrix} \in R^{n\times n}, \tag{1.14.15}$$

$$\bar{\mathbf{b}} = \mathbf{L}^{-1}\mathbf{b} = \begin{bmatrix} 1 \\ 0 \\ \vdots \\ 0 \end{bmatrix} \in R^n, \tag{1.14.16}$$

where $a_0, a_1 \dots, a_{n-1}$ *are coefficients of the characteristic polynomial of the matrix* **A**,

$$\det(s\mathbf{I}_n - \mathbf{A}) = s^n + a_{n-1}s^{n-1} + \dots + a_1 s + a_0. \tag{1.14.17}$$

COROLLARY 1.14.2 (Kaczorek (1983a)). *If the pair* (**A**, **b**) *is controllable, then there exists a nonsingular* $(n\times n)$*-dimensional transformation matrix* **S** *such that the pair* $(\bar{\bar{\mathbf{A}}}, \bar{\bar{\mathbf{b}}}) = (\mathbf{S}\,\mathbf{A}\,\mathbf{S}^{-1}, \mathbf{S}\mathbf{b})$ *is in the second Frobenius canonical form, i.e.*

$$\bar{\mathbf{A}} = \mathbf{SAS}^{-1} = \begin{bmatrix} 0 & 1 & 0 & \dots & 0 \\ 0 & 0 & 1 & \dots & 0 \\ \dots & \dots & \dots & \dots & \dots \\ 0 & 0 & 0 & \dots & 1 \\ -a_0 & -a_1 & -a_2 & \dots & -a_{n-1} \end{bmatrix} \in R^{n\times n}, \tag{1.14.18}$$

$$\bar{\mathbf{b}} = \mathbf{Sb} = \begin{bmatrix} 0 \\ 0 \\ \vdots \\ 0 \\ 1 \end{bmatrix} \in R^n. \tag{1.14.19}$$

The transformation matrices **L** and **S** which are mentioned in the Corollaries 1.14.1 and 1.14.2, can be directly obtained by using formulae (1.14.2) and (1.14.8) and taking into consideration the dependence $m = 1$. Hence

$$\mathbf{L} = [\mathbf{b}\ \ \mathbf{Ab}\ \ \mathbf{A}^2\mathbf{b} \dots\ \mathbf{Ab} \dots\ \mathbf{A}^{n-1}\mathbf{b}]$$

and

$$\mathbf{S} = \begin{bmatrix} \mathbf{s} \\ \mathbf{sA} \\ \mathbf{sA}^2 \\ \vdots \\ \mathbf{sA}^i \\ \vdots \\ \mathbf{sA}^{n-1} \end{bmatrix},$$

where $\mathbf{s} \in R^n$ is a row of the matrix **L** chosen so that $\det\mathbf{S} \neq 0$.

An interesting theoretical problem is to study those properties of the controllable pair of the matrices **(A, B)**, which do not change during an arbitrary nonsingular transformation of the pair into the form $(\mathbf{TAT}^{-1}, \mathbf{TB})$, where $\det\mathbf{T} \neq 0$. These properties are called the *invariants of homological transformations*. The invariants of homological transformations are strictly connected with the Brunovsky–Luenberger and Frobenius canonical forms. The following theorem shows this connection.

THEOREM 1.14.3 (Kaczorek (1983a, Ch. 4), Popov (1972)). *The controllability indices d_j, $j = 1, 2, \dots, m$ and the coefficients a_{ijk} given in the Brunovsky–Luenberger canonical forms are invariants of the homological transformations, i.e., they are identical for the controllable pair of matrices* **(A, B)** *and for an arbitary pair of the matrices* $(\mathbf{TAT}^{-1}, \mathbf{TB})$, *where* **T** *is any nonsingular matrix. Moreover, the set of these invariants is independent (for every system of invariants the appropriate pair of the matrices* **(A, B)** *can be obtained) and complete (the pairs of the matrices with the same invariants are connected by the homological transformation).*

The invariants of the homological transformations are frequently used in theoretical research concerning the structure of linear dynamical systems. They can also be used in the so-called structure theorem (Kaczorek (1983a)), representing the structure of the transfer function matrices of the controllable and observable linear, time-invariant dynamical system. Moreover, with the help of the transformation invariants we can form many different canonical forms of the dynamical systems (Kaczorek (1983a, Ch. 4)), such as for example the Fossard canonical form or the Johnson canonical form (Kaczorek (1983a)).

Canonical forms can also be defined for time-varying linear dynamical systems (Kaczorek (1983a)) and even for time-varying, nonlinear dynamical systems of the special form (Kaczorek (1983a)). However, this requires introduction of several additional notions and symbols.

Various other problems related to invariants, invariant subspaces, canonical forms and their connections with the controllability and observability of the dynamical systems are considered, among other things in the following publications: Armentano (1985), Artstein (1982a), Bhattacharyya and Gomes (1973), Basile and Marro (1973), Bittanti, Guardabassi, Maffezoni and Silverman (1978), Brunovsky (1969), Chen (1970), Dickinson (1973), Emre (1980), Fessas (1979, 1981), Heymann (1976), Johnson (1974), Kalman, Falb and Arbib (1969), Karcianas (1979a, 1979b), Koussiouris (1981, 1982), Molinari (1976a, 1976b), Moore (1981), Moore and Loub (1978), Morse (1971, 1973), Porter and Bradshaw (1972), Korobov, Marinich and Podolski (1975), Kreindler and Saradzik (1964), Lee and Arapostathis (1987), Lion (1987), Levan (1980), Luckel and Müller (1975), Murthy (1986), Petersen (1987), Rastovic (1979), Petersen and Barmish (1983b), Schmitendorf and Elenbogen (1982), Sharma (1986), Silvermann (1969), Sundarashan (1979), Takahashi (1984), Volovich (1968), Yousif (1975).

1.15 Remarks and comments

The dynamical system (1.2.1) is a fairly general model of finite-dimensional linear processes; however, it should be mentioned that there exist slightly more general classes of dynamical systems, for which the assumptions about matrices $\mathbf{A}(t)$ and $\mathbf{B}(t)$ are less restrictive than those given in Section 1.2. In the paper of Dauer (1971) was investigated controllability of a dynamical system with matrix $\mathbf{B}(t)$ whose elements are integrable functions. In this case we cannot define the controllability matrix $\mathbf{W}(t_0, t_1)$ of the form (1.3.8) and use Theorem 1.3.1. Moreover, paper of Dłotko, Klamka and Ligęza (1976) contains the necessary and sufficient conditions for controllability in $[t_0, t_1]$ of dynamical system (1.2.1) in the case where the elements of matrix $\mathbf{A}(t)$ are generalized measures, i.e., the distributional derivatives of functions with bounded variation in the time interval $[t_0, t_1]$.

The results concerning the controllability of the dynamical system (1.2.1) with stochastic inputs are presented in the paper of Kaczorek (1971).

The notion of controllability of dynamical systems (1.2.1) is closely connected with the notion of the so-called "invariants" of a dynamical system, and with the stabilizability problem. The relations between controllability and stabilizability for finite-dimensional dynamical systems are considered in the following papers and books: Athans and Falb (1966), Górecki, Fuksa, Korytowski and Mitkowski (1983), Chu (1988), Lee and Markus (1968), Ogata (1967), Pearson and Kwon (1976), Stanford and Connor (1980), Wolovich (1974), Wonham (1978).

Controllability for singularly perturbed linear, continuous-time dynamical systems is extensively investigated in papers of Kokotovic and Haddad (1975), Sannuti (1977, 1978), and some conditions for special kinds of controllability are formulated in them.

The attainable sets for dynamical systems (1.2.1) are analysed for example in publications Artstein (1973, 1980), Bianchini (1983a, 1983b), Pecsvaradi and Narenda (1971), and some qualitative properties of those sets are presented in them. Moreover, in paper of Artstein (1980) the bang-bang principle for dynamical systems (1.2.1) is discussed.

Problems concerning various aspects of controllability of dynamical systems (1.2.1) are also considered in the following publications: Anichini (1983), Bernhard (1980), Brammer (1975), Bittanti, Colaneri and Guardabassi (1984), Clayton and Kuo (1971), Culllen (1986), Gabasov and Kirilova (1971), Furi, Nistri, Pera and Zezza (1985a), Guardabassi, Locatelli and Rinaldi (1974), Clarke and Loewen (1986), Coxson and Shapiro (1987), Molinari (1976a, 1976b), Silvermann (1969), Troch (1971), Veliov and Krastanov (1986), Warren and Eckberg (1975), Weiss (1973), Willems (1986), Wolovich (1968), Wolovich and Hwang (1974), Yip and Sincovec (1981), Yorke (1972), Zwart (1988), Casti and Wood (1984), Kaczorek (1981).

CHAPTER 2

Controllability of Discrete-Time Dynamical Systems

2.1 Introduction

Controllability theory for continuous-time dynamical systems has been developed side by side with controllability theory for discrete-time dynamical systems described by linear difference equations (Kaczorek (1974, 1977, 1980, 1985), Sorenson (1968), Weiss (1972)).

For a discrete-time dynamical system obtained by the discretization of a continuous-time dynamical system defined in a finite-dimensional space, the controllability of the continuous system does not always imply the controllability of the discrete system. This essentially depends on the eigenvalues of the state matrix in the continuous-time dynamical system and on the period of discretization (Barr-Ness (1975), Chen (1970, Appendix C), Evans and Murthy (1977b)). Hence controllability theory for continuous and for discrete dynamical systems should be treated separately.

Recently, besides discrete dynamical systems with one independent variable, also discrete dynamical systems with two or, in general, M independent variables, i.e., the so-called 2-D or M-D dynamical systems have been considered in the literature (Fornassini and Marchesini (1976, 1978, 1979, 1982), Kaczorek (1985, 1986a, 1987a, 1987b), Klamka (1983a, 1983b, 1983c, 1984a, 1986b)). These multidimensional discrete dynamical systems will also be considered in this chapter.

A separate class of discrete dynamical systems which will be investigated in this chapter are bilinear discrete dynamical systems (Goka, Tarn and Zaborszky (1979), Tarn, Elliot and Goka (1973)), i.e., the discrete dynamical systems which are linear separately with respect to the state variables and controls, and contain the product of the state vector and the control.

Generally, in this chapter we shall formulate and prove several necessary and sufficient conditions for various kinds of controllability of linear finite-dimensional dynamical systems. M-D dynamical systems and bilinear dynamical systems will also be considered and some results for them will be presented. The dependence of controllability on the change of system parameters will also be analysed, especially for the discretization problem.

Moreover, similarly as in the case of continuous-time dynamical systems, the minimum energy control problem for discrete-time dynamical systems will be presented and effectively solved, i.e., we shall give an analytic formula for optimal control and for minimal value of performance index corresponding to optimal control. The same will be done also for M-D dynamical systems.

2.2 System description and the basic definitions

In this section we shall present the basic definitions of controllability for linear discrete nonstationary finite-dimensional dynamical systems described by the following difference equation:

$$\mathbf{x}(k+1) = \mathbf{A}(k)\mathbf{x}(k)+\mathbf{B}(k)\mathbf{u}(k), \quad k \geqslant k_0, \tag{2.2.1}$$

where $\mathbf{x}(k) \in \boldsymbol{R}^n$ is the state vector, $\mathbf{u}(k) \in \boldsymbol{R}^m$ is the control vector, $\mathbf{A}(k)$ is an $(n\times n)$-dimensional matrix, $\mathbf{B}(k)$ is an $(n \times m)$-dimensional matrix, $k \in \boldsymbol{Z}$ is the set of integer numbers.

For a given initial condition $\mathbf{x}(k_0) \in \boldsymbol{R}^n$ and control sequence $\{\mathbf{u}(k)\colon k \geqslant k_0\}$, there exists a unique solution $\mathbf{x}(k, \mathbf{x}(k_0), \mathbf{u})$ for any $k \geqslant k_0$ of the difference equation (2.2.1) expressed by the following formula (Kalman, Falb and Arbib (1969), Sorenson (1968), Zadeh and Desoer (1963)):

$$\mathbf{x}(k, \mathbf{x}(k_0), \mathbf{u}) = \mathbf{F}(k, k_0)\mathbf{x}(k_0)+\sum_{j=k_0}^{j=k-1} \mathbf{F}(k,j+1)\mathbf{B}(j)\mathbf{u}(j), \quad k \geqslant k_0, \tag{2.2.2}$$

where $\mathbf{F}(k, j)$ is the $(n\times n)$-dimensional *transition matrix* for the dynamical system (2.2.1) defined for all $k \geqslant j$ in the following manner:

$$\mathbf{F}(k, k) = \mathbf{I}_{n\times n} \quad \text{for } k \in \boldsymbol{Z},$$

$$\mathbf{F}(k,j) = \mathbf{F}(k,j+1)\mathbf{A}(j) = \mathbf{A}(k-1)\mathbf{A}(k-2)\ldots\mathbf{A}(j+1)\mathbf{A}(j) \quad \text{for } k>j.$$

In the special case where the matrices $\mathbf{A}(k)$ are nonsingular for all $k \in \boldsymbol{Z}$, the transition matrix $\mathbf{F}(k, j)$ is defined also for every $k < j$ as follows:

$$\mathbf{F}(k,j) = \mathbf{A}^{-1}(k)\mathbf{F}(k+1,j) = \mathbf{A}^{-1}(k)\mathbf{A}^{-1}(k+1)\ldots\mathbf{A}^{-1}(j-2)\mathbf{A}^{-1}(j-1),$$
$$k < j.$$

Hence in this case the transition matrix has some additional properties, namely:

$$\mathbf{F}(k,j) \text{ is a nonsingular matrix for all } k,j \in \boldsymbol{Z},$$

$$\mathbf{F}^{-1}(k,j) = \mathbf{F}(j,k) \quad \text{for all } k,j \in \boldsymbol{Z},$$

$$\mathbf{F}(k,j) = \mathbf{F}(k,i)\mathbf{F}(i,j) \quad \text{for all } i,j,k \in \boldsymbol{Z},$$

In the case where $\mathbf{A}(k) = \mathbf{A}$ for all $k \in \boldsymbol{Z}$, the transition matrix depends only on the difference $k-j$ in the following manner:

$$\mathbf{F}(k,j) = \mathbf{A}^{k-j} \quad \text{for } k \geqslant j.$$

Moreover, if the matrix $\mathbf{A}$ is nonsingular, then

$$\mathbf{F}(k,j) = \mathbf{A}^{k-j} \quad \text{for all } k,j \in \boldsymbol{Z}.$$

For linear, nonstationary, discrete dynamical systems (2.2.1) we can formulate several basic definitions of controllability (just as was done for continuous dynamical systems in Section 1.2).

DEFINITION 2.2.1. The dynamical system (2.2.1) is said to be *controllable in the interval* $[k_0, k_1] = \{k_0, k_0+1, \ldots, k_1-1, k_1\}$ if, for any initial state $\mathbf{x}(k_0) \in \boldsymbol{R}^n$, and any vector $\mathbf{x}_1 \in \boldsymbol{R}^n$, there exists a control $\mathbf{u}(k)$ defined for $k \in [k_0, k_1) = \{k_0, k_0+1, \ldots, k_1-2, k_1-1\}$ such that the corresponding trajectory $\mathbf{x}(k, \mathbf{x}(k_0), \mathbf{u})$ of the dynamical system (2.2.1) satisfies the following condition:

$$\mathbf{x}(k_1, \mathbf{x}(k_0), \mathbf{u}) = \mathbf{x}_1.$$

DEFINITION 2.2.2. The dynamical system (2.2.1) is said to be *controllable at* k_0, if there exists a $k_1 > k_0$ such that it is controllable in $[k_0, k_1]$.

Regular controllability and *selective controllability* of the dynamical system (2.2.1) are defined in the same way as in the case of continuous-time dynamical systems. Hence the two definitions will be omitted here.

For discrete dynamical systems it is convenient to introduce the notion o the so-called "controllability index" (Sorenson (1968)), which is understood as the number of steps ensuring the controllability of the dynamical system (2.2.1).

DEFINITION 2.2.3. *Controllability index* $\text{ind}(k_0)$ *at* k_0 *for the dynamical system* (2.2.1) is the minimal integer l such that the dynamical system (2.2.1) is controllable in $[k_0, k_0+l]$.

For a stationary discrete dynamical system

$$\mathbf{x}(k+1) = \mathbf{A}\mathbf{x}(k)+\mathbf{B}\mathbf{u}(k), \quad k \geqslant 0 \tag{2.2.3}$$

the controllability index ind_0 is independent of k_0 and satisfies the relation

$$1 \leqslant \text{ind}(k_0) = \text{ind}_0 \leqslant n.$$

Finally let us define the last type of controllability.

DEFINITION 2.2.4. The dynamical system (2.2.1) is said to be *controllable* if it is controllable at any time $k_0 \in \mathbf{Z}$.

For discrete dynamical systems (2.2.1) we can define the *attainable set from the zero initial state* as follows:

$$K(k_0, k_1) = \left\{\mathbf{x} \in \mathbf{R}^n \colon \mathbf{x} = \sum_{j=k_0}^{k_1-1} \mathbf{F}(k_1, j+1)\mathbf{B}(j)\mathbf{u}(j) \colon \mathbf{u}(j) \in \mathbf{R}^m, \ j \in \mathbf{Z}\right\}; \tag{2.2.4}$$

therefore the controllability in $[k_0, k_1]$ of the dynamical system (2.2.1) is equivalent to the condition $K(k_0, k_1) = \mathbf{R}^n$. In a similar way we can express the remaining types of controllability (using the notion of the attainable set $K(k_0, k_1)$).

2.3 Controllability conditions for nonstationary dynamical systems

In this section we shall formulate the necessary and sufficient conditions for controllability in $[k_0, k_1]$ of the discrete dynamical system (2.2.1). In order to do that, let us introduce the so-called *controllability matrix* defined as follows:

$$\mathbf{W}(k_0, k_1) = \sum_{j=k_0}^{k_1-1} \mathbf{F}(k_1, j+1)\mathbf{B}(j)\mathbf{B}^{\mathrm{T}}(j)\mathbf{F}^{\mathrm{T}}(k_1, j+1). \tag{2.3.1}$$

The controllability matrix $\mathbf{W}(k_0, k_1)$ is a symmetric $(n \times n)$-dimensional matrix and is the discrete version of the continuous controllability matrix given by (1.3.8).

THEOREM 2.3.1 (Sorenson (1968)). *The dynamical system* (2.2.1) *is controllable in* $[k_0, k_1]$ *if and only if*

$$\operatorname{rank}\mathbf{W}(k_0, k_1) = n. \tag{2.3.2}$$

Proof:

Sufficiency. Suppose that condition (2.3.2) is satisfied. Then for any initial state $\mathbf{x}(k_0)$ and any vector $\mathbf{x}_1 \in \mathbf{R}^n$ we can define a control $\mathbf{u}(k)$, $k \in [k_0, k_1-1]$ which steers the dynamical system (2.2.1) from the initial state $\mathbf{x}(k_0)$ to the final state $\mathbf{x}(k_1) = \mathbf{x}_1$ at k_1. This control has the following form:

$$\mathbf{u}(k) = \mathbf{B}^{\mathrm{T}}(k)\mathbf{F}^{\mathrm{T}}(k_1, k+1)\mathbf{W}^{-1}(k_0, k_1)\big(\mathbf{x}_1 - \mathbf{F}(k_1, k_0)\mathbf{x}(k_0)\big), \tag{2.3.3}$$
$$k \in [k_0, k_1-1].$$

In order to prove that the control (2.3.3) steers the dynamical sytem (2.2.1) from $\mathbf{x}(k_0)$ to $\mathbf{x}_1$ at k_1 we substitute (2.3.3) into (2.2.2) and for $k = k_1$ we have

$$\mathbf{x}\big(k_1, \mathbf{x}(k_0), \mathbf{u}\big) = \mathbf{F}(k_1, k_0)\mathbf{x}(k_0) + \sum_{j=k_0}^{k_1-1} \mathbf{F}(k_1, j+1)\mathbf{B}(j)\mathbf{u}(j)$$

$$= \mathbf{F}(k_1, k_0)\mathbf{x}(k_0) + \sum_{j=k_0}^{k_1-1} \mathbf{F}(k_1, j+1)\mathbf{B}(j)\mathbf{B}^{\mathrm{T}}(j)\mathbf{F}^{\mathrm{T}}(k_1, j+1)\mathbf{W}^{-1}(k_0, k_1) \times$$

$$\times\left(\mathbf{x}_1-\mathbf{F}(k_1,k_0)\mathbf{x}(k_0)\right)=\mathbf{F}(k_1,k_0)\mathbf{x}(k_0)+\mathbf{W}(k_0,k_1)\mathbf{W}^{-1}(k_0,k_1)\left(\mathbf{x}_1-\right.$$
$$\left.-\mathbf{F}(k_1,k_0)\mathbf{x}(k_0)\right)=\mathbf{x}_1. \tag{2.3.4}$$

Therefore, by Definition 2.2.1, the dynamical system is controllable in $[k_0, k_1]$.

Necessity. By contradiction. Suppose, that the dynamical system (2.2.1) is controllable in $[k_0, k_1]$, but the condition (2.3.2) does not hold. Then there exists a nonzero vector $\mathbf{g} \in R^n$ such that

$$\mathbf{g}^{\mathrm{T}}\mathbf{W}(k_0,k_1)\mathbf{g}=0. \tag{2.3.5}$$

Using (2.3.1) and the properties of the scalar product in the space R^n, we obtain by (2.3.5) the following equalities:

$$\mathbf{g}^{\mathrm{T}}\mathbf{W}(k_0,k_1)\mathbf{g}=\mathbf{g}^{\mathrm{T}}\left(\sum_{j=k_0}^{k_1-1}\mathbf{F}(k_1,j+1)\mathbf{B}(j)\mathbf{B}^{\mathrm{T}}(j)\mathbf{F}^{\mathrm{T}}(k_1,j+1)\right)\mathbf{g}$$
$$=\sum_{j=k_0}^{k_1-1}\left(\left(\mathbf{F}(k_1,j+1)\mathbf{B}(j)\right)^{\mathrm{T}}\mathbf{g}\right)\left(\left(\mathbf{F}(k_1,j+1)\mathbf{B}(j)\right)^{\mathrm{T}}\mathbf{g}\right)$$
$$=\sum_{j=k_0}^{k_1-1}\|\mathbf{B}^{\mathrm{T}}(j)\mathbf{F}^{\mathrm{T}}(k_1,j+1)\mathbf{g}\|^2=0. \tag{2.3.6}$$

Since the vector $\mathbf{g} \in R^n$ is nonzero, relation (2.3.6) implies the following equalities:

$$\mathbf{B}^{\mathrm{T}}(j)\mathbf{F}^{\mathrm{T}}(k_1,j+1)\mathbf{g}=0 \quad \text{for } j\in[k_0,k_1-1]. \tag{2.3.7}$$

The dynamical system (2.2.1) is by assumption controllable in $[k_0, k_1]$. Hence we can choose the initial state $\mathbf{x}(k_0) \in R^n$ and the final vector $\mathbf{x}_1 \in R^n$ in such a way, that the following equality holds:

$$\mathbf{g}=\mathbf{x}_1-\mathbf{F}(k_1,k_0)\mathbf{x}(k_0). \tag{2.3.8}$$

By assumption the dynamical system (2.2.1) is controllable in $[k_0, k_1]$. Hence taking into account the relations (2.2.2) and (2.3.8), we obtain the equality

$$\mathbf{g}=\mathbf{x}_1-\mathbf{F}(k_1,k_0)\mathbf{x}(k_0)=\sum_{j=k_0}^{k_1-1}\mathbf{F}(k_1,j+1)\mathbf{B}(j)\mathbf{u}(j). \tag{2.3.9}$$

Thus by (2.3.7) and (2.3.9) we conclude that

$$\mathbf{g}^{\mathrm{T}}\mathbf{g}=\left(\sum_{j=k_0}^{k_1-1}\mathbf{F}(k_1,j+1)\mathbf{B}(j)\mathbf{u}(j)\right)^{\mathrm{T}}\left(\sum_{j=k_0}^{k_1-1}\mathbf{F}(k_1,j+1)\mathbf{B}(j)\mathbf{u}(j)\right)$$
$$=\mathbf{g}^{\mathrm{T}}\left(\sum_{j=k_0}^{k_1-1}\mathbf{F}(k_1,j+1)\mathbf{B}(j)\mathbf{u}(j)\right)=\sum_{j=k_0}^{k_1-1}\mathbf{g}^{\mathrm{T}}\mathbf{F}(k_1,j+1)\mathbf{B}(j)\mathbf{u}(j)=0. \tag{2.3.10}$$

Equality (2.3.10) implies that the vector $\mathbf{g}$ is equal to $\mathbf{0}$, which contradicts the assumption that there exists a nonzero vector in the space R^n such that equality

(2.3.5) holds. Therefore our hypothesis that $\operatorname{rank}\mathbf{W}(k_0, k_1) < n$ is false, and thus from the assumption that the dynamical system (2.2.1) is controllable in $[k_0, k_1]$ follows the nonsingularity of the controllability matrix $\mathbf{W}(k_0, k_1)$. Hence the necessary condition is proved and our Theorem 2.3.1 follows. ∎

The proof of Theorem 2.3.1 given above is based entirely on linear algebra, on the properties of scalar product in the space $\boldsymbol{R}^n$ and on the transition matrix $\mathbf{F}(k, j)$. It should be pointed out that the proof of Theorem 2.3.1 gives us also some additional information about the steering controls. Indeed, formula (2.3.3) presents the control $\mathbf{u}(k)$, $k \in [k_0, k_1-1]$ which steers the dynamical system (2.2.1) from the initial state $\mathbf{x}(k_0) \in \boldsymbol{R}^n$ to the required final vector $\mathbf{x}_1 \in \boldsymbol{R}^n$ at k_1.

Theorem 2.3.1 may also be proved in another way, by using the properties of linear bounded operators in Hilbert spaces and some well-known theorems from functional analysis (see, e.g., Dunford and Schwartz (1958)). However, this method of proof does not give any additional information about the steering controls. On the other hand, it should be stressed, that the version of proof based on the operator theory is shorter and more general.

The second version of the proof of Theorem 2.3.1. For the dynamical system (2.2.1) let us define the so-called *controllability operator* $P(k_0, k_1)$: $\boldsymbol{R}^{m(k_1-k_0)} \to \boldsymbol{R}^n$ as follows:

$$P(k_0, k_1)[\mathbf{u}(k_0) \;\; \mathbf{u}(k_0+1) \;\; \dots \;\; \mathbf{u}(k_1-2) \;\; \mathbf{u}(k_1-1)]^{\mathrm{T}}$$
$$= \sum_{j=k_0}^{k_1-1} \mathbf{F}(k_1, j+1)\mathbf{B}(j)\mathbf{u}(j). \tag{2.3.11}$$

The controllability operator $P(k_0, k_1)$ is the linear, bounded operator from Hilbert space $\boldsymbol{R}^{m(k_1-k_0)}$ into Hilbert space $\boldsymbol{R}^n$. Hence its adjoint operator $P^*(k_0, k_1)$: $\boldsymbol{R}^n \to \boldsymbol{R}^{m(k_1-k_0)}$ is of the following form (see, e.g., Sorenson (1968)):

$$P^*(k_0, k_1)\mathbf{x} = [\mathbf{B}^{\mathrm{T}}(k_0)\mathbf{F}^{\mathrm{T}}(k_1, k_0+1)\mathbf{x} \;\; \mathbf{B}^{\mathrm{T}}(k_0+1)\mathbf{F}^{\mathrm{T}}(k_1, k_0+2)\mathbf{x} \;\; \dots$$
$$\dots \;\; \mathbf{B}^{\mathrm{T}}(j)\mathbf{F}^{\mathrm{T}}(k_1, j+1)\mathbf{x} \;\; \dots$$
$$\dots \;\; \mathbf{B}^{\mathrm{T}}(k_1-2)\mathbf{F}^{\mathrm{T}}(k_1, k_1-1)\mathbf{x} \;\; \mathbf{B}^{\mathrm{T}}(k_1-1)\mathbf{F}^{\mathrm{T}}(k_1, k_1)\mathbf{x}]^{\mathrm{T}}. \tag{2.3.12}$$

By (2.2.4) the range of the controllability operator $P(k_0, k_1)$ is equal to the attainable set $K(t_0, t_1)$. Therefore, the controllability in $[k_0, k_1]$ of the dynamical system (2.2.1) is equivalent to the statement that the range of the controllability operator $P(k_0, k_1)$ is the whole space $\boldsymbol{R}^n$. On the other hand the last statement is equivalent to the invertibility of the selfadjoint operator $P(k_0, k_1)P^*(k_0, k_1)$ (Dunford and Schwartz (1958)). By (2.3.1), (2.3.11) and (2.3.12) the selfadjoint operator $P(k_0, k_1)P^*(k_0, k_1)$: $\boldsymbol{R}^n \to \boldsymbol{R}^n$ is equal to the controllability matrix $\mathbf{W}(k_0, k_1)$ and its invertibility is equivalent to the nonsingularity of the controllability matrix $\mathbf{W}(k_0, k_1)$. Hence our theorem follows. ∎

The method of proof used in the second version can also be extended to cover the case of infinite-dimensional linear discrete dynamical systems defined in general Hilbert spaces.

COROLLARY 2.3.1. *The dynamical system* (2.2.1) *is controllable in* $[k_0, k_1]$ *if and only if*

$$\begin{aligned}\operatorname{rank}[\mathbf{F}(k_1, k_1)\mathbf{B}(k_1-1) \vdots \mathbf{F}(k_1, k_1-1)\mathbf{B}(k_1-2) \vdots \dots \vdots \mathbf{F}(k_1, j+1)\mathbf{B}(j) \vdots \dots \\ \dots \vdots \mathbf{F}(k_1, k_0+2)\mathbf{B}(k_0+1) \vdots \mathbf{F}(k_1, k_0+1)\mathbf{B}(k_0)] = n. \qquad (2.3.13)\end{aligned}$$

Proof:

Let us denote by $\mathbf{Q}(k_0, k_1)$ the $(n \times m(k_1-k_0))$-dimensional block matrix in the formula (2.3.13). By (2.3.1) we have

$$\mathbf{W}(k_0, k_1) = \mathbf{Q}(k_0, k_1)\mathbf{Q}^{\mathrm{T}}(k_0, k_1). \qquad (2.3.14)$$

Thus by Theorem 2.3.1 we obtain Corollary 2.3.1. It should also be stressed that in the relation (2.3.13) we may substitute $\mathbf{F}(k_1, k_1) = \mathbf{I}_{n\times n}$. ∎

Now let us consider stationary dynamical systems (2.2.3). Since for stationary dynamical systems the transition matrix is $\mathbf{F}(k, j) = \mathbf{A}^{k-j}$, by Corollary 2.3.1 we can derive the necessary and sufficient conditions for controllability in $[0, k_1]$.

COROLLARY 2.3.2. *The dynamical system* (2.2.3) *is controllable in* $[0, k_1]$ *if and only if*

$$\operatorname{rank}[\mathbf{B} \vdots \mathbf{AB} \vdots \mathbf{A}^2\mathbf{B} \vdots \dots \vdots \mathbf{A}^{k_1-1}\mathbf{B}] = n. \qquad (2.3.15)$$

Proof:

Substituting $\mathbf{B}(k) = \mathbf{B}, \mathbf{F}(k, j) = \mathbf{A}^{k-j}$ and $k_0 = 0$ in the formula (2.3.13), by Corollary 2.3.1 we directly obtain Corollary 2.3.2. ∎

COROLLARY 2.3.3. *The dynamical system* (2.2.3) *is controllable in* $[0, k_1], n \leq k_1$ *if and only if it is controllable in* $[0, n]$.

Proof:

By the Cayley–Hamilton theorem (See, e.g. Chen (1970, p. 53), Zadeh and Desoer (1963, p. 506)) the columns of the matrix $\mathbf{A}^{k_1}$ for $k_1 \geq n$ are linear combinations of the columns of the matrices $\mathbf{A}^l$, $l < n$. Thus we have

$$\operatorname{rank}[\mathbf{B} \vdots \mathbf{AB} \vdots \mathbf{A}^2\mathbf{B} \vdots \dots \vdots \mathbf{A}^{k_1-1}\mathbf{B}] = \operatorname{rank}[\mathbf{B} \vdots \mathbf{AB} \vdots \mathbf{A}^2\mathbf{B} \vdots \dots \vdots \mathbf{A}^{n-1}\mathbf{B}] \quad \text{for } n \leq k_1; \qquad (2.3.16)$$

therefore by Corollary 2.3.2 we derive Corollary 2.3.3. ∎

For stationary, discrete dynamical systems (2.2.3) the minimal natural number $\bar{k}$, such that the dynamical system (2.2.3) is controllable in $[0, \bar{k}]$ is, by the considerations given in Section 2.2, equal to the controllability index ind_0. By Corollary 2.3.3 the controllability index ind_0 satisfies the following inequalities:

$$1 \leqslant \mathrm{ind}_0 \leqslant n.$$

It should be pointed out, that $\mathrm{ind}_0 = 1$ if and only if we have $\mathrm{rank}\,\mathbf{B} = n$ and hence $m = n$.

For stationary discrete dynamical systems we can also use the so-called Jordan canonical form to derive the conditions for various types of controllability. The results obtained are almost the same as the results concerning continuous dynamical systems given in Section 1.5. Hence they will be omitted here.

In many publications instead of the statement "controllability in $[0, k_1]$", for stationary discrete dynamical systems (2.2.3) the single term "controllability" is used. This means that there exists a natural number k_1 such that the dynamical system (2.2.3) is controllable in $[0, k_1]$. Moreover, this also means, that

$$\mathrm{rank}[\mathbf{B} \vdots \mathbf{AB} \vdots \mathbf{A}^2\mathbf{B} \vdots \dots \vdots \mathbf{A}^{n-1}\mathbf{B}] = n, \tag{2.3.17}$$

hence the pair of matrices $(\mathbf{A}, \mathbf{B})$ is a controllable pair.

Now let us consider an intersting special case, namely null-controllability $(\mathbf{x}_1 = 0)$ of the dynamical system (2.2.3) with a singular matrix $\mathbf{A}$ $(\det\mathbf{A} = 0)$.

COROLLARY 2.3.4. *The dynamical system* (2.2.3) *is null-controllable in* $[0, k_1]$ *if and only if*

$$\mathrm{Im}\,\mathbf{A}^{k_1} \subset \langle \mathbf{A} | \mathrm{Im}\,\mathbf{B} \rangle_{k_1}, \tag{2.3.18}$$

where $\mathrm{Im}\,\mathbf{M}$ *is the image of the linear transformation represented by the matrix* $\mathbf{M}$ *and*

$$\langle \mathbf{A} | \mathrm{Im}\,\mathbf{B} \rangle_{k_1} = \sum_{j=0}^{k_1-1} \mathbf{A}^j \,\mathrm{Im}\,\mathbf{B}. \tag{2.3.19}$$

Proof:

From formula (2.2.2) it follows that the solution $\mathbf{x}(k_1, \mathbf{x}(0), \mathbf{u})$ of the dynamical system (2.2.3) is of the following form:

$$\mathbf{x}(k_1, \mathbf{x}(0), \mathbf{u}) = \mathbf{A}^{k_1}\mathbf{x}(0) + \sum_{j=0}^{k_1-1} \mathbf{A}^{k_1-j-1}\mathbf{B}\mathbf{u}(j). \tag{2.3.20}$$

Substituting $\mathbf{x}(k_1, \mathbf{x}(0), \mathbf{u}) = \mathbf{x}_1 = 0$ in (2.3.20) we have the following equality:

$$0 = \mathbf{A}^{k_1}\mathbf{x}(0) + \sum_{j=0}^{k_1-1} \mathbf{A}^{k_1-j-1}\mathbf{B}\mathbf{u}(j). \tag{2.3.21}$$

The dynamical system (2.2.3) is null-controllable in $[0, k_1]$ if and only if, for every initial state $\mathbf{x}(0) \in R^n$ there exists the sequence of controls $\{\mathbf{u}(0), \mathbf{u}(1), \dots, \mathbf{u}(k_1-2), \mathbf{u}(k_1-1)\}$ such that equality (2.3.21) is satisfied. Since

$$\sum_{j=0}^{k_1-1} \mathbf{A}^{k_1-j-1}\,\mathrm{Im}\,\mathbf{B} = \sum_{j=0}^{k_1-1} \mathbf{A}^{j}\,\mathrm{Im}\,\mathbf{B}$$

the null-controllability in $[0, k_1]$ of the dynamical system (2.2.3) is immediately equivalent to condition (2.3.18). ∎

In the case where $\det \mathbf{A} \neq 0$, condition (2.3.18) is of the form $\langle \mathbf{A}|\mathrm{Im}\,\mathbf{B}\rangle_{k_1} = R^n$ and null-controllability in $[0, k_1]$ of the dynamical system (2.2.3) is by Corollary 2.3.2 equivalent to controllability in $[0, k_1]$ (in the sense of Definition 2.2.1). Generally, controllability in $[0, k_1]$ of the dynamical system (2.2.3) implies null-controllability in $[0, k_1]$. In the case where the matrix $\mathbf{A}$ is singular the null-controllability in $[0, k_1]$ of the dynamical system (2.2.3) is essentially weaker notion than controllability in $[0, k_1]$ (in the sense of Definition 2.2.1). We shall illustrate this phenomenon in Example 2.3.1.

Example 2.3.1
Let us consider the dynamical system (2.2.3) with matrices $\mathbf{A}$ and $\mathbf{B}$ of the following form:

$$\mathbf{A} = \begin{bmatrix} 1 & -1 \\ 1 & -1 \end{bmatrix}, \quad \mathbf{B} = \mathbf{b} = \begin{bmatrix} 1 \\ 1 \end{bmatrix}. \tag{2.3.22}$$

Therefore $n = 2, m = 1$ and $\det \mathbf{A} = 0$. Since $\mathrm{rank}[\mathbf{b}, \mathbf{Ab}] = \mathrm{rank}\left[\begin{array}{c:c} 1 & 0 \\ 1 & 0 \end{array}\right] = 1 < 2 = n$, by Corollary 2.3.3 the dynamical system (2.3.22) is not controllable in each interval $[0, k_1]$ for $k_1 \geqslant 1$. On the other hand, since $\mathbf{A}^{k_1} = \begin{bmatrix} 0 & 0 \\ 0 & 0 \end{bmatrix}$ for $k_1 \geqslant 2$, we have

$$\mathrm{Im}\,\mathbf{A}^{k_1} \subset \langle \mathbf{A}|\mathrm{Im}\,\mathbf{b}\rangle_{k_1} \quad \text{for } k_1 \geqslant 1\,.$$

Thus the dynamical system (2.3.22) is null-controllable in each interval $[0, k_1]$ for $k_1 \geqslant 1$.

2.4 Controllability under the changing system parameters

In this section we shall present theorems and corollaries concerning the influence exerted on controllability by the changes of the parameters of the dynamical system (2.2.3). A similar problem has been considered in the Section 1.6 for nonstationary continuous-time dynamical systems under the assumption that the changes of parameters are arbitrary. In this section it is assumed that the changes of parameters of a dynamical system are not arbitrary, but described by a suitably chosen function. The analysis of this problem will proceed in the same way as in the paper of Yunlong, Laixiang and Huisheng (1982). To begin with, let us introduce some symbols and assumptions.

Let f: $C \to C$ and g: $C \to C$ be analytic functions represented by the Taylor series

$$f(s) = \sum_{i=0}^{\infty} a_i s^i, \quad g(s) = \sum_{i=0}^{\infty} b_i s^i \tag{2.4.1}$$

with convergence radii p_f and p_g, respectively.

For the $(n \times n)$-dimensional matrix $\mathbf{A}$ with eigenvalues $s_1, s_2, \ldots, s_l$ $(l \leqslant n)$ lying in the convergence domains of the series (2.4.1), we define the matrix functions $f(\mathbf{A})$ and $g(\mathbf{A})$ in the following way (Chen (1970, Ch. 2.7), Zadeh and Desoer (1963, App. D)):

$$f(\mathbf{A}) = \sum_{i=0}^{\infty} a_i \mathbf{A}^i, \quad g(\mathbf{A}) = \sum_{i=0}^{\infty} b_i \mathbf{A}^i. \tag{2.4.2}$$

Several additional properties of the matrix functions are given in the works cited above. As in Section 1.5, let $r(i)$, $i = 1, 2, \ldots, l$ denote, respectively, the multiplicities of the eigenvalues s_i in the characteristic polynomial. Moreover, let e_i, $i = 1, 2, \ldots, l$ denote the multiplicities of the eigenvalues s_i in the minimal polynomial (Chen 1970, p. 50), Zadeh and Desoer (1963, p. 306)).

In addition to the dynamical system (2.2.3) we shall also consider a single-input $(m = 1)$ stationary, discrete dynamical system of the following form:

$$\mathbf{x}(k+1) = \mathbf{A}\mathbf{x}(k) + \mathbf{b}u(k), \quad \mathbf{b} \in R^n, \ k \geqslant 0. \tag{2.4.3}$$

Moreover, we shall examine stationary discrete dynamical systems obtained from the dynamical systems (2.2.3) and (2.4.3) by changing the parameters according to the given functions. These dynamical systems are given, respectively, by the following equations:

$$\mathbf{x}(k+1) = f(\mathbf{A})\mathbf{x}(k) + g(\mathbf{A})\mathbf{B}\mathbf{u}(k), \quad k \geqslant 0 \tag{2.4.4}$$

and

$$\mathbf{x}(k+1) = f(\mathbf{A})\mathbf{x}(k) + g(\mathbf{A})\mathbf{b}u(k), \quad k \geqslant 0. \tag{2.4.5}$$

THEOREM 2.4.1 (Yunlong, Laixiang and Huisheng (1982)). *Suppose*

$$\operatorname{rank}[\mathbf{b} \,\vdots\, \mathbf{A}\mathbf{b} \,\vdots\, \mathbf{A}^2\mathbf{b} \,\vdots\, \ldots \,\vdots\, \mathbf{A}^{n-1}\mathbf{b}] = n. \tag{2.4.6}$$

Then the dynamical system (2.4.5) *is controllable if and only if*

$$f(s_i) \neq f(s_j) \quad \text{for } i \neq j,$$

$$f^{(1)}(s_i) \neq 0 \quad \text{if } e_i > 1, i = 1, 2, \ldots, l$$

and (2.4.7)

$$g(s_i) \neq 0 \quad \text{for } i = 1, 2, \ldots, l.$$

Proof:

If the relation (2.4.6) is satisfied, then the vectors $\mathbf{b}$, $\mathbf{A}\mathbf{b}$, $\mathbf{A}^2\mathbf{b}$, ..., $\mathbf{A}^{n-1}\mathbf{b}$ are linearly independent in the space R^n, and therefore the minimal polynomial of the matrix **A** is of degree n and $e_1+e_2+\dots+e_l = n$. Since $e_i \leqslant n_i$ and $n_1+n_2+\dots+n_l = n$, we have $e_i = n_i$ for $i = 1, 2, \dots, l$. The analytic function $F(s)$ defined in the domain containing the eigenvalues s_i, $i = 1, 2, \dots, l$ of the matrix **A** can be uniquely extended to the matrix function $F(\mathbf{A})$ as follows (Zadeh and Desoer: (1963, Appendix D)):

$$F(\mathbf{A}) = \sum_{i=0}^{l}\sum_{j=0}^{e_i-1} (F(s_i))^{(j)} Z_i^j(\mathbf{A}), \tag{2.4.8}$$

where $Z_i^j(s)$ are polynomials of the degrees not greater than n and satisfying the relation

$$(Z_i^j(s))^{(k)} = \begin{cases} 1 & \text{for } j = k, \\ 0 & \text{otherwise.} \end{cases}$$

Substituting $F(s) = s^k$, $k = 0, 1, 2, \dots, n-1$, by (2.4.8) we obtain

$$\mathbf{A}^k\mathbf{b} = \sum_{i=0}^{l}\sum_{j=0}^{e_i-1} (s_i^k)^{(j)} Z_i^j(\mathbf{A})b, \quad k = 0, 1, 2, \dots, n-1.$$

Since the vectors $\mathbf{b}, \mathbf{A}\mathbf{b}, \mathbf{A}^2\mathbf{b}, \dots, \mathbf{A}^{n-1}\mathbf{b}$ are linearly independent in the space R^n, also the n vectors

$$\{Z_i^j(\mathbf{A})\mathbf{b}\colon\ i = 1, 2, \dots, l,\ j = 0, 1, 2, \dots, e_i-1\}$$

are linearly independent in the space R^n. These vectors form a basis in the space R^n. Substituting $F(s) = f^k(s)$, $k = 0, 1, 2, \dots, n-1$ by (2.4.8), we have

$$f^k(\mathbf{A})\mathbf{b} = \sum_{i=0}^{l}\sum_{j=0}^{e_i-1} (f^k(s_i))^{(j)} Z_i^j(\mathbf{A})\mathbf{b}.$$

By the results of Yunlong, Laixiang and Huisheng (1982) the vectors $\mathbf{b}$, $f(\mathbf{A})\mathbf{b}$, $f^2(\mathbf{A})\mathbf{b}, \dots, f^{n-1}(\mathbf{A})\mathbf{b}$ are linearly independent in the space R^n if and only if $f(s_i) \neq f(s_j)$ for $i \neq j$ and $f^{(1)}(s_i) \neq 0$ if $e_i > 1$ for $i = 1, 2, \dots, l$.

On the other hand, the dynamical system (2.4.5) is controllable if and only if

$$\operatorname{rank}[g(\mathbf{A})\mathbf{b} \,\vdots\, f(\mathbf{A})g(\mathbf{A})\mathbf{b} \,\vdots\, f^2(\mathbf{A})g(\mathbf{A})\mathbf{b} \,\vdots\, \dots \,\vdots\, f^{n-1}(\mathbf{A})g(\mathbf{A})\mathbf{b}] = n. \tag{2.4.9}$$

Since $g(\mathbf{A})f^k(\mathbf{A}) = f^k(\mathbf{A})g(\mathbf{A})$, we have

$$\begin{aligned}&\operatorname{rank}[g(\mathbf{A})\mathbf{b} \,\vdots\, f(\mathbf{A})g(\mathbf{A})\mathbf{b} \,\vdots\, f^2(\mathbf{A})g(\mathbf{A})\mathbf{b} \,\vdots\, \dots \,\vdots\, F^{n-1}(\mathbf{A})g(\mathbf{A})\mathbf{b}]\\ &= \operatorname{rank}([g(\mathbf{A})][\mathbf{b} \,\vdots\, f(\mathbf{A})\mathbf{b} \,\vdots\, f^2(\mathbf{A})\mathbf{b} \,\vdots\, \dots \,\vdots\, f^{n-1}(\mathbf{A})\mathbf{b}]) = n,\end{aligned}$$

which is equivalent to the statements that $\operatorname{rank} g(\mathbf{A}) = n$ and

$$\operatorname{rank}[\mathbf{b} \,\vdots\, f(\mathbf{A})\mathbf{b} \,\vdots\, f^2(\mathbf{A})\mathbf{b} \,\vdots\, \dots \,\vdots\, f^{n-1}(\mathbf{A})\mathbf{b}] = n. \tag{2.4.10}$$

We have $\operatorname{rank} g(\mathbf{A}) = n$ if and only if $g(s_i) \neq 0$ for $i = 1, 2, \dots, l$. Moreover, by the first part of the proof the relation (2.4.10) holds if and only if $f(s_i) \neq f(s_j)$ for $i \neq j$ and $f^{(1)}(s_i) \neq 0$ if $e_i > 1$ for $i = 1$ $2, \dots, l$. Hence our theorem follows. ■

For multiinput dynamical systems ($m > 1$), instead of a necessary and sufficient condition of controllability for the dynamical system (2.4.4) we can only obtain a sufficient condition of controllability. (Theorem 2.4.1 is no longer true.)

THEOREM 2.4.2 (Yunlong, Laixiang and Huisheng (1982)). *Suppose that*

$$\operatorname{rank}[\mathbf{B} \vdots \mathbf{AB} \vdots \mathbf{A}^2\mathbf{B} \vdots \ldots \vdots \mathbf{A}^{n-1}\mathbf{B}] = n \tag{2.4.11}$$

and the functions f and g satify the same assumptions as in Theorem 2.4.1. *Then the dynamical system* (2.4.4) *is controllable.*

Proof:

Let $\mathbf{B} = [\mathbf{b}_1 \vdots \mathbf{b}_2 \vdots \ldots \vdots \mathbf{b}_n \vdots \ldots \vdots \mathbf{b}_m]$. Then by (2.4.8) and subsequent relations we have

$$\mathbf{A}^k\mathbf{b}_p = \sum_{i=1}^{l}\sum_{j=0}^{e_i-1} (s_i^k)^{(j)} Z_i^j(\mathbf{A})\,\mathbf{b}_p, \quad p = 1, 2, \ldots, m, \; k = 0, 1, 2, \ldots, n-1. \tag{2.4.12}$$

By (2.4.11) and (2.4.12) it is always possible to choose n linearly independent vectors is the space $\boldsymbol{R}^n$ from the set of the following vectors:

$$\{Z_i^j(\mathbf{A})\mathbf{b}_p\colon\; i = 1, 2, \ldots, l,\; j = 1, 2, \ldots, e_i-1, p = 1, 2, \ldots, m\}. \tag{2.4.13}$$

As in the proof of Theorem 2.4.1, we can obtain

$$f^k(\mathbf{A})\mathbf{b}_p = \sum_{i=0}^{l}\sum_{j=0}^{e_i-1} (f^k(s_i))^{(j)} Z_i^j(\mathbf{A})\mathbf{b}_p,$$

$$p = 1, 2, \ldots, m, \; k = 0, 1, 2, \ldots, n-1.$$

Suppose that the functions f and g satisfy the conditions of Theorem 2.4.1. Then by the results of Yunlong, Laixiang and Huisheng (1982) and the fact that in the set (2.4.13) there exist n linearly independent vectors we conlcude that

$$\operatorname{rank}[g(\mathbf{A})\mathbf{B} \vdots f(\mathbf{A})g(\mathbf{A})\mathbf{B} \vdots f^2(\mathbf{A})g(\mathbf{A})\mathbf{B} \vdots \ldots \vdots f^{n-1}(\mathbf{A})g(\mathbf{A})\mathbf{B}] = n \tag{2.4.14}$$

which means that the dynamical system (2.4.4) is controllable. ■

Since the controllability conditions for stationary continuous and discrete dynamical systems are purely algebraic and have the same structure, all the results given in Theorems 2.4.1 and 2.4.2 are also true for continuous dynamical systems (1.2.8).

Theorems 2.4.1 and 2.4.2 can also be used to determine controllability conditions for a discrete dynamical system obtained by discretization of the continuous dynamical system (1.2.8). Controllability after the introduction of sampling of the dynamical system (1.2.8) has been considered in Bar-Ness and Langholtz (1975),

Chen (1970, Appendix C), Yunlong, Laixiang and Huisheng (1982), where several conditions for controllability are presented.

Suppose we are given a linear, continuous stationary dynamical system of the form (1.2.8). We consider now the case in which the input vector $\mathbf{u}(t)$ is piecewise constant; that is, the control $\mathbf{u}(t)$ changes value only at a discrete instant of time $t_k = kT$, $k = 0, 1, 2, \ldots$, where $T > 0$ is the so-called *sampling period.* Inputs of this type occur in sampled-data systems or in systems in which digital computers are used to generate the control $\mathbf{u}(t)$. A piecewise-constant control is often generated by a sampler and a filter called a *zero-order hold.* The control $\mathbf{u}(t)$, $t \geq 0$ is then of the following form:

$$\mathbf{u}(t) = \mathbf{u}(k) \in R^m \quad \text{for } t \in [kT, (k+1)T) = [t_k, t_{k+1}), \; k = 0, 1, 2, \ldots \tag{2.4.15}$$

By (1.2.2) and (1.2.10) we have

$$\begin{aligned}\mathbf{x}(t_{k+1}, \mathbf{x}(0), \mathbf{u}) &= \exp(\mathbf{A}(k+1)T)\mathbf{x}(0) + \int_0^{(k+1)T} \exp(\mathbf{A}(kT+T-t))\mathbf{B}\mathbf{u}(t)\,dt \\ &= \exp(\mathbf{A}T)\Big(\exp(\mathbf{A}kT)\mathbf{x}(0) + \int_0^{kT} \exp(\mathbf{A}(kT-t))\mathbf{B}\mathbf{u}(t)\,dt\Big) + \\ &\quad + \int_{kT}^{(k+1)T} \exp(\mathbf{A}(kT+T-t))\mathbf{B}\mathbf{u}(t)\,dt.\end{aligned} \tag{2.4.16}$$

Since $\mathbf{x}(t_k, \mathbf{x}(0), \mathbf{u}) = \exp(\mathbf{A}kT)\mathbf{x}(0) + \int_0^{kT} \exp(\mathbf{A}(kT-t))\mathbf{B}\mathbf{u}(t)dt$ and the control $\mathbf{u}(t)$ equals $\mathbf{u}(k)$ for $t \in [kT, (k+1)T)$, then changing the variable of integration $\tau = kT+T-t$, by (2.4.16), we obtain the following equality:

$$\mathbf{x}(t_{k+1}, \mathbf{x}(0), \mathbf{u}) = \exp(\mathbf{A}T)\mathbf{x}(t_k, \mathbf{x}(0), \mathbf{u}) + \Big(\int_0^T \exp(\mathbf{A}\tau)\,d\tau\Big)\mathbf{B}\mathbf{u}(k). \tag{2.4.17}$$

Substituting $\mathbf{x}(t_k, \mathbf{x}(0), \mathbf{u}) = \mathbf{x}(k)$, $k = 0, 1, 2, \ldots$, $\tilde{\mathbf{A}} = \exp(\mathbf{A}T)$, $\tilde{\mathbf{B}} = (\int_0^T \exp(\mathbf{A}\tau)d\tau)\mathbf{B}$ we can transform the relation (2.4.17) into the state linear difference equation

$$\mathbf{x}(k+1) = \tilde{\mathbf{A}}\mathbf{x}(k) + \tilde{\mathbf{B}}\mathbf{u}(k), \quad k \geq 0. \tag{2.4.18}$$

It is easy to verify, that equation (2.4.18) is the same as equation (2.2.3), which represents the linear discrete stationary dynamical system. The dynamical system (2.4.18) is the discrete version of the continuous dynamical system (2.2.8). It should be pointed out that the matrix $\tilde{\mathbf{A}} = \exp(\mathbf{A}T)$ is always nonsingular for any sampling period $T > 0$ and any matrix $\mathbf{A}$ (the matrix $\mathbf{A}$ may be even the zero matrix, and hence

$\tilde{\mathbf{A}} = \mathbf{I}_{n\times n}$). To investigate the relations between the controllability of the continuous dynamical system (1.2.8) and its discrete version (2.4.18) Theorems 2.4.1 and 2.4.2 can be used.

COROLLARY 2.4.1 (Bar-Ness and Langholtz (1975), Chen (1970, App. C), Yunlong, Laixiang and Huisheng (1982)). *Suppose the dynamical system* (1.2.8) *is controllable. Then the dynamical system* (2.4.18) *is controllable if*

$$\mathrm{Im}(s_i - s_j) \neq \frac{2\pi q}{T} \quad \text{for } q = \pm 1, \pm 2, \pm 3, \ldots \tag{2.4.19}$$

whenever

$$\mathrm{Re}(s_i - s_j) = 0 \quad \text{for } i, j = 1, 2, \ldots, l.$$

Moreover, if $m = 1$, i.e., the dynamical system is a single input one, then condition (2.4.19) *is the necessary and sufficient condition for the controllability of the dynamical system* (2.4.18).

Proof:

Putting $f(s) = \exp(sT)$ and $g(s) = \int_0^T \exp(st)dt$, we obtain from Theorem 2.4.2 the sufficient condition of controllability and from Theorem 2.4.1 the necessary and sufficient condition of controllability for single input dynamical system (2.4.18). ∎

Example 2.4.1 (Chen (1970, p. 408))
Let us consider the controllable dynamical system (1.2.8) with matrices **A** and **B** of the following form:

$$\mathbf{A} = \begin{bmatrix} -1 & 0 & 0 \\ 0 & -1-2i & 0 \\ 0 & 0 & -1+2i \end{bmatrix}, \quad \mathbf{B} = \mathbf{b} = \begin{bmatrix} 1 \\ 1 \\ 1 \end{bmatrix}. \tag{2.4.20}$$

Therefore $n = 3$, $m = 1$, $s_1 = -1$, $s_2 = -1-2i$, $s_3 = -1+2i$. Hence $\mathrm{Re}(s_1 - s_2) = 0$, $\mathrm{Re}(s_1 - s_3) = 0$ and $\mathrm{Re}(s_2 - s_3) = 0$. From Corollary 2.4.1 it follows that the discrete dynamical system (2.4.18) obtained by discretization of the continuous dynamical system (2.4.20) is controllable if and only if all the following conditions are fulfilled:

$$\begin{gathered} \mathrm{Im}(s_1 - s_2) = 2 \neq \frac{2\pi q}{T}, \quad \mathrm{Im}(s_1 - s_3) = -2 \neq \frac{2\pi q}{T}, \\ \mathrm{Im}(s_2 - s_3) = -4 \neq \frac{2\pi q}{T} \quad \text{for } q = \pm 1, \pm 2, \pm 3, \ldots \end{gathered} \tag{2.4.21}$$

Thus the dynamical system (2.4.18) is controllable if and only if the sampling period T satisfies the followng inequality:

$$T \neq \tfrac{1}{2}\pi q \quad \text{for} \quad q = \pm 1, \pm 2, \pm 3, \ldots \tag{2.4.22}$$

It is easy to verify, that in this case the matrices $\tilde{\mathbf{A}}$ and $\tilde{\mathbf{B}} = \tilde{\mathbf{b}}$ have the form

$$\tilde{\mathbf{A}} = \begin{bmatrix} \exp(-T) & 0 & 0 \\ 0 & \exp(-T-2iT) & 0 \\ 0 & 0 & \exp(-T+2iT) \end{bmatrix},$$

$$\tilde{\mathbf{B}} = \tilde{\mathbf{b}} = \begin{bmatrix} 1-\exp(-T) \\ (1-\exp(-T-2iT))/(1+2i) \\ (1-\exp(-T+2iT))/(1-2i) \end{bmatrix}.$$

Knowing the matrices $\tilde{\mathbf{A}}$ and $\tilde{\mathbf{B}}$, we can derive the controllability condition (2.4.21) directly from the equality $\text{rank}[\tilde{\mathbf{B}} \vdots \tilde{\mathbf{A}}\tilde{\mathbf{B}} \vdots \tilde{\mathbf{A}}^2\tilde{\mathbf{B}}] = 3$. The result is of course the same.

In practice Corollary 2.4.1 gives us the information about the sampling period, which guarantees the controllability of discrete version dynamical system. It is obvious that the set of admissible controls is essentially smaller for the discrete version than for the continuous version of a dynamical system.

Example 2.4.2

Let us consider the dynamical system (1.2.8) with the following matrices **A** and **B**:

$$\mathbf{A} = \begin{bmatrix} 0 & 1 \\ -w_0^2 & 0 \end{bmatrix}, \quad \mathbf{B} = \mathbf{b} = \begin{bmatrix} 0 \\ w_0^2 \end{bmatrix}, \quad w_0 \neq 0. \tag{2.4.23}$$

Since

$$\text{rank}[\mathbf{b} \vdots \mathbf{Ab}] = \text{rank}\begin{bmatrix} 0 & w_0^2 \\ w_0^2 & 0 \end{bmatrix} = 2 = n,$$

then the dynamical system (2.4.23) is controllable.

The eigenvalues s_1 and s_2 of the matrix **A**, obtained from the characteristic equation

$$\det(s\mathbf{I}-\mathbf{A}) = \det\begin{bmatrix} s & -1 \\ w_0^2 & s \end{bmatrix} - s^2 + w_0^2 = 0$$

are as follows: $s_1 = iw_0$ and $s_2 = -iw_0$. Hence $\text{Re}(s_1 - s_2) = 0$ and $\text{Im}(s_1 - s_2) = 2w_0$. Thus, by Corollary 2.4.1, the discrete dynamical system (2.4.18) obtained by discretization of the continuous dynamical system (2.4.23) is controllable if and only if

$$T \neq \frac{\pi q}{w_0} \quad \text{for } q = \pm 1, \pm 2, \pm 3, \dots \tag{2.4.24}$$

It is easily verified, that the matrices $\tilde{\mathbf{A}}$ and $\tilde{\mathbf{B}}$ are of the following form:

$$\tilde{\mathbf{A}} = \begin{bmatrix} \cos(w_0 T) & \sin(w_0 T)/w_0 \\ -w_0 \sin(w_0 T) & \cos(w_0 T) \end{bmatrix}, \quad \tilde{\mathbf{B}} = \tilde{\mathbf{b}} = \begin{bmatrix} 0 \\ 1 \end{bmatrix}.$$

Hence

$$\text{rank}[\tilde{\mathbf{b}} \vdots \tilde{\mathbf{A}}\tilde{\mathbf{b}}] = \text{rank}\begin{bmatrix} 0 & \sin(w_0 T)/w_0 \\ 1 & \cos(w_0 T) \end{bmatrix}.$$

Therefore we have rank$[\tilde{\mathbf{b}} \vdots \tilde{\mathbf{A}}\tilde{\mathbf{b}}] = 2 = n$ if and only if the relation (2.4.24) holds. This example shows how to compute a sampling period, which guarantees the controllability of the discrete version of a dynamical system.

In the book of Chen (1970), the Jordan canonical form of a continuous dynamical system is used to prove theorems analogous to Corollary 2.4.1. Using the dual principle we can solve the problem of observability of the discrete version of a dynamical system. The methods of proofs and the results obtained are quite similar to the controllability results. It is a consequence of the pure algebraic nature of our considerations and of the duality principle. Finally, it should be stressed that in the case where the sampling period is not constant, the obtained results are no longer valid.

2.5 Minimum energy control

For linear, discrete dynamical systems (2.2.1), just as for continuous dynamical systems (1.2.1) we can formulate the so-called *minimum energy control* problem. Under the assumption that the dynamical system (2.2.1) is controllable in $[k_0, k_1]$, there generally exist many different controls $\mathbf{u}(k)$, $k \in [k_0, k_1-1]$, which can transfer the initial state $\mathbf{x}(k_0) \in \boldsymbol{R}^n$ of the dynamical system (2.2.1) to the desired final state $\mathbf{x}_1 \in \boldsymbol{R}^n$ at the moment k_1. Among these possible controls which achieve the same transfer we may ask which is the optimal one according to the following cost function (performance index):

$$J(\mathbf{u}) = \sum_{j=k_0}^{k_1-1} ||\mathbf{u}(k)||^2_{\mathbf{R}(k)}, \tag{2.5.1}$$

where $\mathbf{R}(k)$, $k \in [k_0, k_1-1]$ is the $(n \times n)$-dimensional, nonsingular, symmetric, positive definite weighting matrix. Moreover, we denote:

$$||\mathbf{u}(k)||^2_{\mathbf{R}(k)} = \langle \mathbf{u}(k), \mathbf{R}(k)\mathbf{u}(k)\rangle_{R^m} = \mathbf{u}^{\mathrm{T}}(k)\mathbf{R}(k)\mathbf{u}(k).$$

Generally the cost function $J(\mathbf{u})$ denotes the control energy in $[k_0, k_1-1]$ with a certain weight. In the special case where $\mathbf{R}(k) = \mathbf{I}_{m \times m}$, for $k \in [k_0, k_1-1]$ the performance index $J(\mathbf{u})$ denotes directly the energy of the control $\mathbf{u}(k)$ in the interval $[k_0, k_1-1]$. The performance index $J(\mathbf{u})$ given by (2.5.1) is the special case of the general quadratic cost function, and so the existence of the solution of the optimization problem follows from the general theory (Sorenson (1968)) of quadratic optimization.

In this section, using a similar method to that applied in Section 1.7, we shall solve the minimum energy control problem for the discrete dynamical system (2.2.1). Namely, we shall prove a theorem which gives an analytic formula for the optimal control function and, moreover, the corresponding minimal value of the cost function (2.5.1).

For abbreviation, let us introduce the following notation:

$$\mathbf{W}_\mathbf{R}(k_0, k_1) = \sum_{j=k_0}^{k_1-1} \mathbf{F}(k_1, j+1)\mathbf{B}(j)\mathbf{R}^{-1}(j)\mathbf{B}^\mathrm{T}(j)\mathbf{F}^\mathrm{T}(k_1, j+1). \tag{2.5.2}$$

Hence $\mathbf{W}_\mathbf{R}(k_0, k_1)$ is an $(n \times n)$-dimensional, symmetric matrix.

Under the assumption, that the dynamical system (2.2.1) is controllable in $[k_0, k_1]$, we conclude by Theorem 2.3.1 that there exists an inverse matrix $\mathbf{W}^{-1}(k_0, k_1)$ and therefore by (2.5.2) that there exists a matrix $\mathbf{W}_\mathbf{R}^{-1}(k_0, k_1)$. Using the matrix $\mathbf{W}_\mathbf{R}^{-1}(k_0, k_1)$, we define the control $\mathbf{u}^0(k)$ for $k \in [k_0, k_1 - 1]$ as follows:

$$\mathbf{u}^0(k) = \mathbf{R}^{-1}(k)\mathbf{B}^\mathrm{T}(k)\mathbf{F}^\mathrm{T}(k_1, k)\mathbf{W}^{-1}(k_0, k_1)\big(\mathbf{x}_1 - \mathbf{F}(k_1, k_0)\mathbf{x}(k_0)\big),$$
$$k \in [k_0, k_1 - 1]. \tag{2.5.3}$$

THEOREM 2.5.1 (Klamka (1977d), Sorenson (1968)). *Suppose the dynamical system* (2.2.1) *is controllable in* $[k_0, k_1]$ *and let* $\mathbf{u}'(k)$, $k \in [k_0, k_1 - 1]$ *be any control which steers that system* (2.2.1) *from the initial state* $\mathbf{x}(k_0) \in \mathbf{R}^n$ *to the desired final state* $\mathbf{x}_1 \in \mathbf{R}^n$ *at the moment* k_1. *Then the control* $\mathbf{u}^0(k)$, $k \in [k_0, k_1 - 1]$ *given by formula* (2.5.3) *also steers the dynamical system* (2.2.1) *from the initial state* $\mathbf{x}(k_0)$ *to the final state* $\mathbf{x}_1$ *at* k_1 *and additionally*

$$\sum_{j=k_0}^{k_1-1} \|\mathbf{u}'(j)\|^2_{\mathbf{R}(j)} \geqslant \sum_{j=k_0}^{k_1-1} \|\mathbf{u}^0(j)\|^2_{\mathbf{R}(j)}. \tag{2.5.4}$$

Moreover, the minimal value of the performance index $J(\mathbf{u}^0)$ *is given by the formula*

$$J(\mathbf{u}^0) = \sum_{j=k_0}^{k_1-1} \|\mathbf{u}^0(j)\|^2_{\mathbf{R}(j)} = \|\mathbf{x}_1 - \mathbf{F}(k_1, k_0)\mathbf{x}(k_0)\|^2_{\mathbf{W}_\mathbf{R}^{-1}(k_0, k_1)}. \tag{2.5.5}$$

Proof:

The proof of Theorem 2.5.1 is divided into two parts. In the first part we shall show that the control $\mathbf{u}^0(k)$, $k \in [k_0, k_1 - 1]$ given by formula (2.5.3) steers the dynamical system (2.2.1) from the initial state $\mathbf{x}(k_0) \in \mathbf{R}^n$ to the desired final state $\mathbf{x}_1 \in \mathbf{R}^n$ at k_1. From the controllability assumption it follows that there exists an inverse matrix $\mathbf{W}_\mathbf{R}^{-1}(k_0, k_1)$. Therefore the control $\mathbf{u}^0(k)$ given by formula (2.5.3) is well defined. Substituting (2.5.3) in (2.2.2), for $k = k_1$ we have

$$\mathbf{x}\big(k_1, \mathbf{x}(k_0), \mathbf{u}^0\big) = \mathbf{F}(k_1, k_0)\mathbf{x}(k_0) + \sum_{j=k_0}^{k_1-1} \mathbf{F}(k_1, j+1)\mathbf{B}(j) \times$$
$$\times \mathbf{R}^{-1}(j)\mathbf{B}^\mathrm{T}(j)\mathbf{F}^\mathrm{T}(k_1, j)\mathbf{W}_\mathbf{R}^{-1}(k_0, k_1)\big(\mathbf{x}_1 - \mathbf{F}(k_1, k_0)\mathbf{x}(k_0)\big)$$
$$= \mathbf{F}(k_1, k_0)\mathbf{x}(k_0) + \mathbf{W}_\mathbf{R}(k_0, k_1)\mathbf{W}_\mathbf{R}^{-1}(k_0, k_1)\big(\mathbf{x}_1 - \mathbf{F}(k_1, k_0)\mathbf{x}(k_0)\big) = \mathbf{x}_1. \tag{2.5.6}$$

Formula (2.5.6) directly implies, that the control $\mathbf{u}^0(k)$ given by (2.5.3) steers the dynamical system (2.2.1) from any initial state $\mathbf{x}(k_0)$ to any desired final state

$\mathbf{x}_1$ at k_1. Hence, by Defintion 2.2.1, the dynamical system (2.2.1) is controllable in $[k_0, k_1]$.

In the next step of the proof, we shall prove that inequality (2.5.4) is satisfied. In order to do that, we shall use the fundamental properties of the norm and the scalar product in the space $\boldsymbol{R}^n$. Let us asume that $\mathbf{u}'(k)$, $k \in [k_0, k_1 - 1]$ is any control which steers the dynamical system (2.2.1) from a given initial state $\mathbf{x}(k_0)$ to the desired final state $\mathbf{x}_1$ at k_1. Hence the following equality holds:

$$\sum_{j=k_0}^{k_1-1} \mathbf{F}(k_1, j+1)\mathbf{B}(j)\mathbf{u}'(j) = \sum_{j=k_0}^{k_1-1} \mathbf{F}(k_1, j+1)\mathbf{B}(j)\mathbf{u}^0(j). \tag{2.5.7}$$

Subtracting by both sides, we obtain the following equality:

$$\sum_{j=k_0}^{k_1-1} \mathbf{F}(k_1, j+1)\mathbf{B}(j)\left(\mathbf{u}'(j) - \mathbf{u}^0(j)\right) = 0. \tag{2.5.8}$$

Therefore, by (2.5.8) also the following inequality is satisfied:

$$\left\langle \sum_{j=k_0}^{k_1-1} \mathbf{F}(k_1, j+1)\mathbf{B}(j)\left(\mathbf{u}'(j) - \mathbf{u}^0(j)\right), \mathbf{W}_{\mathbf{R}}^{-1}(k_0, k_1)\left(\mathbf{x}_1 - \mathbf{F}(k_1, k_0)\mathbf{x}(k_0)\right)\right\rangle_{\boldsymbol{R}^n} = 0. \tag{2.5.9}$$

Taking into account the properties of the scalar product in the space $\boldsymbol{R}^n$, we can transfer inequality (2.5.9) into the following form:

$$\begin{aligned}
&\sum_{j=k_0}^{k_1-1} \langle \mathbf{F}(k_1, j+1)\mathbf{B}(j)\left(\mathbf{u}'(j) - \mathbf{u}^0(j)\right), \mathbf{W}_{\mathbf{R}}^{-1}(k_0, k_1)(\mathbf{x}_1 - \mathbf{F}(k_1, k_0)\mathbf{x}(k_0))\rangle_{\boldsymbol{R}^n} \\
&= \sum_{j=k_0}^{k_1-1} \langle \mathbf{u}'(j) - \mathbf{u}^0(j), \mathbf{B}^{\mathrm{T}}(j)\mathbf{F}^{\mathrm{T}}(k_1, j+1)\mathbf{W}_{\mathbf{R}}^{-1}(k_0, k_1)(\mathbf{x}_1 - \mathbf{F}(k_1, k_0)\mathbf{x}(k_0))\rangle_{\boldsymbol{R}^n} \\
&= \sum_{j=k_0}^{k_1-1} \langle \mathbf{u}'(j) - \mathbf{u}^0(j), \mathbf{R}(j)\mathbf{R}^{-1}(j)\mathbf{B}^{\mathrm{T}}(j)\mathbf{F}^{\mathrm{T}}(k_1, j+1)\mathbf{W}_{\mathbf{R}}^{-1}(k_0, k_1)(\mathbf{x}_1 - \\
&\qquad - \mathbf{F}(k_1, k_0)\rangle_{\boldsymbol{R}^n} = \sum_{j=k_0}^{k_1-1} \langle \mathbf{u}'(j) - \mathbf{u}^0(j), \mathbf{R}(j)\mathbf{u}^0(j)\rangle_{\boldsymbol{R}^n} = 0.
\end{aligned} \tag{2.5.10}$$

Next, using once again the properties of a scalar product and the norm in the space $\boldsymbol{R}^n$ we can transform equality (2.5.10) and obtain the following relations:

$$\begin{aligned}
&\sum_{j=k_0}^{k_1-1} \|\mathbf{u}'(j)\|^2_{\mathbf{R}(j)} = \sum_{j=k_0}^{k_1-1} \langle \mathbf{u}'(j), \mathbf{R}(j)\mathbf{u}'(j)\rangle_{\boldsymbol{R}^n} \\
&= \sum_{j=k_0}^{k_1-1} \langle \mathbf{u}'(j) - \mathbf{u}^0(j) + \mathbf{u}^0(j), \mathbf{R}(j)\left(\mathbf{u}'(j) - \mathbf{u}^0(j) + \mathbf{u}^0(j)\right)\rangle_{\boldsymbol{R}^n}
\end{aligned}$$

$$= \sum_{j=k_0}^{k_1-1} \langle \mathbf{u}'(j)-\mathbf{u}^0(j), \mathbf{R}(j)(\mathbf{u}'(j)-\mathbf{u}^0(j))\rangle_{R^n} +$$
$$+ \sum_{j=k_0}^{k_1-1} \langle \mathbf{u}'(j), \mathbf{R}(j)(\mathbf{u}'(j)-\mathbf{u}^0(j))\rangle_{R^n} +$$
$$+ \sum_{j=k_0}^{k_1-1} \langle \mathbf{u}'(j)-\mathbf{u}^0(j), \mathbf{R}(j)\mathbf{u}^0(j)\rangle_{R^n} + \sum_{j=k_0}^{k_1-1} \langle \mathbf{u}^0(j), \mathbf{R}(j)\mathbf{u}^0(j)\rangle_{R^n}$$
$$= \sum_{j=k_0}^{k_1-1} \langle \mathbf{u}'(j)-\mathbf{u}^0(j), \mathbf{R}(j)(\mathbf{u}'(j)-\mathbf{u}^0(j))\rangle_{R^n} + \sum_{j=k_0}^{k_1-1} \langle \mathbf{u}^0(j), \mathbf{R}(j)\mathbf{u}^0(j)\rangle_{R^n}$$
$$= \sum_{j=k_0}^{k_1-1} ||\mathbf{u}'(j)-\mathbf{u}^0(j)||^2_{\mathbf{R}(j)} + \sum_{j=k_0}^{k_1-1} ||\mathbf{u}^0(j)||^2_{\mathbf{R}(j)}. \tag{2.5.11}$$

Since we have the term $\sum_{j=k_0}^{k_1-1} ||\mathbf{u}'(j)-\mathbf{u}^0(j)||^2_{\mathbf{R}(j)} \geqslant 0$, by using relation (2.5.11) we immediately obtain inequality (2.5.4).

In the last step of the proof we shall derive formula (2.5.5), which defines the minimal value of the performance index $J(\mathbf{u}^0)$, corresponding to the optimal control $\mathbf{u}^0(k)$ given by equality (2.5.3). In order to do that, let us substitute formula (2.5.3) in relation (2.5.1), which leads to the following equalities:

$$J(\mathbf{u}^0) = \sum_{j=k_0}^{k_1-1} ||\mathbf{u}^0(j)||^2_{\mathbf{R}(j)}$$
$$= \sum_{j=k_0}^{k_1-1} ||\mathbf{R}^{-1}(j)\mathbf{B}^{\mathrm{T}}(j)\mathbf{F}^{\mathrm{T}}(k_1, j+1)\mathbf{W}_{\mathbf{R}}^{-1}(k_0, k_1)(\mathbf{x}_1-\mathbf{F}(k_1, k_0)\mathbf{x}(k_0)||^2_{\mathbf{R}(j)}$$
$$= \sum_{j=k_0}^{k_1-1} \langle \mathbf{R}^{-1}(j)\mathbf{B}^{\mathrm{T}}(j)\mathbf{F}^{\mathrm{T}}(k_1, j+1)\mathbf{W}_{\mathbf{R}}^{-1}(k_0, k_1)(\mathbf{x}_1-\mathbf{F}(k_1, k_0)\mathbf{x}(k_0)),$$
$$\mathbf{B}^{\mathrm{T}}(j)\mathbf{F}^{\mathrm{T}}(k_1, j+1)\mathbf{W}_{\mathbf{R}}^{-1}(k_0, k_1)(\mathbf{x}_1-\mathbf{F}(k_1, k_0)\mathbf{x}(k_0))\rangle_{R^n}. \tag{2.5.12}$$

Since the matrix $\mathbf{W}_{\mathbf{R}}(k_0, k_1)$ is symmetric, by (2.5.2) and (2.5.3) the performance index $J(\mathbf{u}^0)$ can be expressed in the following form:

$$J(\mathbf{u}^0) = \sum_{j=k_0}^{k_1-1} \langle (\mathbf{x}_1-\mathbf{F}(k_1, k_0)\mathbf{x}(k_0)), \mathbf{W}_{\mathbf{R}}^{-1}(k_0, k_1)\mathbf{F}(k_1, k_0)\mathbf{B}(j)\mathbf{R}^{-1}(j) \times$$
$$\times \mathbf{B}^{\mathrm{T}}(j)\mathbf{F}^{\mathrm{T}}(k_1, j+1)\mathbf{W}_{\mathbf{R}}^{-1}(k_0, k_1)(\mathbf{x}_1-\mathbf{F}(k_1, k_0)\mathbf{x}(k_0))\rangle_{R^n}$$
$$= \Big\langle (\mathbf{x}_1-\mathbf{F}(k_1, k_0)\mathbf{x}(k_0)), \mathbf{W}_{\mathbf{R}}^{-1}(k_0, k_1) \sum_{j=k_0}^{k_1-1} \mathbf{F}(k_1, j+1)\mathbf{B}(j)\mathbf{R}^{-1}(j) \times$$
$$\times \mathbf{B}^{\mathrm{T}}(j)\mathbf{F}^{\mathrm{T}}(k_1, j+1)\mathbf{W}_{\mathbf{R}}^{-1}(k_0, k_1)(\mathbf{x}_1-\mathbf{F}(k_1, k_0)\mathbf{x}(k_0))\Big\rangle_{R^n}$$
$$= \langle (\mathbf{x}_1-\mathbf{F}(k_1, k_0)\mathbf{x}(k_0)), \mathbf{W}_{\mathbf{R}}^{-1}(k_0, k_1)\mathbf{W}_{\mathbf{R}}(k_0, k_1)\mathbf{W}_{\mathbf{R}}^{-1}(k_0, k_1)(\mathbf{x}_1 -$$

$$-\mathbf{F}(k_1, k_0)\mathbf{x}(k_0))\rangle_{R^n}$$
$$= \langle(\mathbf{x}_1 - \mathbf{F}(k_1, k_0)\mathbf{x}(k_0)), \mathbf{W}_\mathbf{R}^{-1}(k_0, k_1)(\mathbf{x}_1 - \mathbf{F}(k_1, k_0)\mathbf{x}(k_0))\rangle_{R^n}$$
$$= \|\mathbf{x}_1 - \mathbf{F}(k_1, k_0)\mathbf{x}(k_0)\|^2_{\mathbf{W}_\mathbf{R}^{-1}(k_0, k_1)}.$$

Hence formula (2.5.5) is proved and our theorem follows. ■

Theorem 2.5.1 has been proved by using only the fundamental properties of the norm and scalar product in the space R^n. We did not use any theorems from control theory. This was possible for the following reasons:

(1) the linearity of the discrete state equation of the dynamical system,

(2) the lack of constraints on control,

(3) the special form of the performance index, namely the lack of the state variables $\mathbf{x}(k)$ in it,

(4) the quadratic performance index,

(5) controllability in $[k_0, k_1]$ of the dynamical system (2.2.1)

All the above reasons together make the solution of the minimum energy control problem very easily obtainable.

2.6 Controllability of M-D dynamical systems

In the last few years great attention has been paid to so-called *M-D systems*, i.e., the discrete, linear, dynamical systems with M independent variables. The rapidly growing interest in M-D system techniques is motivated by their applications in the image processes and in the military area. The problems of M-D systems and especially 2-D and 3-D systems have been recently considered in the following papers: Morf, Levy and Kung (1977), Ciftibassi and Yuksel (1983), Eising (1978, 1979a, 1979b, 1979c), Eising and Hautus (1981), Fornassini and Marchesini (1976, 1978, 1979, 1982), Kaczorek (1983b, 1985, 1986a, 1986b, 1986c, 1987a, 1987b), Kaczorek and Klamka (1987, 1988), Klamka (1983a, 1983b, 1983c, 1984a, 1984b, 1986c, 1988a, 1988b), Klamka and Kaczorek (1986), Kung, Levy, Morf and Kailath (1977), Kurek (1987), Roesser (1975), Sebek, Bisiacco and Fornasini (1988), Theodorou and Tzafestas (1984), Lin, Kawamata and Higuchi (1987).

In this section we shall give a description of the M-D dynamical systems, define the so-called transition matrix for M-D systems and then, basing ourselves on the appropriate definitions, we shall formulate and prove sufficient and necessary conditions for local controllability in a prescribed rectangular. Finally the minimum energy control problem for M-D systems will be considered and effectively solved. Particular attention will be paid to 2-D dynamical systems, i.e., discrete, linear systems with two independent variables.

For brevity let us introduce the following notation:

$I = \{0, 1, 2, \ldots\}$ — the set of nonnegative integer numbers,

$I^m = \underbrace{I \times I \times \ldots \times I}_{m \text{ times}}$ — the Cartesian product of M sets I.

For $(i_1, i_2, \dots, i_M) \in I^M$ and $(k_1, k_2, \dots, k_M) \in I^M$ we shall introduce ordering relation:

$$(i_1, i_2, \dots, i_M) \leqslant (k_1, k_2, \dots, k_M) \Leftrightarrow i_j \leqslant k_j \quad \text{for } j = 1, 2, \dots, M,$$
$$(i_1, i_2, \dots, i_M) = (k_1, k_2, \dots, k_M) \Leftrightarrow i_j = k_j \quad \text{for } j = 1, 2, \dots, M,$$
$$(i_1, i_2, \dots, i_M) < (k_1, k_2, \dots, k_M) \Leftrightarrow (i_1, i_2, \dots, i_M) \leqslant (k_1, k_2, \dots, k_M)$$

and

$$(i_1, i_2, \dots, i_M) \neq (k_1, k_2, \dots, k_M).$$

Moreover, for $(i_1, i_2, \dots, i_M) < (k_1, k_2, \dots, k_M)$ we define an M-dimensional rectangular prism

$$\begin{aligned}&[(i_1, i_2, \dots, i_M) < (k_1, k_2, \dots, k_M)]\\ &= \{(s_1, s_2, \dots, s_M) \in I^M\colon (i_1, i_2, \dots, i_M) \leqslant (s_1, s_2, \dots, s_M)\\ &\qquad \leqslant (k_1, k_2, \dots, k_M)\}.\end{aligned}$$

The notation $(i_1, i_2, \dots, 0_j, \dots, i_M)$ denotes an ordered M-tuple of nonnegative integer numbers with zero on the j-th position. Similarly $(0, 0, \dots, i_j, \dots, 0)$ denotes an ordered M-tuple containing a nonzero integer on the j-th position and zeros elsewhere.

The symbols introduced above allow us to simplify the formulas and equalities concerning M-D systems. However, it should be mentioned that the descriptions and symbols for M-D systems are very complicated and obscure.

Let us consider a linear, stationary, discrete dynamical M-D system described by the following set of difference equations:

$$\begin{aligned}\mathbf{x}_j(i_1, \dots, i_j+1, \dots, i_M) &= \sum_{k=1}^{M} \mathbf{A}_{jk}\mathbf{x}_k(i_1, \dots, i_j, \dots, i_M) +\\ &+\mathbf{B}_j\mathbf{u}(i_1, \dots, i_j, \dots, i_M), \quad j = 1, 2, \dots, M,\end{aligned} \tag{2.6.1}$$

$$\mathbf{y}(i_1, \dots, i_j, \dots, i_M) = \sum_{k=1}^{M} \mathbf{C}_k\mathbf{x}_k(i_1, \dots, i_j, \dots, i_M) + \mathbf{D}\mathbf{u}(i_1, \dots, i_j, \dots, i_M), \tag{2.6.2}$$

where $\mathbf{x}_j(i_1, \dots, i_j, \dots, i_M) \in \boldsymbol{R}^{n_j}$ for $j = 1, 2, \dots, M$, $\mathbf{u}(i_1, \dots, i_p, \dots, i_M) \in \boldsymbol{R}^p$ is the control vector, $\mathbf{y}(i_1, \dots, i_j, \dots, i_M) \in \boldsymbol{R}^q$ is the output vector, $\mathbf{A}_{jk}$, $j, k = 1, 2, \dots, M$ are $(n_j \times n_k)$-dimensional matrices with constant coefficients, $\mathbf{B}_j$, $j = 1, 2, \dots, M$ are $(n_j \times p)$-dimensional matrices with constant coefficients, $\mathbf{C}_k$, $k = 1, 2, \dots, M$ are $(q \times n_k)$-dimensional matrices with constant coefficients, $\mathbf{D}$ is a $(q \times p)$-dimensional matrix with constant coefficients.

In order to solve the set of difference equations (2.6.1), let us assume, that we are given the boundary conditions $\mathbf{x}_j(i_1, \dots, 0_j, \dots, i_M)$ for $i_1 \in \boldsymbol{Z}^+, \dots, i_{j-1} \in \boldsymbol{Z}^+$, $i_{j+1} \in \boldsymbol{Z}^+, \dots, i_M \in \boldsymbol{Z}^+, j = 1, 2, \dots, M$, where $\boldsymbol{Z}^+$ denotes the set of nonnegative integers.

In order to present the solution of the set of difference equations (2.6.1) in a compact, legible form we define the so-called transition matrix for an M-D system.

DEFINITION 2.6.1. For the block matrix $\mathbf{A} = [\mathbf{A}_{jk}]$, $j, k = 1, 2, \ldots, M$ we define the *transition matrix* $\mathbf{A}^{i_1,\ldots,i_j,\ldots,i_M}$ as follows:

(1) $\mathbf{A}^{0,0,\ldots,0,\ldots,0} = \mathbf{I}_{n\times n}$, where $n = \sum_{j=1}^{M} n_j$,

(2) $$\mathbf{A}^{0,0,\ldots,1_j,\ldots,0} = \begin{bmatrix} 0 & 0 & \ldots & 0 \\ \cdots & \cdots & \cdots & \cdots \\ \mathbf{A}_{j1} & \mathbf{A}_{j2} & \ldots & \mathbf{A}_{jM} \\ \cdots & \cdots & \cdots & \cdots \\ 0 & 0 & \ldots & 0 \end{bmatrix},$$

(3) $\mathbf{A}^{i_1,i_2,\ldots,i_j,\ldots,i_M} = \sum_{j=1}^{M} \mathbf{A}^{0,\ldots,1_j,\ldots,0}\mathbf{A}^{i_1,\ldots,i_j-1,\ldots,i_M}$,

(4) $\mathbf{A}^{i_1,i_2,\ldots,-i_j,\ldots,i_M} = 0$ for $j = 1, 2, \ldots, M$.

Definition 2.6.1 immediately implies the following properties of the transition matrix:

(1) $\mathbf{A} = \sum_{j=1}^{M} \mathbf{A}^{0,0,\ldots,1_j,\ldots,0}$,

(2) $\mathbf{A}^{0,0,\ldots,i_j,\ldots,0} = (\mathbf{A}^{0,0,\ldots,1_j,\ldots,0})^i$ for $j = 1, 2, \ldots, M$,

(3) $\mathbf{A}^{0,0,\ldots,1_j,\ldots,0} = \mathbf{I}^{0,0,\ldots,1_j,\ldots,0}\mathbf{A}$ for $j = 1, 2, \ldots, M$,

(4) $\mathbf{I}^{0,0,\ldots,1_k,\ldots,0}\mathbf{A}^{0,0,\ldots,1_j,\ldots,0} = 0$ for $j \neq k$.

Using the transition matrix $\mathbf{A}^{i_1,i_2,\ldots,i_j,\ldots,i_M}$ and introducing the notation $\mathbf{x} = (\mathbf{x}_1^{\mathrm{T}}, \mathbf{x}_2^{\mathrm{T}}, \ldots, \mathbf{x}_j^{\mathrm{T}}, \ldots, \mathbf{x}_M^{\mathrm{T}})^{\mathrm{T}} \in \boldsymbol{R}^n$, we can prove (Klamka (1983a)), that the solution of the set of difference equation (2.6.1) with a given boundary conditions has the following compact form:

$$\mathbf{x}(i_1, i_2, \ldots, i_M)$$

$$= \sum_{j=1}^{M} \sum_{\substack{0,\ldots,0,\ldots,0 \\ \leqslant (k_1,\ldots,0_j,\ldots,k_M) \\ \leqslant (i_1,\ldots,0_j,\ldots,i_M)}} \mathbf{A}^{i_1-k_1,\ldots,i_j,\ldots,i_M-k_M} \begin{bmatrix} 0 \\ \cdots \\ \mathbf{x}_j(k_1, \ldots, 0_j, \ldots, k_M) \\ \cdots \\ 0 \end{bmatrix} +$$

$$+ \sum_{\substack{(0,\ldots,0,\ldots,0) \\ \leqslant (k_1,k_2,\ldots,k_M) \\ < (i_1,i_2,\ldots,i_M)}} \sum_{j=1}^{M} \mathbf{A}^{i_1-k_1,\ldots,i_j-k_j-1,\ldots,i_M-k_M}\mathbf{B}^{0,0,\ldots,1_j,\ldots,0}\mathbf{u}(k_1, k_2, \ldots, k_M). \tag{2.6.3}$$

By (2.6.2) and (2.6.3) we can easily compute the output vector $\mathbf{y}(i_1, i_2, \ldots, i_M)$, i.e.,

$$\mathbf{y}(i_1, i_2, \ldots, i_M) = [\mathbf{C}_1 \vdots \mathbf{C}_2 \vdots \ldots \vdots \mathbf{C}_M]\mathbf{x}(i_1, i_2, \ldots, i_M)+\mathbf{D}\mathbf{u}(i_1, i_2, \ldots, i_M). \tag{2.6.4}$$

For the special case of 2-D systems equations (2.6.1) and (2.6.2) have the following simple form (Roesser (1975)):

$$\begin{aligned} \mathbf{x}_1(i_1+1, i_2) &= \mathbf{A}_{11}\mathbf{x}_1(i_1, i_2)+\mathbf{A}_{12}\mathbf{x}_2(i_1, i_2)+\mathbf{B}_1\mathbf{u}(i_1, i_1), \\ \mathbf{x}_2(i_1, i_2+1) &= \mathbf{A}_{21}\mathbf{x}_1(i_1, i_2)+\mathbf{A}_{22}\mathbf{x}_2(i_1, i_2)+\mathbf{B}_2\mathbf{u}(i_1, i_2), \end{aligned} \tag{2.6.5}$$

and

$$\mathbf{y}(i_1, i_2) = \mathbf{C}_1\mathbf{x}_1(i_1, i_2)+\mathbf{C}_2\mathbf{x}_2(i_1, i_2)+\mathbf{D}\mathbf{u}(i_1, i_2). \tag{2.6.6}$$

For a given boundary conditions $\mathbf{x}_1(0, i_2)$, $i_2 \in \boldsymbol{Z}^+$ and $\mathbf{x}_2(i_1, 0)$, $i_1 \in \boldsymbol{Z}^+$, by (2.6.3) the solution of the 2-D system (2.6.5) can be expressed by using the transition matrix $\mathbf{A}^{i_1,i_2}$ as follows (Roesser (1975)):

$$\mathbf{x}(i_1, i_2) = \begin{bmatrix}\mathbf{x}_1(i_1, i_2)\\ \mathbf{x}_2(i_1, i_2)\end{bmatrix} = \sum_{k_2=0}^{i_2} \mathbf{A}^{i_1,i_2-k_2}\begin{bmatrix}\mathbf{x}_1(0, k_2)\\ 0\end{bmatrix}+$$

$$+\sum_{k_1=0}^{i_1} \mathbf{A}^{i_1-k_1,i_2}\begin{bmatrix}0\\ \mathbf{x}_2(k_1, 0)\end{bmatrix}+\sum_{(0,0)\leqslant(k_1,k_2)<(i_1,i_2)} +\mathbf{A}^{i_1-k_1-1-i_2-k_2}\mathbf{B}^{1,0}+$$

$$+\mathbf{A}^{i_1-k_1,i_2-k_2-1}\mathbf{B}^{0,1}\mathbf{u}(k_1, k_2). \tag{2.6.7}$$

Similarly, by (2.6.4), we obtain the formula for the output vector $\mathbf{y}(i_1, i_2)$ as follows:

$$\mathbf{y}(i_1, i_2) = [\mathbf{C}_1 \vdots \mathbf{C}_2]\begin{bmatrix}\mathbf{x}_1(i_1, i_2)\\ \mathbf{x}_2(i_1, i_2)\end{bmatrix}+\mathbf{D}\mathbf{u}(i_1, i_2). \tag{2.6.8}$$

In the paper of Kaczorek (1983b), the author considers a mathematical model for 3-D systems and presents the general form of solution.

Now we shall formulate the necessary and sufficient conditions for local controllability in a rectangular prism of the dynamical system M-D.

DEFINITION 2.6.2. The dynamical system M-D is said to be *locally controllable in an M-dimensional rectangular prism* $[(0, 0, \ldots, 0), (r_1, r_2, \ldots, r_M)]$ if for any boundary conditions $\mathbf{x}_j(i_1, \ldots, 0_j, \ldots, i_M)$, $i_1 \in \boldsymbol{Z}^+, \ldots, i_{j-1} \in \boldsymbol{Z}^+$, $i_{j+1} \in \boldsymbol{Z}^+, \ldots, i_M \in \boldsymbol{Z}^+$, $j = 1, 2, \ldots, M$ and any vector $\tilde{\mathbf{x}} \in \boldsymbol{R}^n$, there exists a sequence of controls $\{\mathbf{u}(i_1, i_2, \ldots, i_M)\colon (0, 0, \ldots, 0) \leqslant (i_1, i_2, \ldots, i_M) < (r_1, r_2, \ldots, r_M)\}$ such that the corresponding trajectory of the M-D system satisfies the following condition:

$$\mathbf{x}(r_1, r_2, \ldots, r_M) = \tilde{\mathbf{x}}. \tag{2.6.9}$$

In order to obtain verifiable criteria for the local controllability of M-D systems we define the so-called *local controllability matrix* (Klamka (1983a)) given by

$$\mathbf{Q}(r_1, r_2, \ldots, r_M) = [\mathbf{M}(i_1, i_2, \ldots, i_M)], \tag{2.6.10}$$

where $(0, 0, \ldots, 0) \leqslant (i_1, i_2, \ldots, i_M) < (r_1, r_2, \ldots, r_M)$ and the $(n\times p)$-dimensional matrices $\mathbf{M}(i_1, i_2, \ldots, i_M)$ are defined as follows:

$$\mathbf{M}(i_1, i_2, \dots, i_M) = \sum_{j=1}^{M} \mathbf{A}^{i_1, i_2, \dots, i_j - 1, \dots, i_M} \mathbf{B}^{0,0,\dots,1_j,\dots,0}. \tag{2.6.11}$$

Using the local controllability matrix $Q(r_1, r_2, \dots, r_M)$, we can obtain the necessary and sufficient condition for the local controllability of the dynamical system M-D.

THEOREM 2.6.1 (Klamka (1983a)). *The M-D dynamical system* (2.6.1) *is locally controllable in an M-dimensional rectangular prism* $[(0, 0, \dots, 0), (r_1, r_2, \dots, r_M)]$ *if and only if*

$$\operatorname{rank} \mathbf{Q}(r_1, r_2, \dots, r_M) = n. \tag{2.6.12}$$

Proof:

Since the M-D system is linear, we can assume without loss of generality, that all the boundary conditions are equal to zero. Therefore by (2.6.3) we have

$$\begin{aligned}
&\mathbf{x}(r_1, r_2, \dots, r_M) \\
&= \sum_{\substack{(0,0,\dots,0) \\ \leqslant (k_1,k_2,\dots,k_M) \\ < (r_1,r_2,\dots,r_M)}} \sum_{j=1}^{M} \mathbf{A}^{r_1-k_1, r_2-k_2, \dots, r_j-k_j-1, \dots, r_M-k_M} \mathbf{B}^{0,0,\dots,1_j,\dots,0} \mathbf{u}(k_1, k_2, \dots, k_M) \\
&= \sum_{\substack{(0,0,\dots,0) \\ \leqslant (k_1,k_2,\dots,k_M) \\ < (r_1,r_2,\dots,r_M)}} \mathbf{M}(r_1-k_1, r_2-k_2, \dots, r_M-k_M) \mathbf{u}(k_1, k_2, \dots, k_M) \\
&= \mathbf{Q}(r_1, r_2, \dots, r_M) \mathbf{U}(r_1, r_2, \dots, r_M),
\end{aligned} \tag{2.6.13}$$

where

$$\begin{aligned}
&\mathbf{U}(r_1, r_2, \dots, r_M) \\
&= [\mathbf{u}^{\mathrm{T}}(0, 0, \dots, 0) \;\; \mathbf{u}^{\mathrm{T}}(1, 0, \dots, 0) \;\; \dots \;\; \mathbf{u}^{\mathrm{T}}(r_1, r_2, \dots, r_M - 1)]^{\mathrm{T}}.
\end{aligned}$$

The symbol $\mathbf{U}(r_1, r_2, \dots, r_M)$ represents the controls in the rectangular prism $[(0, 0, \dots, 0), (r_1, r_2, \dots, r_M)]$.

Using an analogous method to that used in the proof of Theorem 2.3.1, after some easy computations and taking into account (2.6.13) and (2.6.10) we obtain Theorem 2.6.1. ∎

COROLLARY 2.6.1 (Klamka (1983a)). *Suppose the dynamical system M-D is locally controllable in an M-dimensional rectangular prism* $[(0, 0, \dots, 0), (i_1, i_2, \dots, i_M)]$. *Then this system is also locally controllable in every M-dimensional rectangular prism* $[(0, 0, \dots, 0), (r_1, r_2, \dots, r_M)]$, *where* $(i_1, i_2, \dots, i_M) \leqslant (r_1, r_2, \dots, r_M)$.

Proof:

From (2.6.10) and (2.6.11) it follows that

$$\operatorname{rank}\mathbf{Q}(i_1, i_2, \dots, i_M) \leqslant \operatorname{rank}\mathbf{Q}(r_1, r_2, \dots, r_M)$$
$$\text{for } (i_1, i_2, \dots, i_M) \leqslant (r_1, r_2, \dots, r_M).$$

This implies our corollary. ■

For the special case of 2-D dynamical systems the local controllability matrix $\mathbf{Q}(r_1, r_2)$ is by (2.6.10) and (2.6.11) of the following simple form (Klamka (1983a)):

$$\mathbf{Q}(r_1, r_2) = [\mathbf{M}(0,0) \vdots \mathbf{M}(1,0) \vdots \mathbf{M}(0,1) \vdots \mathbf{M}(1,1) \vdots \dots \vdots \mathbf{M}(i_1, i_2) \vdots \dots \vdots \mathbf{M}(r_1, r_2)], \tag{2.6.14}$$

where

$$\mathbf{M}(i_1, i_2) = \mathbf{A}^{i_1-1,i_2}\mathbf{B}^{1,0} + \mathbf{A}^{i_1,i_2-1}\mathbf{B}^{0,1} \quad \text{for } (0,0) \leqslant (i_1, i_2) < (r_1, r_2). \tag{2.6.15}$$

From Theorem 2.6.1 and Corollary 2.6.1 we easy derive the local controllability conditions for the 2-D dynamical system defined in the rectangular prism $[(0,0), (r_1, r_2)]$.

COROLLARY 2.6.2 (Kaczorek (1985), Klamka (1984a)). *A 2-D dynamical system (2.6.5) is locally controllable in the rectangular prism* $[(0,0), (r_1, r_2)]$ *if and only if*

$$\operatorname{rank}\mathbf{Q}(r_1, r_2) = n_1 + n_2. \tag{2.6.16}$$

Proof:

Putting $M = 2$ in Theorem 2.6.1, we directly obtain our corollary. ■

COROLLARY 2.6.3 (Klamka (1984a)). *Suppose 2-D dynamical system is locally controllable in rectangular prism* $[(0,0), (i_1, i_2)]$. *Then it is also locally controllable in every rectangular prism* $[(0,0), (r_1, r_2)]$, *where* $(i_1, i_2) \leqslant (r_1, r_2)$.

Proof:

Putting $M = 2$ in Corollary 2.6.1, we directly obtain our corollary. ■

Similar results for 3-D dynamical systems can be found in the excellent monograph of Kaczorek (1985) or in the papers of Kaczorek (1983b), Theodorou and Tzafestas (1984). It should also be stressed that in the book of Kaczorek (1985) we find several different mathematical models of 2-D dynamical systems and generally of M-D dynamical systems, presented and investigated. Moreover, the interrelations between different types of 2-D dynamical systems are considered and explained. The transformations which transform one 2-D system into another one are also presented. Pure algebraic methods for investigating various kinds of controllability of 2-D systems are extensively studied in the paper of Eising and Hautus (1981), Kung, Levy, Morf and Kailath (1977), Sebek, Bisiacco and Fornasini (1978).

Just as in the case of 1-D dynamical systems, also for M-D dynamical systems it is possible, under the assumption of local controllability in the M-dimensional rectangular prism $[(0, 0, \ldots, 0), (r_1, r_2, \ldots, r_M)]$, to solve analytically the problem of minimum energy control. For M-D dynamical systems locally controllable in the rectangular prism $[(0, 0, \ldots, 0), (r_1, r_2, \ldots, r_M)]$ there generally exist many different sequences of controls $\{\mathbf{u}(i_1, i_2, \ldots, i_M): (i_1, i_2, \ldots, i_M) \in [(0, 0, \ldots, 0), (r_1, r_2, \ldots, r_M))\}$ which steers the M-D system from given initial conditions to the desired final local state $\tilde{\mathbf{x}} \in \boldsymbol{R}^n$ at the point $(r_1, r_2, \ldots, r_M)$. Among the controls which achieve the same mission we may look for the one which is the optimal according to some given performance index $J(\mathbf{u})$:

$$J(\mathbf{u}) = \sum_{\substack{(0,0,\ldots,0) \\ \leqslant (i_1,i_2,\ldots,i_M) \\ < (r_1,r_2,\ldots,r_M)}} \mathbf{u}^{\mathrm{T}}(i_1, i_2, \ldots, i_M)\mathbf{R}(i_1, i_2, \ldots, i_M)\mathbf{u}(i_1, i_2, \ldots, i_M), \tag{2.6.17}$$

where $\mathbf{R}(i_1, i_2, \ldots, i_M)$ are $(p \times p)$-dimensional, symmetric, positive definite matrices.

The interpretation of the matrices $\mathbf{R}(i_1, i_2, \ldots, i_M)$ and the performance index $J(\mathbf{u})$ is the same as in the case of the 1-D dynamical systems considered in Section 2.5.

To shorten the notation, let us denote:

$$\begin{aligned}&\mathbf{Q}(r_1, r_2, \ldots, r_M)\\ &= \sum_{\substack{(0,0,\ldots,0) \\ \leqslant (i_1,i_2,\ldots,i_M) \\ < (r_1,r_2,\ldots,r_M)}} \mathbf{M}(r_1-i_1, r_2-i_2, \ldots, r_M-i_M)\mathbf{R}^{-1}(i_1, i_2, \ldots, i_M) \times\\ &\qquad\times \mathbf{M}^{\mathrm{T}}(r_1-i_1, r_2-i_2, \ldots, r_M-i_M).\end{aligned} \tag{2.6.18}$$

Therefore, $\mathbf{Q}(r_1, r_2, \ldots, r_M)$ is an $(n \times n)$-dimensional, symmetric matrix.

Local controllability of an M-D dynamical system in the M-dimensional rectangular prism $[(0, 0, \ldots, 0), (r_1, r_2, \ldots, r_M)]$ implies by Theorem 2.6.1 and formula (2.6.18) that there exists an inverse matrix $\mathbf{Q}_{\mathbf{R}}^{-1}(r_1, r_2, \ldots, r_M)$. This matrix is used in defining the controls $\mathbf{u}^0(i_1, i_2, \ldots, i_M)$ for $(i_1, i_2, \ldots, i_M) \in [(0, 0, \ldots, 0), (r_1, r_2, \ldots, r_M))$

$$\begin{aligned}&\mathbf{u}^0(i_1, i_2, \ldots, i_M)\\ &= \mathbf{R}^{-1}(i_1, i_2, \ldots, i_M)\mathbf{M}^{\mathrm{T}}(r_1-i_1, r_2-i_2, \ldots, r_M-i_M)\mathbf{Q}_{\mathbf{R}}^{-1}(r_1, r_2, \ldots, r_M) \times\\ &\times \mathbf{q}(r_1, r_2, \ldots, r_M),\end{aligned} \tag{2.6.19}$$

where

$$\begin{aligned}&\boldsymbol{R}^n \ni \mathbf{q}(r_1, r_2, \ldots, r_M)\\ &= \tilde{\mathbf{x}} - \sum_{j=1}^{M} \sum_{\substack{(0,0,\ldots,0) \\ \leqslant (k_1,\ldots,0_j,\ldots,k_M) \\ \leqslant (i_1,\ldots,0_j,\ldots,i_M)}} \mathbf{A}^{i_1-k_1, i_2-k_2, \ldots, i_j, \ldots, i_M-k_M} \begin{bmatrix} 0 \\ \cdots\cdots\cdots \\ \mathbf{x}_j(k_1, \ldots, 0_j, \ldots, k_M) \\ \cdots\cdots\cdots \\ 0 \end{bmatrix}.\end{aligned} \tag{2.6.20}$$

Using (2.6.18), (2.6.20) and the method presented in Section 2.5, we may effectively solve the minimum energy control problem for an M-D dynamical system, and give the analytic formula for the optimal control sequence. Moreover, the minimum value of the performance index can also be computed.

THEOREM 2.6.2 (Klamka (1983a)). *Suppose the M-D dynamical system* (2.6.1) *is locally controllable in the M-dimensional rectangular prism* $[(0, 0, \dots, 0), (r_1, r_2, \dots, r_M)]$, *and let* $\tilde{\mathbf{u}}(i_1, i_2, \dots, i_M)$ *for* $(i_1, i_2, \dots, i_M) \in [(0, 0, \dots, 0), (r_1, r_2, \dots, r_M))$ *be any control sequence which steers the M-D system from the given boundary conditions* $\mathbf{x}_j(i_1, \dots, 0_j, \dots, i_M)$, $i_1 \in \boldsymbol{Z}^+, \dots, i_{j-1} \in \boldsymbol{Z}^+, i_{j+1} \in \boldsymbol{Z}^+, \dots, i_M \in \boldsymbol{Z}^+, j = 1, 2, \dots, M$, *to the desired final local state* $\mathbf{x} \in \boldsymbol{R}^n$ *at the point* $(r_1, r_2, \dots, r_M)$. *Then the control sequence* $\mathbf{u}^0(i_1, i_2, \dots, i_M)$ *given by* (2.6.19) *accomplishes the same transfer and additionally*:

$$J(\mathbf{u}^0) \leqslant J(\tilde{\mathbf{u}}). \tag{2.6.21}$$

Moreover, the minimum value of the performance index $J(\mathbf{u}^0)$ *is given by the formula*

$$J(\mathbf{u}^0) = \mathbf{q}^{\mathrm{T}}(r_1, r_2, \dots, r_M)\mathbf{Q}_{\mathrm{R}}^{-1}(r_1, r_2, \dots, r_M)\mathbf{q}(r_1, r_2, \dots, r_M). \tag{2.6.22}$$

Proof:

Substituting (2.6.19) in (2.6.3) and using relations (2.6.18) and (2.6.20), we can show that the sequence of controls $\mathbf{u}^0(i_1, i_2, \dots, i_M)$, for $(i_1, i_2, \dots, i_M) \in [(0, 0, \dots, 0), (r_1, r_2, \dots, r_M))$ steers the M-D dynamical system from given boundary conditions to the desired final local state $\mathbf{x} \in \boldsymbol{R}^n$ at the point $(r_1, r_2, \dots, r_M)$. In the next part of the proof, just as in the proof of Theorem 2.5.1, we can prove inequality (2.6.21) and finally compute formula (2.6.22). It should be pointed out that the proof of Theorem 2.6.2 is quite similar to the proof of Theorem 2.5.1 but requires very long and tedious computations (see Klamka (1983a) for details). ■

The modification of Theorem 2.6.2 for 2-D dynamical systems is obvious and the details are presented in Kaczorek (1985), Klamka (1984a).

A similar problem for 3-D dynamical system has been considered in the paper of Kaczorek (1983b), where its complete solution is presented.

It should be stressed that in the proof of Theorem 2.6.2 only the fundamental properties of the norm and scalar product in the space $\boldsymbol{R}^n$ are used. The reasons for doing that are the same as those mentioned in Sections 1.7 and 2.5.

The results presented in this section can quite easily be extended to cover the case of nonstationary, linear M-D dynamical systems or even the M-D dynamical systems defined in the infinite-dimensional Hilbert space. In the latter case instead of matrices we have linear, bounded operators in the state difference equation (2.6.1). However, it should be stressed that there exists no maximum principle for M-D dynamical systems. On the other hand, some satisfactiory results concerning optimal problems can be derived by using mathematical programming methods.

Example 2.6.1
Let us consider an M-D dynamical system with the following matrices:

$$\mathbf{A} = \begin{bmatrix} \mathbf{A}_{11} & \mathbf{A}_{12} \\ \mathbf{A}_{21} & \mathbf{A}_{22} \end{bmatrix} = \left[\begin{array}{c:cc} 0 & 0 & 1 \\ \hdashline 0 & 0 & 1 \\ 1 & 0 & 0 \end{array}\right], \quad \mathbf{B} = \begin{bmatrix} \mathbf{B}_1 \\ \mathbf{B}_2 \end{bmatrix} = \begin{bmatrix} 1 \\ 1 \\ 0 \end{bmatrix}. \tag{2.6.23}$$

It is assumed that the M-D system is defined in the rectangular prism $[(0, 0), (1, 2)] = [(0, 0), (r_1, r_2)]$. Hence we have $n_1 = 1$, $n_2 = 2$, $p = 1$ and $n = n_1 + n_2 = 3$. By (2.6.11) and the formulas given in Definition 2.6.1 we compute the matrices $\mathbf{M}(i_1, i_2)$ for $(i_1, i_2) \in ((0, 0), (1, 2)]$, namely,

$$\mathbf{M}(0, 1) = \mathbf{A}^{-1,1}\mathbf{B}^{1,0} + \mathbf{A}^{0,0}\mathbf{B}^{0,1} = \begin{bmatrix} \mathbf{I}_{n_1} & 0 \\ 0 & \mathbf{I}_{n_2} \end{bmatrix}\begin{bmatrix} 0 \\ \mathbf{B}_2 \end{bmatrix} = \begin{bmatrix} 0 \\ 1 \\ 0 \end{bmatrix},$$

$$\mathbf{M}(1, 0) = \mathbf{A}^{0,0}\mathbf{B}^{1,0} + \mathbf{A}^{1,-1}\mathbf{B}^{1,0} = \begin{bmatrix} \mathbf{I}_{n_1} & 0 \\ 0 & \mathbf{I}_{n_2} \end{bmatrix}\begin{bmatrix} \mathbf{B}_1 \\ 0 \end{bmatrix} = \begin{bmatrix} 1 \\ 0 \\ 0 \end{bmatrix},$$

$$\mathbf{M}(1, 1) = \mathbf{A}^{0,1}\mathbf{B}^{1,0} + \mathbf{A}^{1,0}\mathbf{B}^{0,1} = \begin{bmatrix} 0 & 0 \\ \mathbf{A}_{21} & \mathbf{A}_{22} \end{bmatrix}\begin{bmatrix} \mathbf{B}_1 \\ 0 \end{bmatrix} + \begin{bmatrix} \mathbf{A}_{11} & \mathbf{A}_{12} \\ 0 & 0 \end{bmatrix}\begin{bmatrix} 0 \\ \mathbf{B}_2 \end{bmatrix}$$

$$= \begin{bmatrix} 0 \\ 0 \\ 1 \end{bmatrix},$$

$$\mathbf{M}(0, 2) = \mathbf{A}^{-1,2}\mathbf{B}^{1,0} + \mathbf{A}^{0,1}\mathbf{B}^{0,1} = \begin{bmatrix} 0 & 0 \\ \mathbf{A}_{21} & \mathbf{A}_{22} \end{bmatrix}\begin{bmatrix} 0 \\ \mathbf{B}_2 \end{bmatrix} = \begin{bmatrix} 0 \\ 0 \\ 0 \end{bmatrix},$$

$$\mathbf{M}(1,2) = \mathbf{A}^{0,2}\mathbf{B}^{1,0} + \mathbf{A}^{1,1}\mathbf{B}^{0,1} = \mathbf{A}^{0,1}\mathbf{A}^{0,1}\mathbf{B}^{1,0} + (\mathbf{A}^{1,0}\mathbf{A}^{0,1} + \mathbf{A}^{0,1}\mathbf{A}^{1,0})\mathbf{B}^{0,1}$$

$$= \begin{bmatrix} 0 & 0 \\ \mathbf{A}_{21} & \mathbf{A}_{22} \end{bmatrix}\begin{bmatrix} 0 & 0 \\ \mathbf{A}_{21} & \mathbf{A}_{22} \end{bmatrix}\begin{bmatrix} \mathbf{B}_1 \\ 0 \end{bmatrix} + \left(\begin{bmatrix} \mathbf{A}_{11} & \mathbf{A}_{12} \\ 0 & 0 \end{bmatrix}\begin{bmatrix} 0 & 0 \\ \mathbf{A}_{21} & \mathbf{A}_{22} \end{bmatrix} + \right.$$

$$\left. + \begin{bmatrix} 0 & 0 \\ \mathbf{A}_{21} & \mathbf{A}_{22} \end{bmatrix}\begin{bmatrix} \mathbf{A}_{11} & \mathbf{A}_{12} \\ 0 & 0 \end{bmatrix}\right)\begin{bmatrix} 0 \\ \mathbf{B}_2 \end{bmatrix} = \begin{bmatrix} 0 \\ 1 \\ 0 \end{bmatrix}.$$

Therefore, by (2.6.10), the local controllability matrix $\mathbf{Q}(1, 2)$ for the dynamical system (2.6.23) is given by the following equality:

$$\mathbf{Q}(1, 2) = [\mathbf{M}(0, 1) \vdots \mathbf{M}(1, 0) \vdots \mathbf{M}(1, 1) \vdots \mathbf{M}(0, 2) \vdots \mathbf{M}(1, 2)] = \left[\begin{array}{c:c:c:c:c} 0 & 1 & 0 & 0 & 0 \\ 1 & 0 & 0 & 0 & 1 \\ 0 & 0 & 1 & 0 & 0 \end{array}\right].$$

Since $\operatorname{rank}\mathbf{Q}(1, 2) = 3 = n = n_1 + n_2$, hence by Theorem 2.6.1 the 2-D dynamical system (2.6.23) is locally controllable in the two-dimensional rectangular prism $[0, 0), (1, 2)]$.

In the next part of our example we shall consider in detail the minimum energy control problem for the 2-D dynamical system (2.6.23) in the rectangular prism $[(0, 0), (1, 2)]$.

Let $\mathbf{R}(i_1, i_2) = 1$ for $(i_1, i_2) \in [(0, 0), (1, 2))$. Since the dynamical system (2.6.23) is locally controllable in the rectangular prism $[(0, 0), (1, 2)]$, then all the assumptions of Theorem 2.6.2 are satisfied, and therefore we may use this theorem to solve our minimum energy control problem. Moreover, we shall be able to compute the minimum value of the performance index.

Let us assume that the initial boundary conditions for the system (2.6.23) are

$$\mathbf{x}_1(0, 0) = 1, \quad \mathbf{x}_1(0, 1) = 0, \quad \mathbf{x}_1(0, 2) = 0, \quad \mathbf{x}_2(0, 0) = \begin{bmatrix} 5 \\ 3 \end{bmatrix},$$

$$\mathbf{x}_2(1, 0) = \begin{bmatrix} 0 \\ 0 \end{bmatrix}.$$

Moreover let us assume, that the local final state at the point $(1, 2)$ is also given and $\mathbf{x}(1, 2) = \begin{bmatrix} 3 \\ 2 \\ 1 \end{bmatrix}$. We want to steer the dynamical system (6.2.23) from the above boundary conditions to the given final local state with minimum control energy. In order to solve this minimum energy control problem, we shall use Theorem 2.6.2, whose assumptions are all satisfied.

Since $\mathbf{R} = 1$, by (2.6.18) we have $\mathbf{Q_R}(1, 2) = \mathbf{Q}(1, 2) = \mathbf{M}(0, 1)\mathbf{M}^T(0, 1) +$
$+\mathbf{M}(1,0)\mathbf{M}^T(1,0)+\mathbf{M}(1, 1)\mathbf{M}^T(1, 1)+\mathbf{M}(0, 2)\mathbf{M}^T(0, 2)+\mathbf{M}(1, 2)\mathbf{M}^T(1, 2) = \begin{bmatrix} 1 & 0 & 1 \\ 0 & 2 & 0 \\ 0 & 0 & 1 \end{bmatrix}$.

By (2.6.20) the vector $\mathbf{q}(1, 2) \in R^3$ is given by the following formula:

$$\mathbf{q}(1, 2) = \mathbf{x}(1, 2) - \mathbf{A}^{1,2} \begin{bmatrix} \mathbf{x}_1(0, 0) \\ 0 \\ 0 \end{bmatrix} - \mathbf{A}^{1,2} \begin{bmatrix} 0 \\ \mathbf{x}_2(0, 0) \end{bmatrix}$$

$$= \mathbf{x}(1, 2) - \mathbf{A}^{1,2} \begin{bmatrix} \mathbf{x}_1(0, 0) \\ \mathbf{x}_2(0, 0) \end{bmatrix} = \begin{bmatrix} 3 \\ 2 \\ 0 \end{bmatrix}.$$

By (2.6.19) the optimal control sequence is given by the following equalities:

$$\begin{aligned}
\mathbf{u}(0, 0) &= \mathbf{M}^T(1, 2)\mathbf{Q}^{-1}(1, 2)\mathbf{q}(1, 2) = 0, \\
\mathbf{u}(1, 0) &= \mathbf{M}^T(0, 2)\mathbf{Q}^{-1}(1, 2)\mathbf{q}(1, 2) = 0, \\
\mathbf{u}(0, 1) &= \mathbf{M}^T(1, 1)\mathbf{Q}^{-1}(1, 2)\mathbf{q}(1, 2) = 1, \\
\mathbf{u}(1, 1) &= \mathbf{M}^T(0, 1)\mathbf{Q}^{-1}(1, 2)\mathbf{q}(1, 2) = 3, \\
\mathbf{u}(0, 2) &= \mathbf{M}^T(1, 0)\mathbf{Q}^{-1}(1, 2)\mathbf{q}(1, 2) = 1.
\end{aligned}$$

The minimum value of the performance index $J(\mathbf{u}^0)$ found from (2.6.22) is

$$J(\mathbf{u}^0) = \mathbf{q}^T(1, 2)\mathbf{Q}^{-1}(1, 2)\mathbf{q}(1, 2) = [3 \ \ 2 \ \ 0]\begin{bmatrix} 1 & 0 & 0 \\ 0 & 0.5 & 0 \\ 0 & 0 & 1 \end{bmatrix}\begin{bmatrix} 3 \\ 2 \\ 0 \end{bmatrix} = 11.$$

Hence the minimum energy control problem has been solved.

Example 2.6.2
Let us consider a 2-D dynamical system with the matrices

$$\mathbf{A} = \begin{bmatrix} \mathbf{A}_{11} & \mathbf{A}_{12} \\ \mathbf{A}_{21} & \mathbf{A}_{22} \end{bmatrix} = \left[\begin{array}{c:cc} 1 & 1 & 0 \\ \hdashline 1 & 1 & 0 \\ 1 & 1 & 1 \end{array}\right], \qquad B = \begin{bmatrix} \mathbf{B}_1 \\ \mathbf{B}_2 \end{bmatrix} = \begin{bmatrix} 1 \\ 1 \\ 1 \end{bmatrix} \tag{2.6.24}$$

defined in the rectangular prism $[(0, 0), (r_1, r_2)] = [(0, 0), (1, 2)]$. Hence $n_1 = 1$, $n_2 = 2$, $n = n_1+n_2 = 3$ and $p = 1$. By (2.6.11) and by the formulas given in Definition 2.6.1 the matrices $\mathbf{M}(i_1, i_2)$ for $(i_1, i_2) \in ((0, 0), (1, 2)]$ are of the following form:

$$\mathbf{M}(0, 1) = \begin{bmatrix} 1 \\ 1 \\ 1 \end{bmatrix}, \quad \mathbf{M}(1, 0) = \begin{bmatrix} 1 \\ 1 \\ 1 \end{bmatrix}, \quad \mathbf{M}(1, 1) = \begin{bmatrix} 1 \\ 1 \\ 2 \end{bmatrix},$$

$$\mathbf{M}(0, 2) = \begin{bmatrix} 2 \\ 2 \\ 3 \end{bmatrix}, \quad \mathbf{M}(1, 2) = \begin{bmatrix} 1 \\ 1 \\ 3 \end{bmatrix}.$$

Since $\operatorname{rank}\mathbf{Q}(1, 2) = \operatorname{rank}[\mathbf{M}(0, 1) \vdots \mathbf{M}(1, 0) \vdots \mathbf{M}(1, 1) \vdots \mathbf{M}(0, 2) \vdots \mathbf{M}(1, 2)] = 2 < 3 = n = n_1+n_2$ then by Theorem 2.6.1 the dynamical system (2.6.24) is not locally controllable in the rectangular prism $[(0, 0), (1, 2)]$.

2.7 Controllability of discrete bilinear dynamical systems

The study of bilinear dynamical systems has been developed fast for the last few years. In the literature there are many items concerning control theory for discrete and continuous bilinear dynamical systems (see. e.g. Ball, Marsden and Slemrod (1982), Brockett (1976), Chang, Tarn and Elliot (1975), Elliot, Tarn and Goka (1973), Evans and Murthy (1977a, 1978), Frazho (1980, 1981), Gauthier and Bornard (1982a), Mohler (1973), Rink and Mohler (1968), Tarn, Elliot and Goka (1973), Wei and Pearson (1978a), Kern (1978)).

Continuous-time bilinear dynamical systems arise as models for chemical and biological processes (Mohler (1973, Ch. 2)). Discrete-time bilinear dynamical systems are very natural models in economics, controlled Markov processes and astronautics (Wei and Pearson (1978b, 1978c)).

The purpose of this section is to investigate the question of controllability of time invariant discrete single-input bilinear dynamical systems. The proposed method is purely algebraic in contrast to that followed with regard continuous bilinear dynamical systems, where extensive use is made of Lie algebras and vector fields (Brockett (1972, 1976), Chang, Tarn and Elliot (1975)), and of some methods of mathematical analysis (Wei and Pearson (1978a, 1978b, 1978c)). For discrete-time bilinear dynamical systems several necessary and sufficient controllability conditions have been recently formulated and proved for homogeneous systems as well as for no nhomogeneous ones (Evans and Murthy (1978), Hohler (1973)). It should be mentioned that we may also use the Jordan canonical form to obtain simple criteria for the controllability of discrete-time bilinear dynamical systems (Klamka (1974b)). Minimum energy control problem for bilinear dynamical systems have been considered and solved in the papers of Wei and Pearson (1978b, 1978c).

In the theory of bilinear dynamical systems we generally distinguish two types of systems: homogeneous and nonhomogeneous ones. A discrete-time, homogeneous dynamical system is described by the following difference equation:

$$\mathbf{x}(k+1) = \mathbf{A}\Big(\mathbf{I}+\sum_{j=1}^{m} u_j(k)\mathbf{B}_j\Big)\mathbf{x}(k), \quad k = 0, 1, 2, \ldots, \tag{2.7.1}$$

where $\mathbf{A}$ is an $(n\times n)$-dimensional nonsingular matrix, $\mathbf{I}$ is an $(n\times n)$-dimensional identity matrix, $\mathbf{B}_j$ are for $j = 1, 2, \ldots, m$, $(n\times n)$-dimensional matrices, $u_j(k) \in \boldsymbol{R}$, $k = 0, 1, 2, \ldots$, $j = 1, 2, \ldots, m$ are scalar controls.

Bilinear, discrete-time nonhomogeneous dynamical systems are defined by the following difference equation (Evans and Murthy (1977a, 1978)):

$$\mathbf{x}(k+1) = \mathbf{A}\Big(\mathbf{I}+\sum_{j=1}^{m} u_j(k)\mathbf{B}_j\Big)\mathbf{x}(k)+\mathbf{B}_0\mathbf{u}(k), \quad k = 0, 1, 2, \ldots, \tag{2.7.2}$$

where $\mathbf{B}_0$ is $(n\times n)$-dimensional matrix, $\mathbf{u}(k) = [u_1(k)\ u_2(k)\ \ldots\ u_j(k)\ \ldots\ u_m(k)]^{\mathrm{T}} \in \boldsymbol{R}^m$ is m-dimensional control vector, the remainder notations are the same as in the equation (2.7.1).

For discrete bilinear dynamical systems (homogeneous and nonhomogeneous ones) we introduce the following definitions of controllability (Evans and Murthy (1977a, 1978), Goka, Tarn and Zaborszky (1973)).

DEFINITION 2.7.1. A discrete bilinear dynamical system is said to be:

(I) *controllable from the set* $D_0 \subset \boldsymbol{R}^n$ *to the set* $D_1 \subset \boldsymbol{R}^n$ *in the interval* $[0, k_1]$ if for any initial state $\mathbf{x}(0) \in D_0$ and any final state $\mathbf{x}_1 \in D_1$ there exist a $k_2 \in [0, k_1]$ and a sequence of controls $\{\mathbf{u}(0), \mathbf{u}(1), \ldots, \mathbf{u}(k_2-1)\}$, such that $\mathbf{x}(k_2, \mathbf{x}(0), \mathbf{u}) = \mathbf{x}_1$,

(II) *uniformly controllable from the set* $D_0 \subset \boldsymbol{R}^n$ *to the set* $D_1 \subset \boldsymbol{R}^n$ if there exists a k_1 such that the bilinear system is controllable from D_0 to D_1 in the interval $[0, k_1]$,

(III) *controllable from the set* $D_0 \subset \boldsymbol{R}^n$ *to the set* $D_1 \subset \boldsymbol{R}^n$ if for any initial state $\mathbf{x}(0) \in D_0$ and any final state $\mathbf{x}_1 \in D_1$ there exist a k_1 and a sequence of controls $\{\mathbf{u}(0), \mathbf{u}(1), \ldots, \mathbf{u}(k_1-1)\}$ (both depending on $\mathbf{x}(0)$ and $\mathbf{x}_1$) such that $\mathbf{x}(k_1, \mathbf{x}(0, \mathbf{u}) = \mathbf{x}_1$.

Definition 2.7.1 implies that the following implications hold: (I) $\Rightarrow$ (II) $\Rightarrow$ (III). The converse implications are not always true.

If we put in Definition 2.7.1 $D_0 = D_1 = D \subset \boldsymbol{R}^n$, then instead "controllability from D_0 to D_1" we shortly say "controllability in the set D". Moreover, if $D_0 = \boldsymbol{R}^n$ and $D_1 = \{0\}$, we speak about "controllability to zero". In the case where $D_0 = D_1 = \boldsymbol{R}^n$ we speak about "controllability" of bilinear system.

We now restrict our attention to single-input, homogeneous bilinear systems

$$\mathbf{x}(k+1) = (\mathbf{A}+u(k)\mathbf{B})\mathbf{x}(k), \quad k = 0, 1, 2, \ldots, \tag{2.7.3}$$

where $u(k) \in \boldsymbol{R}$, $k = 0, 1, 2, \ldots$ are the scalar controls, $\mathbf{A}$ is an $(n \times n)$-dimensional, nonsingular matrix, $\mathbf{B}$ is an $(n \times n)$-dimensional matrix such that $\operatorname{rank}\mathbf{B} = 1$.

For the bilinear dynamical system (2.7.3) we can formulate simple necessary (but not sufficient) or sufficient (but not necessary) controllability conditions.

THEOREM 2.7.1 (Elliot, Tarn and Goka (1973), Evans and Murthy (1977a, 1978), Tarn, Elliot and Goka (1976)). *Suppose the pair* (**A**, **B**) *is controllable and observable. Then*

(I) *if the bilinear system* (2.7.3) *is controllable to zero, then*

$$\mathbf{B}\mathbf{A}^{-1}\mathbf{B} \neq 0, \tag{2.7.4}$$

(II) *the bilinear system* (2.7.3) *is controllable to zero in* $[0, n]$ *if and only if*

$$\mathbf{B}\mathbf{A}^{-i}\mathbf{B} \neq 0 \quad \textit{for } i = 2, 3, 4, \ldots, n, \tag{2.7.5}$$

(III) *if* $\mathbf{B}\mathbf{A}^{-1}\mathbf{B} \neq 0$ *and the bilinear system* (2.7.3) *is controllable in the set* $\boldsymbol{R}^n - \{0\}$, *then it is also controllable to zero.*

Detailed proofs of each part of Theorem 2.7.1 can be found in the literature mentioned above.

For a nonsingular matrix $\mathbf{A}$ and $\operatorname{rank}\mathbf{B} = 1$ we can transform the bilinear, homogeneous system (2.7.3) into an equivalent, more convenient form:

$$\mathbf{x}(k+1) = \mathbf{A}(\mathbf{1}+\mathbf{u}(k)\mathbf{c}\mathbf{d}^{\mathrm{T}})\mathbf{x}(k), \quad k = 0, 1, 2, \ldots, \tag{2.7.6}$$

where $\mathbf{c} \in \boldsymbol{R}^n$, $\mathbf{d} \in \boldsymbol{R}^n$ and $\mathbf{A}^{-1}\mathbf{B} = \mathbf{c}\mathbf{d}^{\mathrm{T}}$.

THEOREM 2.7.2 (Elliot, Tarn and Goka (1973), Evans and Murthy (1977a)). *Let* r *be the greatest common divisor of the set of natural numbers* $J = \{j \in \boldsymbol{N}\colon \mathbf{d}^{\mathrm{T}}\mathbf{A}^j\mathbf{c} \neq 0,\ 0 < j \leqslant n^2\}$. *Then bilinear system* (2.7.6) *is controllable in the set* $\boldsymbol{R}^n - \{0\}$ *if and only if*

$$\operatorname{rank}[\mathbf{c} \,\vdots\, \mathbf{A}\mathbf{c} \,\vdots\, \mathbf{A}^2\mathbf{c} \,\vdots\, \ldots \,\vdots\, \mathbf{A}^{n-1}\mathbf{c}] = n, \tag{2.7.7}$$

$$\operatorname{rank}[\mathbf{d} \vdots \mathbf{A}^{\mathrm{T}}\mathbf{d} \vdots (\mathbf{A}^{\mathrm{T}})^2\mathbf{d} \vdots \ldots \vdots (\mathbf{A}^{\mathrm{T}})^{n-1}\mathbf{d}] = n, \tag{2.7.8}$$

$$r = 1.$$

Basing oneselves on the controllability results concerning homogeneous bilinear systems we can formulate necessary and sufficient conditions for the controllability in $\boldsymbol{R}^n$ of nonhonogeneous bilinear systems described by the following difference equation:

$$\mathbf{x}(k+1) = \mathbf{A}(\mathbf{I}+\mathbf{u}(k)\mathbf{c}\mathbf{d}^{\mathrm{T}})\mathbf{x}(k)+\mathbf{c}\mathbf{u}(k), \quad k = 0, 1, 2, \ldots \tag{2.7.9}$$

These conditions can be obtained by transforming the nonhomogeneous bilinear system (2.7.9) into a homogeneous bilinear system with greater dimensionality.

THEOREM 2.7.3 (Evans and Murthy (1977a, 1978)). *Bilinear, discrete-time, nonhomogeneous dynamical system* (2.7.9) *is controllable if and only if the condition* (2.7.7) *is satisfied and moreover*

$$\operatorname{rank}\begin{bmatrix} 1 & \mathbf{d}^{\mathrm{T}} \\ 1 & \mathbf{d}^{\mathrm{T}}\mathbf{A} \\ 1 & \mathbf{d}^{\mathrm{T}}\mathbf{A}^{2r} \\ \cdots & \cdots \\ 1 & \mathbf{d}^{\mathrm{T}}\mathbf{A}^{ar} \end{bmatrix} = a+1, \tag{2.7.10}$$

where

$$a = m/r \quad \textit{and} \quad m = \operatorname{rank}[\mathbf{d} \vdots \mathbf{A}^{\mathrm{T}}\mathbf{d} \vdots (\mathbf{A}^{\mathrm{T}})^2\mathbf{d} \vdots \ldots \vdots (\mathbf{A}^{\mathrm{T}})^{n-1}\mathbf{d}], \quad m \leqslant n. \tag{2.7.11}$$

In the paper of Evans and Murthy (1978) it is proved, that the number a in formula (2.7.11) is always a natural number, and hence equality (2.7.10) is well posed.

The proofs of Theorems 2.7.1, 2.7.2 and 2.7.3 are based on the matrix calculus and some theorems from linear algebra and the theory of polynomials. Moreover we can construct an algorithm for checking controllability to zero, basing ourselves on the decision methods. It is interesting to point out, that the necessary and sufficient conditions for the controllability of bilinear dynamical systems have been formulated by the successive weakening of the sufficient conditions or the successive strengthening of the necessary conditions. Papers of Chang, Tarn and Elliot (1975) and Mohler (1973) contain some results concerning the controllability of bilinear systems with control contraints.

Now let us consider more exactly the noncontrollable, bilinear homogeneous dynamical systems (2.7.6). Theorem 2.7.2 implies that the dynamical system (2.7.6) is not controllable in $\boldsymbol{R}^n - \{0\}$ in the following three cases:

(I) condition (2.7.7) is satisfied but condition (2.7.8) is not,

(II) condition (2.7.8) is satisfied but condition (2.7.7) is not,

(III) neither condition (2.7.2) nor conditin (2.7.8) is satisfied.

In case (I), using the nonsingular transformation we can convert the dynamical system (2.7.6) into the following equivalent form (Hollis and Murthy (1982)):

$$\begin{bmatrix} \mathbf{x}_1(k+1) \\ \mathbf{x}_2(k+1) \end{bmatrix} = \begin{bmatrix} \mathbf{A}_{11} & 0 \\ \mathbf{A}_{21} & \mathbf{A}_{22} \end{bmatrix} \begin{bmatrix} \mathbf{x}_1(k) \\ \mathbf{x}_2(k) \end{bmatrix} + \begin{bmatrix} \mathbf{c}_1 \\ \mathbf{c}_2 \end{bmatrix} [\mathbf{d}_1^{\mathrm{T}} \ 0] \begin{bmatrix} \mathbf{x}_1(k) \\ \mathbf{x}_2(k) \end{bmatrix} \mathbf{u}(k), \tag{2.7.12}$$
$$k = 0, 1, 2, \ldots,$$

where $\mathbf{x}_1(k) \in \boldsymbol{R}^{n_1}$, $\mathbf{x}_2(k) \in \boldsymbol{R}^{n_2}$ $(n = n_1+n_2)$, $k = 0, 1, \ldots, n_1 = \operatorname{rank}[\mathbf{d} \vdots \mathbf{A}^{\mathrm{T}}\mathbf{d} \vdots (\mathbf{A}^{\mathrm{T}})^2\mathbf{d} \vdots \ldots \ldots \vdots (\mathbf{A}^{\mathrm{T}})^{n-1}\mathbf{d}]$.

From (2.7.12) it follows that if $\mathbf{x}_1(k) = 0$, then $\mathbf{x}_1(j) = 0$ and $\mathbf{x}_2(j) = \mathbf{A}_{22}\mathbf{x}_2(j-1)$ for all $j > k$. Therefore, in fact, the dynamical system (2.7.12) is not controllable in $\boldsymbol{R}^n - \{0\}$. If we want to steer only the state vector $\mathbf{x}_1(k)$, then such partial controllability can be checked by using the following bilinear state equation:

$$\mathbf{x}_1(k+1) = \mathbf{A}_{11}\mathbf{x}_1(k) + \mathbf{c}_1\mathbf{d}_1^{\mathrm{T}}\mathbf{x}_1(k)\mathbf{u}(k), \quad k = 0, 1, 2, \ldots, \tag{2.7.13}$$

which can be directly obtained from equation (2.7.12). Controllability conditions in $\boldsymbol{R}^n - \{0\}$ for the dynamical system (2.7.13) can be derived directly from Theorem 2.7.2 by putting $\mathbf{A} = \mathbf{A}_{11}$, $\mathbf{c} = \mathbf{c}_1$, $\mathbf{d} = \mathbf{d}_1$ and $n = n_1$. If the greatest common divisor for the set $J_1 = \{j \in N\colon \mathbf{d}_1^{\mathrm{T}}\mathbf{A}_{11}^{j-1}\mathbf{c}_1 \neq 0, 0 < j \leqslant 2n_1\}$ is equal to r_1, then the controllability domain for the dynamical system (2.7.6) consists of r_1 subspaces, each of dimension p_i, $i = 1, 2, \ldots, r_1$, $p_1+p_2+ \ldots +p_i+ \ldots +p_{r_1} = n_1$. In every controllability subspace the projection of the n_1-dimensional state vector is nonzero.

In some cases the state vector $\mathbf{x}_2(k)$ can be controllable by the vector $\mathbf{x}_1(k)$ (Hollis and Murthy (1982)). Let X_0 be the kernel of the linear transformation defined by the $(n \times n)$-dimensional matrix $[\mathbf{d} \vdots \mathbf{A}^{\mathrm{T}}\mathbf{d} \vdots (\mathbf{A}^{\mathrm{T}})^2\mathbf{d} \vdots \ldots \vdots (\mathbf{A}^{\mathrm{T}})^{n-1}\mathbf{d}]$. Let us consider the subspace $\boldsymbol{R}^n - X_0$, whose dimensionality is greater than n_1, and which contains controllability subspace defined for the state vector $\mathbf{x}_1(k)$. The necessary and sufficient conditions for the controllability in $\boldsymbol{R}^n - X_0$ of the dynamical system (2.7.12) are given in Theorem 2.7.4.

THEOREM 2.7.4 (Hollis and Murthy (1982)). *The dynamical system* (2.7.12) (*or the quivalent dynamical system* (2.7.6) *in case* (I)) *is controllable in* $\boldsymbol{R}^n - X_0$ *if and only if the greatest common divisor of the set* $J_2 = \{j \in N\colon \mathbf{d}^{\mathrm{T}}\mathbf{A}^{j-1}\mathbf{c} \neq 0,\ 0 < j \leqslant 2n_1\}$ *is equal to zero.*

Since the greatest common divisor for the set J_2 is the same as for the set J_1, we have

$$\mathbf{d}^{\mathrm{T}}\mathbf{A}^{j-1}\mathbf{c} = \mathbf{d}_1^{\mathrm{T}}\mathbf{A}_{11}^{j-1}\mathbf{c}_1.$$

In the case where the greatest common divisor $r_1 > 1$ we have $r_1 a_1 = n_1$ for some natural number a_1 (Hollis and Murthy (1982)).

Now, let us consider the so-called adjoint system for system (2.7.6) given by the following difference equation (Hollis and Murthy (1982)):

$$\mathbf{y}(k+1) = (\mathbf{I} + \mathbf{A}^{-k}\bar{\mathbf{c}}\mathbf{d}^{\mathrm{T}}\mathbf{A}^{k}\mathbf{u}(k))\mathbf{y}(k), \quad k = 0, 1, 2, \ldots \tag{2.7.14}$$

with the initial condition $\mathbf{y}(0) = \mathbf{x}(0)$, where $\mathbf{x}(k) = \mathbf{A}^k\mathbf{y}(k)$ and $\bar{\mathbf{c}} = \mathbf{A}^{-1}\mathbf{c}$.

Let $P_i \subset R^n$, be the linear subspaces spanned by the columns of the matrices $\mathbf{L}_i = \mathbf{A}\,[\mathbf{c} \vdots \mathbf{A}^{r_1}\mathbf{c} \vdots \mathbf{A}^{2r_1}\mathbf{c} \vdots \dots \vdots \mathbf{A}^{(n-1)r_1}\mathbf{c}]$ for $i = 1, 2, \dots, r_1-1$. The first a_1 columns in every matrix $\mathbf{L}_i$ are linearly independent. Since $n_1 < n$, the (n_1+1)-th column is linearly independent of the first n_1 columns. Therefore the dimension q of every subspace P_i satisfies the inequalities

$$r_1 q \geqslant n > r_1(q-1) \tag{2.7.15}$$

and, moreover, is the same for all $P_i = \mathbf{A}^i P_0$.

Controllability in $R^n - X_0$ is characterized by the following theorem.

THEOREM 2.7.5 (Hollis and Murthy (1982)). *Suppose condition* (2.7.7) *is satisfied and* $\mathrm{rank}[\mathbf{d} \vdots \mathbf{A}^T\mathbf{d} \vdots (\mathbf{A}^T)^2\mathbf{d} \vdots \dots \vdots (\mathbf{A}^T)^{n-1}\mathbf{d}] = n_1 < n$. *Moreover, let r_1 be the greatest common divisor of the set J_2. Then an arbitraty initial state of the adjoint dynamical system* (2.7.14) *can be expressed in the following form:*

$$\mathbf{y}(0) = \mathbf{x}(0) = \sum_{i=1}^{r_1-1} \mathbf{x}_i, \qquad \textit{where } \mathbf{x}_i \in P_i \textit{ for } i = 0, 1, 2, \dots, r_1-1. \tag{2.7.16}$$

The adjoint dynamical system (2.7.14) *is not controllable in $R^n - X_0$ if for some index i we have $\mathbf{x}_i \in X_0$. Moreover, the adjoint dynamical system* (2.7.14) *is controllable in $R^n - X_0$ if $\mathbf{x}_i \notin X_0$ for all $i \in [0, r_1-1]$.*

Hence the controllability domain for the adjoint dynamical system (2.7.14) *has the form $P - X_0$, where P is a linear subspace in R^n given by the following formula*

$$P = \bigcup_{i=0}^{r_1-1} P_i. \tag{2.7.17}$$

Since the dynamical system (2.7.6) is connected with the ajoint system (2.7.14) by nonsingular transformation, we have the subspace equality: $\mathbf{A}^{r_1+j}P = \mathbf{A}^j P$. Denoting $Q^j = \mathbf{A}^j P$ for $j \in [0, r_1-1]$, we obtain r_1 different subspaces Q^j. It can be shown (Hollis and Murthy (1982)), that for a given initial state $\mathbf{x}(0)$ of the dynamical system (2.7.6) all final states $\mathbf{x}_1 \in Q^j - X_0, j = 0, 1, 2, \dots, r_1-1$, are attainable.

Now we shall consider case (II), for which the condition (2.7.8) is satisfied and, moreover, $\mathrm{rank}[\mathbf{c} \vdots \mathbf{A}\mathbf{c} \vdots \mathbf{A}^2\mathbf{c} \vdots \dots \vdots \mathbf{A}^{n-1}\mathbf{c}] = n_1 < n$.

THEOREM 2.7.6. *Suppose condition* (2.7.8) *is satisfied and, moreover,* $\mathrm{rank}[\mathbf{c} \vdots \mathbf{A}\mathbf{c} \vdots \mathbf{A}^2\mathbf{c} \vdots \dots \vdots \mathbf{A}^{n-1}\mathbf{c}] = n_1 < n$, *and let r_1 be the greatest common divisor of the set J_2. Then the dynamical system* (2.7.6) *is controllable in subspace R^{n_1}, i.e., the initial state $\mathbf{x}(0) \in R^{n_1}$ can be steered to any final state $\mathbf{x}_{1f} \in R^{n_1}$ if there exists a natural number $p \geqslant n^2$ for which $\mathbf{x}_{2f} = \mathbf{A}_{22}^p \mathbf{x}_2(0)$, $\mathbf{x}_2(0) \in R^{n_2}$, $\mathbf{x}_{2f} \in R^{n_2}$, $n_1 + n_2 = n$, and there exist natural numbers $l_j \in L$, $m_j \in M$ such that for all $j = 0, 1, 2, \dots, r_1-1$ we have*

$$\mathbf{A}^{l_j}\mathbf{c} \in V_j \quad \textit{and} \quad \mathbf{A}^{m_j}\mathbf{c} \in V_j,$$

where V_j, $j = 0, 1, 2, \ldots, r_1 - 1$ *are linear subspaces spanned by the columns of the matrix* $\mathbf{A}^j[\mathbf{c}|\mathbf{A}^{r_1}\mathbf{c}|\mathbf{A}^{2r_1}\mathbf{c}\vdots\ldots\vdots\mathbf{A}^{r_1(n-1)}\mathbf{c}]$, *and the sets L and M are defined as follows:*

$$L = \{l \in N\colon\ \mathbf{d}^{\mathrm{T}}\mathbf{A}^l\mathbf{x}(0) \neq 0\}, \qquad M = \{m \in N\colon\ \mathbf{d}^{\mathrm{T}}\mathbf{A}^{-(m+1)}\mathbf{x}_f \neq 0\},$$

where $\mathbf{x}^{\mathrm{T}}(0) = [\mathbf{x}_1(0)\ \mathbf{x}_2(0)]^{\mathrm{T}}$ *and* $x_f^{\mathrm{T}}[\mathbf{x}_{1f}\ \mathbf{x}_{2f}]^{\mathrm{T}}$.

Finally, let us consider case (III). In order to do that, let us use a nonsingular transformation, converting the dynamical system (2.7.6) into the form (Hollis and Murthy (1982))

$$\begin{bmatrix} \mathbf{x}_1(k+1) \\ \mathbf{x}_2(k+1) \\ \mathbf{x}_3(k+1) \\ \mathbf{x}_4(k+1) \end{bmatrix} = \begin{bmatrix} \mathbf{A}_{11} & \mathbf{A}_{12} & 0 & 0 \\ 0 & \mathbf{A}_{22} & 0 & 0 \\ \mathbf{A}_{31} & \mathbf{A}_{32} & \mathbf{A}_{33} & \mathbf{A}_{34} \\ 0 & \mathbf{A}_{42} & 0 & \mathbf{A}_{44} \end{bmatrix} \begin{bmatrix} \mathbf{x}_1(k) \\ \mathbf{x}_2(k) \\ \mathbf{x}_3(k) \\ \mathbf{x}_4(k) \end{bmatrix} +$$

$$+ \begin{bmatrix} \mathbf{c}_1 \\ 0 \\ \mathbf{c}_2 \\ 0 \end{bmatrix} [\mathbf{d}_1^{\mathrm{T}}\ \ \mathbf{d}_2^{\mathrm{T}}\ \ 0\ \ 0] \begin{bmatrix} \mathbf{x}_1(k) \\ \mathbf{x}_2(k) \\ \mathbf{x}_3(k) \\ \mathbf{x}_4(k) \end{bmatrix} \mathbf{u}(\mathbf{k}), \qquad k = 0, 1, 2, \ldots \tag{2.7.18}$$

where $\mathbf{x}_i(k) \in R^{n_i}$, for $i = 1, 2, 3, 4$, $\sum_{i=0}^{4} n_i = n$ and $\operatorname{rank}[\mathbf{c}\vdots\mathbf{Ac}\vdots\mathbf{A}^2\mathbf{c}\vdots\ldots\vdots\mathbf{A}^{n-1}\mathbf{c}] = n_1 + n_3 < n$, $\operatorname{rank}[\mathbf{d}\vdots\mathbf{A}^{\mathrm{T}}\mathbf{d}\vdots(\mathbf{A}^{\mathrm{T}})^2\mathbf{d}\vdots\ldots\vdots(\mathbf{A}^{\mathrm{T}})^{n-1}\mathbf{d}] = n_1 + n_2 < n$. The state vectors $\mathbf{x}_2(k)$ and $\mathbf{x}_4(k)$ are not controllable, however, the sufficient conditions for the controllability of the vectors $\mathbf{x}_1(k)$ and $\mathbf{x}_3(k)$ are given in Theorem 2.7.7.

THEOREM 2.7.7 (Hollis and Murthy (1982)). *The state vectors* $\mathbf{x}_1(k)$ *and* $\mathbf{x}_3(k)$ *are controllable if there exists a natural number* $p \geqslant n^2$ *for which* $\mathbf{x}_{2f} = \mathbf{A}_{22}^p\mathbf{x}_2(0)$, *and there exist natural numbers* $l_j \in L$ *and* $m_j \in M$ *such that for for all* $j = 0, 1, 2, \ldots, r-1$ *we have*

$$\mathbf{A}^{l_j}\mathbf{c} \in W_j \quad \text{and} \quad \mathbf{A}^{m_j} \in W_j,$$

where W_j, $j = 0, 1, 2, \ldots, r-1$ *are linear subspaces spanned by the columns of the matrix* $\mathbf{A}^j[\mathbf{c}\vdots\mathbf{A}^r\mathbf{c}\vdots\mathbf{A}^{2r}\mathbf{c}\vdots\ldots\vdots\mathbf{A}^{r(n-1)}\mathbf{c}]$, *and the sets L and M are the same as in Theorem* 2.7.6.

The proofs of Theorems 2.7.5, 2.7.6, and 2.7.7, based on the matrix calculus and the geometrical properties of the invariant subspaces, are presented in detail in the paper of Hollis and Murthy (1982).

Example 2.7.1 (Goka, Tarn and Zaborszky (1979))
Let us consider a bilinear discrete homogeneous dynamical system of the form (2.7.9) with the following matrices **A** and **B**:

$$\mathbf{A} = \begin{bmatrix} 0 & 1 \\ 1 & 0 \end{bmatrix}, \quad \mathbf{B} = \begin{bmatrix} 1 & 0 \\ 0 & 0 \end{bmatrix}, \quad \mathbf{A}^{-1}\mathbf{B} = \begin{bmatrix} 0 & 0 \\ 1 & 0 \end{bmatrix}, \quad \operatorname{rank}\mathbf{B} = 1. \tag{2.7.19}$$

Vectors **c** and **d** can be chosen for instance as $\mathbf{c} = \begin{bmatrix} 0 \\ 1 \end{bmatrix}$ and $\mathbf{d} = \begin{bmatrix} 1 \\ 0 \end{bmatrix}$. Hence $\operatorname{rank}[\mathbf{c} \vdots \mathbf{Ac}] = \operatorname{rank}\begin{bmatrix} 0 & 1 \\ 1 & 0 \end{bmatrix} = 2 = n$, and $\operatorname{rank}[\mathbf{d} \vdots \mathbf{A}^T\mathbf{d}] = \operatorname{rank}\begin{bmatrix} 1 & 0 \\ 0 & 1 \end{bmatrix} = 2 = n$. In order to obtain the set J we successively compute

$$\mathbf{d}^T\mathbf{Ac} = [1 \;\; 0]\begin{bmatrix} 0 & 1 \\ 1 & 0 \end{bmatrix}\begin{bmatrix} 0 \\ 1 \end{bmatrix} = 1 \neq 0, \quad \mathbf{d}^T\mathbf{A}^2\mathbf{c} = [1 \;\; 0]\begin{bmatrix} 1 & 0 \\ 0 & 1 \end{bmatrix}\begin{bmatrix} 0 \\ 1 \end{bmatrix} = 0,$$

$$\mathbf{d}^T\mathbf{A}^3\mathbf{c} = [1 \;\; 0]\begin{bmatrix} 0 & 1 \\ 1 & 0 \end{bmatrix}\begin{bmatrix} 0 \\ 1 \end{bmatrix} = 1 \neq 0, \quad \mathbf{d}^T\mathbf{A}^4\mathbf{c} = [1 \;\; 0]\begin{bmatrix} 1 & 0 \\ 0 & 1 \end{bmatrix}\begin{bmatrix} 0 \\ 1 \end{bmatrix} = 0.$$

Hence the set $J = \{1, 3\}$, and thus $r = 1$ is the greatest common divisor of the set J. Therefore by Theorem 2.7.2 the dynamical system (2.7.19) is controllable in $R^n - \{0\}$.

Example 2.7.2 (Evans and Murthy (1977a))
Let us consider a bilinear discrete homogeneous dynamical system of the form (2.7.3) with the following matrices **A** and **B**:

$$\mathbf{A} = \begin{bmatrix} 0 & 1 \\ a_{12} & a_{22} \end{bmatrix}, \quad \mathbf{B} = \begin{bmatrix} 0 & 0 \\ 1 & 0 \end{bmatrix}, \quad a_{12} \neq 0, \quad \mathbf{A}^{-1}\mathbf{B} = \begin{bmatrix} 1/a_{12} & 0 \\ 0 & 0 \end{bmatrix},$$

$$\operatorname{rank}\mathbf{B} = 1. \tag{2.7.20}$$

Vectors **c** and **d** can be chosen for instance as $\mathbf{c} = \begin{bmatrix} 1/a_{12} \\ 0 \end{bmatrix}$ and $\mathbf{d} = \begin{bmatrix} 1 \\ 0 \end{bmatrix}$. Hence $\operatorname{rank}[\mathbf{c} \vdots \mathbf{Ac}] = \operatorname{rank}\begin{bmatrix} 1/a_{12} & 0 \\ 0 & 1 \end{bmatrix} = 2 = n$ and $\operatorname{rank}[\mathbf{d} \vdots \mathbf{A}^T\mathbf{d}] = \operatorname{rank}\begin{bmatrix} 1 & 0 \\ 0 & 1 \end{bmatrix} = 2 = n$. In order to obtain the set J we successively compute:

$$\mathbf{d}^T\mathbf{Ac} = 0, \quad \mathbf{d}^T\mathbf{A}^2\mathbf{c} = 1, \quad \mathbf{d}^T\mathbf{A}^3\mathbf{c} = a_{22}, \quad \mathbf{d}^T\mathbf{A}^4\mathbf{c} = a_{12} + a_{22}^2.$$

Thus the set J is of the following form:

(1) $J = \{2, 3, 4\}$ if $a_{22} \neq 0$ and then $r = 1$,
(2) $J = \{2, 4\}$ if $a_{22} = 0$ and then $r = 2$.

In case (1), i.e., if $a_{22} \neq 0$, the dynamical system (2.7.20) is controllable in $R^n - \{0\}$. In case (2), i.e., if $a_{22} = 0$, the dynamical system (2.7.20) is not controllable in $R^n - \{0\}$. By direct computation we can show that in the case (2) the following initial states are uncontrollable:

$$\mathbf{x}(0) = \begin{bmatrix} x_1(0) \\ 0 \end{bmatrix} \quad \text{or} \quad \mathbf{x}(0) = \begin{bmatrix} 0 \\ x_2(0) \end{bmatrix}.$$

The attainable sets for the above initial states are always one-dimensional and, moreover, Q^k is orthogonal to Q^{k+1}, where Q^k is the attainable set at the step k (Q^k is in fact a linear subspace). Since $\operatorname{rank}[\mathbf{c} \vdots \mathbf{Ac}] = n$ and $\operatorname{rank}[\mathbf{d} \vdots \mathbf{A}^T\mathbf{d}] = n$, in case (2) we cannot directly use Theorems 2.7.4, 2.7.5, 2.7.6 or 2.7.7.

2.8 Remarks and comments

The results presented in Section 2.3 can be extended to the case of discrete linear time-invariant dynamical systems defined in infinite-dimensional linear spaces (mainly in Hilbert or Banach spaces); see e.g. Fuhrmann (1972, 1973, 1974a, 1974b, 1975, 1976), Helton (1974, 1976), Klamka (1978c) for details. Moreover, linear, discrete dynamical systems with the so-called "shift operator" are considered in papers of Fuhrmann (1974a, 1974b, 1975, 1976), in which a great deal of attention has been paid to the problem of realization and its solution with the use of the theory of Hankel operators and invariant subspaces.

Discrete linear dynamical systems with delays in the state variables and in the control can be fairly easily transormed into linear discrete dynamical systems of the form (2.2.1) but with greater dimensionality. This is particularly true for dynamical systems of the form:

$$\mathbf{x}(k+1) = \sum_{i=0}^{N} \mathbf{A}_i(k)\mathbf{x}(k-i) + \sum_{j=0}^{M} \mathbf{B}_j(k)\mathbf{u}(k-j), \qquad k \geqslant k_0, \tag{2.8.1}$$

where $\mathbf{A}_i(k)$, $i = 0, 1, 2, \ldots, N$ are $(n \times n)$-dimensional matrices for $k \geqslant k_0$, $\mathbf{B}_j(k)$, $j = 0, 1, 2, \ldots, M$ are $(n \times m)$-dimensional matrices for $k \geqslant k_0$.

In papers of Busłowicz (1981), Inouye (1982), Kucera (1980), Klamka (1977c, 1977d, 1978c), Mallela (1982), Mullis (1973), Nguyen (1986) several controllability conditions for the dynamical system (2.8.1) and its special cases have been derived. For instance, relative and absolute controllabilities of the dynamical system (2.8.1) have been considered. Moreover, in paper of Klamka (1977d) the minimum energy control problem for dynamical system (2.8.1) has been solved. Paper of Klamka (1978c) contains the criteria for the relative and absolute controllability of dynamical system with delays in control and defined in an infinite-dimensional Hilbert space. Moreover, in the paper of Klamka (1978c) the minimum energy contral problem is also investigated and solved.

Recently, in the analysis of discrete dynamical systems extensive use has been made of the theory of abstract algebra, especially the theory of groups, rings, fields and polynomials of several variables. For instance, several controllability conditions obtained by using pure algebraic and topological methods have been presented in papers of Banaszuk, Kocięcki and Przyłuski (1988), Evans (1986), Kamen (1975a, 1975b, 1976), Mitter and Foulkes (1971), Sontag (1976a, 1976b, 1976c, 1978). The controllability of composite discrete linear dynamical systems has been investigated in the paper of Panda (1970), and the output controllability has been considered in the papers of Mitter and Foulkes (1971, Mullis (1973)). Finally, in the paper of Harris (1978) stochastic controllability of discrete sytems with multiplicative noise has been analysed.

CHAPTER 3

Controllability of Continuous-Time, Infinite-Dimensional Dynamical Systems

3.1 Introduction

In the last few years several works have appeared concerning various kinds of controllability of linear continuous-time dynamical systems, defined in infinite-dimensional spaces (mainly in Hilbert or Banach spaces): Balakrishnan (1977), Baras, Brockett and Fuhrmann (1974), Baras and Brockett (1975), Brogan (1968), Brockett and Fuhrmann (1976), Butkowski (1975, 1977), Dolecki (1976), Dolecki and Russell (1977), Fattorini (1966a, 1966b, 1967, 1974), Feintuch (1978), Fujii and Sakawa(1974), (1974), Fuhrmann (1972, 1974b), Helton (1976), Hiris and Topuzu (1974), Hiris and Megan (1977a, 1977b), Kobayashi (1978, 1980a, 1980b, 1982), Korobov and Rabakh (1979), Korobov and Shon (1980a, 1980b, 1980c), Megan (1974, 1976, 1978a, 1978b), Naito (1987), Narukawa (1981), Reghis (1976), Reghis and Megan (1977), Seidman (1978, 1979, 1987), Sourour (1982), Triggiani (1975a, 1975b, 1975c, 1976, 1977), Yamamoto (1980, 1981), Zhou (1983, 1984), Farlov (1986), Goncalves (1987), Krabs, Leugering and Seidman (1985), Leugering (1985), Littman and Markus (1988).

The most popular class of infinite-dimensional dynamical systems are the systems described by partial differential equations. In order to investigate the controllability of such systems we may use of course the general theorems concerning infinite-dimensional dynamical systems described by abstract differential equations, (Delfour and Mitter (1972b), Fattorini (1966a, 1966b, 1967), Triggiani (1975a, 1976). However, in view of the specific problems arising in partial differential equations, in many papers special methods of controllability investigations have been proposed and extensively used.

Generally those publications can be divided into two different groups: the first group concerns various kinds of controllability of dynamical systems described by partial differential equations of parabolic type with different boundary conditions. In this group we have the following publications: Fattorini (1975a), Fattorini and Russell (1971, 1974), Klamka (1980c, 1981a, 1982b), Lasiecka and Triggiani (1983a, 1983b), Martin (1976, 1977), McGlothin (1974, 1978), MacCamy, Mizel, Seidman (1968, 1969), Nambu (1979a, 1979b), Prabhu and McCausland (1976), Russell (1973), Sakawa (1974, 1975, 1983), Sakawa and Matsushita (1955), Seidman (1977), Triggiani (1978, 1979), Wang (1964, 1972), Weck (1984), White (1980), Zhou (1982), Schmidt (1980a, 1980b).

The second group concerns the various kinds of controllability of dynamical systems described by partial differential equations of hyperbolic type with different boundary conditions: Chen (1979a, 1979b), Fattorini (1975b), Graham and Russell (1975), Lagnese (1977, 1978, 1979), Lasiecka and Triggiani (1983a), Russell (1971, 1973), Triggiani (1978), Weck (1982), Narukawa (1982, 1983, 1984).

The mutual relations between the controllability of dynamical systems of parabolic type and the controllability of dynamical systems of hyperbolic type are explained in the paper of Triggiani (1978) on the basis of the theory of semigroups of bounded linear operators, and in the paper of Russell (1973) with the use of some theorems from harmonic analysis.

Moreover, it should be stressed that in controllability analysis for infinite-dimensional dynamical systems the following factors are very important: the type of control (distributed or boundary control), the shape of the domain in which the partial differential equations are defined, and the type of boundary conditions (Dirichlet, Neumann or mixed boundary conditions). All these factors determine the form of the unbounded linear operator which is connected with the given partial differential equation, and hence also determine the controllability conditions.

3.2 System descripiton and the basic definitions

In this section we shall present the basic definitions of controllability for linear continuous time-invariant infinite-dimensional dynamical systems described by the following abstract differential equation:

$$\dot{\mathbf{x}}(t) = A\mathbf{x}(t)+B\mathbf{u}(t), \quad t \geqslant 0, \tag{3.2.1}$$

where $\mathbf{x}(t) \in X$—Banach space (state space), $\mathbf{u} \in L^2_{\mathrm{loc}}([0, \infty), U)$, U—Banach space (control space), $B \in L(U, X)$, i.e., B: $U \to X$ is a linear and bounded operator, A: $X \supset D(A) \to X$ is the infinitesimal generator of a strongly continuous semigroup of linear, bounded operators $S(t)$: $X \to X$, $t \geqslant 0$ (Fattorini (1983)).

In the special case where the control space U is finite-dimensional, i.e., $U = R^m$, the dynamical system (3.2.1) can be more conveniently written as

$$\dot{\mathbf{x}}(t) = A\mathbf{x}(t) + \sum_{j=0}^{m} \mathbf{b}_j u_j(t), \quad t \geqslant 0, \tag{3.2.2}$$

where $\mathbf{b}_j \in X$ for $j = 1, 2, \ldots, m$

$$u_j \in L^2_{\text{loc}}([0, \infty), R) \quad \text{for} \quad j = 1, 2, \ldots, m.$$

For every initial condition $\mathbf{x}(0) \in X$ and each control $\mathbf{u} \in L^2_{\text{loc}}([0, \infty), U)$, there exists a unique solution, called a "mild" solution, $\mathbf{x}(t, \mathbf{x}(0), \mathbf{u})$ of the abstract differential equation (3.2.1) given by (Fattorini (1974), Triggiani (1975a, 1975b))

$$\mathbf{x}(t, \mathbf{x}(0), \mathbf{u}) = S(t)\mathbf{x}(0) + \int_0^t S(t-\tau) B\mathbf{u}(\tau) d\tau. \tag{3.2.3}$$

In the case of dynamical system (3.2.2) we have

$$\mathbf{x}(t, \mathbf{x}(0), \mathbf{u}) = S(t)\mathbf{x}(0) + \int_0^t S(t-\tau) \sum_{j=1}^{m} \mathbf{b}_j u_j(\tau) d\tau. \tag{3.2.4}$$

Let K_t denote the *attainable set at time* $t > 0$ *from the initial state* $\mathbf{x}(0) = 0$, i.e.

$$K_t = \left\{\mathbf{x} \in X \colon \mathbf{x} = \int_0^t S(t-\tau) B\mathbf{u}(\tau) d\tau \colon \ \mathbf{u} \in L^2_{\text{loc}}([0, t], U)\right\}. \tag{3.2.5}$$

In dynamical systems defined in infinite-dimensional spaces we distinguish two basic types of controllability, namely approximate controllability and exact controllability (Dolecki (1976), Dolecki and Russell (1977), Fattorini (1966a, 1967), Feintuch (1978), Fuhrmann (1972, 1974b), Triggiani (1975a, 1975b)).

This is strongly related to the fact that in infinite-dimensional spaces not every linear subspace is closed (in contrast to the finite-dimensional case, where all the linear subspaces are closed).

Now we shall list fundamental facts concerning semigoups (Balakrishnan (1977, Chapter 4)).

DEFINITION 3.2.1. A family $\{S(t) \colon X \to X, — t \geqslant 0\}$ of bounded linear operators in a Banach space X is called a *strongly continuous semigroup* if

(a) $S(t+\tau) = S(t)S(\tau)$ for all $t, \tau \geqslant 0$,

(b) $S(0) = I$ (identity operator),

(c) for each $\mathbf{x} \in X$, $S(t)\mathbf{x}$ is continuous in t on $[0, \infty)$.

If, in addition, the operators $S(t)$ are compact for every $t > 0$, then the semigroup $\{S(t), t \geqslant 0\}$ is called a *compact semigroup*. Moreover, if for every $t \geqslant 0$, $\|S(t)\|_X \leqslant 1$, then the semigroup is called a *contraction semigroup*. In applications most of the semigroups are compact semigroups.

The another important class of semigroups are analytic semigroups (see e.g. Triggiani (1976)). We call a vector $\mathbf{x}_a \in X$ *analytic for the semigroup* $S(t)$ in the case the map $t \to S(t)\mathbf{x}_a$ is analytic in t for $t > 0$. Let X_a denote the totality of analytic vectors for the semigroup $S(t)$. If $X_a = X$, then the semigroup $S(t)$ is an *analytic samigroup* and, moreover, the map $t \to S(t)$ is infinitely many times continuously differentiable in the uniform operator topology for $t > 0$, and for all integrs $n \geqslant 0$ we have the following formula:

$$\frac{d^n S(t)}{dt^n} = A^n S(t) \in L(X, X), \tag{3.2.6}$$

where A is the infinitesimal generator of the semigroup $S(t)$.

The notion of the infinitesimal generator is strongly related to a strongly continuous semigroup $S(t)$ (see e.g. Balakrishnan (1977, Chapter 4), Dunford and Schwartz (1958, 1963), Fattorini (1983)).

DEFINITION 3.2.2. For $h > 0$ the linear operator A_h is defined by the formula

$$A_h \mathbf{x} = (S(h)\mathbf{x} - \mathbf{x})/h, \quad \mathbf{x} \in X.$$

Let $D(A)$ be the set of all $\mathbf{x}$ in X for which the limit, $\lim_{h \to 0} A_h \mathbf{x}$, exists and define the operator A with domain $D(A)$ by the following formula:

$$A\mathbf{x} = \lim_{h \to 0} A_h \mathbf{x}.$$

The operator A with domain $D(A)$ is called the *infinitesimal generator of the semigroup* $S(t)$.

The set $D(A)$ is a linear manifold everywhere dense in X and A is a linear operator on $D(A)$, which is closed on $D(A)$.

In the case where, in addition, the map $t \to S(t)$ is continuous in the uniform operator topology, the family $\{S(t), t \geqslant 0\}$ is called a *uniformly continuous semigroup*, and then it has a bounded infinitesimal generator. Generally the infinitesimal generator A is an unbounded operator.

More informations concerning the semigroups and infinitesimal generators can be found in the extensive monographs of Dunford and Schwartz (1958, 1963) and Fattorini (1983).

DEFINITION 3.2.3. The dynamical system (3.2.1) is said to be *approximately controllable in the finite interval* $[0, t_1]$ if

$$\overline{K_{t_1}} = X.$$

DEFINITION 3.2.4. The dynamical system (3.2.1) is said to be *approximately controllable in finite time* if

$$\overline{\bigcup_{0 \leqslant t < \infty} K_t} = X.$$

DEFINITION 3.2.5. The dynamical system (3.2.1) is said to be *exactly controllable on the finite interval* $[0, t_1]$ if

$$K_{t_1} = X.$$

DEFINITION 3.2.6. The dynamical system (3.2.1) is said to be *exactly controllable in finite time* if

$$\bigcup_{0 \leqslant t < \infty} K_t = X.$$

Since an affine transformation of dense subspaces gives us again a dense subspace, in the definitions of approximate controllability on the finite interval $[0, t_1]$ and approximate controllability in finite time it is enough to consider the attainable sets from zero initial state.

In the special case, where the semigroup $S(t)$ is an analytic semigroup (Triggiani (1975b)) the approximate (exact) controllability on finite interval $[0, t_1]$ is equivalent to approximate (exact) controllability in the finite time (Triggiani (1975a, 1975b)), and hence also to approximate (exact) controllability in an arbitrary time interval. In this case the closedness of the attainable set $\bar{K}_t$ is constant for all $t > 0$ (Seidman (1979), Triggiani (1978)).

In practice, most of dynamical systems defined in infinite-dimensional spaces are only approximately controllable and not exactly controllable (Triggiani (1975a, 1975b, 1977, 1978), Yamamoto (1980), Zhou (1982, 1983, 1984)). Moreover, it should be pointed out that approximate controllability is a sufficiently strong notion and assumption to solve most problems in optimal control theory.

If the operator A is bounded and so generates a uniformly continuous analytic group $S(t) = \exp(At)$, $-\infty < t < +\infty$, then the conditions for approximate and exact controllability and their proofs have simple forms. Unfortunately most of dynamical systems do not satisfy this assumption. For instance, dynamical systems described by partial differential equations have an unbounded operator A.

3.3 Controllability conditions for time-invariant dynamical systems

This section is devoted entirely to generalizations of the familiar rank condition given in Theorem 1.4.1 to the general dynamical systems (3.2.1) if both the state space X and the control space U are infinite-dimensional Banach spaces and the operator A is only assumed to generate a strongly continuous semigroup (group).

LEMMA 3.3.1 (Triggiani (1975b)). *The dynamical system* (3.2.1) *is approximately controllable on* $[0, t_1]$ *(in finite time) if and only if for all* $x^* \in X^*$

$$x^*(S(t)B) = 0 \text{ for } t \in [0, t_1],\ t > 0 \text{ implies } x^* = 0. \tag{3.3.1}$$

In particular, the dynamical system (3.2.2) *is approximately controllable on* $[0, t_1]$ *(in finite time) if and only if, for all* $x^* \in X^*$,

$$x^*(S(t)\mathbf{b}_j) = 0 \text{ for } j = 1, 2, \ldots, m \text{ and } t \in [0, t_1],\ t > 0 \text{ implies } x^* = 0. \tag{3.3.2}$$

Proof:

By the Hahn–Banach theorem (Triggiani (1975a, 1975b)) we have $\overline{K_{t_1}} = X$ if and only if the zero functional is the only bounded linear functional that vanishes on the attainable set K_{t_1}. Hence by (3.2.5) and the linearity of the integral operation we obtain Lemma 3.3.1. ∎

Using Lemma 3.3.1, we can formulate the necessary and sufficient conditions for the approximate controllability of the dynamical systems (3.2.1) and (3.2.2). For simplicity of considerations it is assumed that the operator A generates the analytic semigroup $S(t)$, $t \in [0, \infty)$. Criteria of approximate controllability for more general dynamical systems are given in Triggiani (1975b). Since the semigroup $S(t)$ is an analytic semigroup, we have

$$\overline{K_{t_1}} = \overline{\bigcup_{0 \leqslant t < \infty} K_t} \quad \text{for } t_1 > 0. \tag{3.3.3}$$

Thus approximate controllability in $[0, t_1]$ is equivalent to approximate controllability in finite time. In order to simplify the notation we shall speak shortly about approximate controllability (without mentioning the time interval).

THEOREM 3.3.1 (Triggiani (1975b)). *The dynamical system* (3.2.1) *is approximately controllable if and only if for an arbitrary* $\tilde{t} > 0$ *we have*

$$\overline{\text{sp}}\{A^n S(\tilde{t})B, n \geqslant 0\} = X. \tag{3.3.4}$$

Proof:

By the analyticity of the semigroup $S(t)$, $t > 0$, we have

$$\frac{d^n S(t)}{dt^n} = A^n S(t) \in L(X, X) \quad \text{for } t > 0,\ n = 0, 1, 2, \ldots \tag{3.3.5}$$

Sufficiency. By contradition. Suppose that equality (3.3.4) holds but the dynamical system (3.2.1) is not approximately controllable. Thus by Lemma 3.3.1 there exists a nonzero functional $x_1^* \in X^*$ such that

$$x_1^*(S(t)B) = 0, \quad t > 0.$$

Differentiating the above equality and using (3.3.5) we obtain

$$x_1^*(A^n S(t)B) = 0, \quad t > 0, \quad n = 0, 1, 2, \ldots$$

Thus by the Hahn–Banach theorem this contradicts formula (3.3.4).

Necessity. By contradiction. Suppose the dynamical system (3.2.1) is approximately controllable, but (3.3.4) does not hold. Thus by (3.3.4) and the Hahn–Banach theorem there exists a nonzero functional $x_1^* \in X^*$ such that

$$x_1^*(A^n S(t)B) = x_1^*(S(t)A^n B) = 0, \quad t > 0, \quad n = 0, 1, 2, \ldots$$

By analyticity of $S(t)B$ for $t > 0$ and by Lemma 3.3.1 we obtain a contradiction of relation (3.3.4). ∎

COROLLARY 3.3.1 (Triggiani (1975b)). *The dynamical system* (3.2.2) *is approximately controllable if and only if for arbitrary* $\tilde{t} > 0$ *we have*

$$\overline{\text{sp}}\{A^n S(\tilde{t}) b_j, n \geqslant 0, j = 1, 2, \ldots, m\} = X. \tag{3.3.6}$$

Proof:

Corollary 3.3.1 immediately follows from Theorem 3.3.1 and $U = \boldsymbol{R}^m$. ■

COROLLARY 3.3.2 (Triggiani (1975b)). *Suppose the operator A generates an analytic group $S(t)$, $-\infty < t < +\infty$, then the dynamical system* (3.2.1) *is approximately controllable if and only if*

$$\overline{\text{sp}}\{A^n B, n \geqslant 0\} = X. \tag{3.3.7}$$

Proof:

If the operator A generates the analytic group $S(t)$, $-\infty < t < +\infty$, then we may put $\tilde{t} = 0$ in formula (3.3.4). Thus $S(\tilde{t}) = S(0) = I$, which immediately leads to the condition (3.3.7). ■

COROLLARY 3.3.3 (Triggiani (1975)). *Suppose the operator A generates an analytic group $S(t)$, $-\infty < t < +\infty$. Then the dynamical system* (3.2.2) *is approximately controllable if and only if*

$$\overline{\text{sp}}\{A^n b_j, n \geqslant 0, j = 1, 2, \ldots, m\} = X. \tag{3.3.8}$$

Proof:

Corollary 3.3.3 immediately follows from Corollaries 3.3.1 and 3.3.2. ■

Since the analyticity of the group $S(t)$ is equivalent to the boundedness of the operator A, Corollaries 3.3.2 and 3.3.3 are the necessary and sufficient conditions for approximate controllability of the dynamical systems (3.2.1) and (3.2.2) with bounded operator A. In the special case $X = \boldsymbol{R}^n$, $U = \boldsymbol{R}^m$ we obtain Theorem 1.4.1.

It should be pointed out, that for concrete forms of the operators A and B, the verification of conditions (3.3.4) and (3.3.7) is commonly very difficult and requires the use of advanced methods of functional analysis. However, under some additional assumptions concerning the operator A, we can obtain simple and easily computable criteria for approximate controllability. Subsequently we shall assume that X is a Hilbert space and A is an infinitesimal generator of the analytic semigroup $S(t)$ with the compact resolvent $R(s, A) \in L(X, X)$ for all $s \in \varrho(A)$, ($\varrho(A)$ is the resolvent set).

Now let us put together some known facts from functional analysis (Triggiani (1975b, 1976, 1978)).

(a) Spectrum $\sigma(A)$ of the operator A is a point spectrum consisting entirely of s_i, $i = 1, 2, \ldots$ distinct isolated, real eigenvalues of the operator A, each with finite multiplicity $r(i)$, $i = 1, 2, \dots$ equal to the dimensionality of the corresponding eigenmanifolds.

(b) There is a corresponding complete orthonormal set $\{\mathbf{x}_{ik}\}$, $i = 1, 2, \dots$, $k = 1, 2, \dots, r(i)$ of eigenvetcors of A.

(c) If $X \ni \mathbf{x} = \sum_{i=1}^{\infty} \sum_{k=1}^{r(i)} \langle \mathbf{x}, \mathbf{x}_{ik} \rangle_X \mathbf{x}_{ik}$, then

$$A^n \mathbf{x} = \sum_{i=1}^{\infty} s_i^n \sum_{k=1}^{r(i)} \langle \mathbf{x}, \mathbf{x}_{ik} \rangle_X \mathbf{x}_{ik}, \quad n = 0, 1, 2, \dots \tag{3.3.9}$$

for $\mathbf{x} \in D(A^n) = \left\{ \mathbf{x} \in X: \sum_{i=1}^{\infty} |s_i^n|^2 \sum_{k=1}^{r(i)} |\langle \mathbf{x}, \mathbf{x}_{ik} \rangle_X|^2 < \infty \right\}$,

$$S(t)\mathbf{x} = \sum_{i=1}^{\infty} \exp(s_i t) \sum_{k=1}^{r(i)} \langle \mathbf{x}, \mathbf{x}_{ik} \rangle_X \mathbf{x}_{ik}, \quad t \geqslant 0, \mathbf{x} \in X, \tag{3.3.10}$$

$$A^n S(t) b_j = \sum_{i=1}^{\infty} \exp(s_i t) s_i^n \sum_{k=1}^{r(i)} \langle \mathbf{b}_j, \mathbf{x}_{ik} \rangle_X \mathbf{x}_{ik}, \tag{3.3.11}$$

$$n = 0, 1, \dots, \quad b_j \in X, \quad j = 1, 2, \dots, m, \quad t \geqslant 0.$$

To simplify the notation let us introduce the $(r(i) \times m)$-dimensional matrices $\mathbf{B}_i$ defined as follows:

$$\mathbf{B}_i = \begin{bmatrix} \langle \mathbf{b}_1, \mathbf{x}_{i1} \rangle_X & \langle \mathbf{b}_2, \mathbf{x}_{i1} \rangle_X & \dots & \langle \mathbf{b}_m, \mathbf{x}_{i1} \rangle_X \\ \langle \mathbf{b}_1, \mathbf{x}_{i2} \rangle_X & \langle \mathbf{b}_2, \mathbf{x}_{i2} \rangle_X & \dots & \langle \mathbf{b}_m, \mathbf{x}_{i2} \rangle_X \\ \dots & \dots & \dots & \dots \\ \langle \mathbf{b}_1, \mathbf{x}_{ir(i)} \rangle_X & \langle \mathbf{b}_2, \mathbf{x}_{ir(i)} \rangle_X & \dots & \langle \mathbf{b}_m, \mathbf{x}_{ir(i)} \rangle_X \end{bmatrix}, \quad i = 1, 2, \dots \tag{3.3.12}$$

In the sequel, we shall have to refer to the following defintion: $\sup_i r(i) = r \leqslant \infty$, i.e., r is the highest multiplicity (if $r < \infty$) of the eigenvalues of A. If $r = 1$, we simply write $\mathbf{x}_i$ instead of $\mathbf{x}_{i1}$. If $r < \infty$, then the point spectrum of A is of finite multiplicity.

THEOREM 3.3.2 (Triggiani (1976)). *Suppose a selfadjoint operator A with a compact resolvent $R(s, A) \in L(X, X)$, X—Hilbert space, is an infinitesimal generator of an analytic semigroup $S(t)$, $t \geqslant 0$. Then the dynamical system* (3.2.2) *is approximately controllable if and only if*

$$\operatorname{rank} \mathbf{B}_i = r(i) \quad \textit{for } i = 1, 2, 3, \dots \tag{3.3.13}$$

Proof:

Necessity. By contradiction. Suppose the dynamical system (3.2.2) is apporximately controllable but the condition (3.3.13) does not hold. Then there exists an index $\bar{i}$ such that $\operatorname{rank} \mathbf{B}_{\bar{i}} < r(\bar{i})$. Thus there exists a nonzero $r(\bar{i})$-dimensional vector

$\mathbf{v}_{\bar{i}} = [x_{\bar{1}}^*(\mathbf{x}_{\bar{i}1})\, x_{\bar{1}}^*(\mathbf{x}_{\bar{i}2}) \;\ldots\; x_{\bar{1}}^*(\mathbf{x}_{\bar{i}r(\bar{i})})]^{\mathrm{T}}$, $0 \neq x_{\bar{1}}^* \in X^*$ and $x_{\bar{1}}^*(\mathbf{x}_{ik}) = 0$ for $i \neq \bar{i}$, $k = 1, 2, \ldots, r(i)$, such that $\mathbf{B}_{\bar{i}}^{\mathrm{T}}\mathbf{v}_{\bar{i}} = 0$. Thus by (3.3.11) we have $x_{\bar{1}}^*(A^n S(\tilde{t})\mathbf{b}_j) = 0$ for $n = 0, 1, 2, \ldots, j = 1, 2, \ldots, m, \tilde{t} > 0$.

Hence, by Corollary 3.3.1 and the Hanhn–Banach theorem, we obtain a contradiction.

Sufficiency. Let $\tilde{t} > 0$ and $\tilde{\mathbf{b}}_j = S(\tilde{t})\mathbf{b}_j$. Then we have $\langle \tilde{\mathbf{b}}_j, \mathbf{x}_{ik}\rangle_X = \exp(s_i t) \times$ $\times \langle \mathbf{b}_j, \mathbf{x}_{ik}\rangle_X$. Let $\tilde{\mathbf{B}}_i$ denote the matrix $\mathbf{B}_i$ consisting of the vectors $\tilde{\mathbf{b}}_j, j = 1, 2, \ldots, m$. Thus $\operatorname{rank}\tilde{\mathbf{B}}_i = \operatorname{rank}\mathbf{B}_i$ for $i = 1, 2, \ldots$ Since the operator $AS(\tilde{t}) \in L(X, X)$, we conclude by Corollary 3.3.3 and result of Triggiani (1975a) that condition (3.3.13) implies approximate controllability of the dynamical system (3.2.2). ■

COROLLARY 3.3.4 (Triggiani (1976)). *Suppose that the assumptioms of Theorem* 3.3.2 *are satisfied. Then the necessary condition for approximate controllability of the dynamical system* (3.3.2) *is*

$$r \leqslant m, \tag{3.3.14}$$

Proof:

If $m < r$, then there exists an index $\bar{i}$ such that $r(\bar{i}) > m$. Thus $\operatorname{rank}\mathbf{B}_{\bar{i}} < r(\bar{i})$ and by (3.3.13) the dynamical system (3.2.2) is not approximately controllable. ■

COROLLARY 3.3.5 (Triggiani (1976)). *Let* $r(i) = 1$ *for* $i = 1, 2, \ldots$ *Suppose that the assumptions of Theorem* 3.3.2 *are satisfied. Then the dynamical system* (3.2.2) *is approximately controllable if and only if*

$$\sum_{j=1}^{\infty} \langle \mathbf{b}_j, \mathbf{x}_i \rangle_X^2 \neq 0 \quad \text{for } i = 1, 2, \ldots \tag{3.3.15}$$

Proof:

If $r(i) = 1$ for $i = 1, 2, \ldots$, then by (3.3.12) matrices $\mathbf{B}_i$ reduce to the row vectors and the condition $\operatorname{rank}\mathbf{B}_1 = 1$ denotes the existence of a nonzero element in this row. Then by (3.3.15) the above statement is equivalent to the condition (3.3.15). ■

Now let us consider the exact controllability of the dynamical systems (3.2.1) and (3.2.2). A direct study of some of the approximately controllable systems shows that there are systems which are not exactly controllable on any finite time interval. In practice, most dynamical systems which are approximately controllable on $[0, t_1]$ or in finite time are not exactly controllable in finite time and particularly in the time interval $[0, t_1]$.

THEOREM 3.3.3 (Triggiani (1975a, 1975c, 1977)). *The dynamical system* (3.2.1) *is not excatly controllable in finite time if the operator* B *is compact or the operator* A *generates a compact semigroup* $S(t)$, $t > 0$.

Proof:

Let $P_{t_1}: L^2([0, t_1], U) \to X$ be a linear, bounded operator given by

$$P_{t_1}\mathbf{u} = \mathbf{x}(t_1, 0, \mathbf{u}) = \int_0^{t_1} S(t_1 - \tau)B\mathbf{u}(\tau)d\tau \tag{3.3.16}$$

and defined for arbitrary, but fixed $t_1 > 0$.

Since for a compact operator B or for a compact semigroup $S(t)$, $t > 0$, the operator $S(t_1 - \tau)B$ is also compact, by (3.3.16) the operator P_{t_1} is compact for arbitrary but fixed $t_1 > 0$. Thus the image $K_{t_1}(n)$ of a ball with radius n and centre at zero under the operator P_{t_1} is a precompact set in the space X. Let $\overline{K_{t_1}(n)}$ be its closure. Since X is infinite-dimensional, $\overline{K_{t_1}(n)}$ cannot contain spheres and hence is nowhere dense in X. Next observe that exact controllability on $[0, t_1]$ demands $X = \bigcup_{n=1}^{\infty} K_{t_1}(n)$ but this is impossible by the Baire category theorem. Consequently the range of the operator P_{t_1} does not fill all of X. Since t_1 is arbitrary, by Definition 3.2.4 the dynamical system (3.2.1) is not exactly controllable in finite time. ∎

COROLLARY 3.3.6 (Triggiani (1975a, 1975c, 1977)). *The dynamical system* (3.2.2) *is not exactly controllable in finite time.*

Proof:

Since the operator B in (3.2.2) is finite-dimensional (in fact m-dimensional), it is compact. Therefore by Theorem 3.3.3 the dynamical system (3.2.2) is not exactly controllable. ∎

Corollary 3.3.6 explains why, in practice, most dynamical systems defined in infinite-dimensional spaces X are not exactly controllable. This follows from the fact that almost all dynamical systems have finite-dimensional control vectors and hence compact operators B.

In order to check the exact or approximate controllability of the dynamical system (3.2.1), where X is a Hilbert space, it is sometimes convenient to use the so-called *controllability operator* W_{t_1} defined as follows (Kobayashi (1978), Megan and Hiris (1975)):

$$L(X, X) \ni W_{t_1} = P_{t_1}P_{t_1}^* = \int_0^{t_1} S(t_1 - t)BB^*S^*(t_1 - t)dt, \qquad t_1 > 0. \tag{3.3.17}$$

From (3.3.17) it immediately follows, that the controllability operator W_{t_1} is selfadjoint and bounded. The controllability operator W_{t_1} generalizes to the infinite-dimensional case the controllability matrix defined by formula (1.3.8).

THEOREM 3.3.4 (Kobyashi (1978)). *The dynamical system* (3.2.1) *is approximately controllable in* $[0, t_1]$ *if and only if the controllability operator* W_{t_1} *is positive definite.*

Proof:

By Definition 3.2.1 the dynamical system (3.2.1) is approximately controllable in $[0, t_1]$ if and only if the operator P_{t_1} has a dense image, which is equivalent to the statement that it is invertible, i.e., that its kernel has only the zero element. Since

$$||P_{t_1}^* \mathbf{x}||_X^2 = \langle P_{t_1} P_{t_1}^* \mathbf{x}, \mathbf{x} \rangle_X = \langle W_{t_1} \mathbf{x}, \mathbf{x} \rangle_X, \quad \mathbf{x} \in X \tag{3.3.18}$$

then the kernel of the operator $R_{t_1}^*$ consists only of the zero element if and only if the controllability operator $W_{t_1} \in L(X, X)$ is positive definite, i.e., if and only if for an arbitrary $\mathbf{x} \in X$ we have

$$\langle W_{t_1} \mathbf{x}, \mathbf{x} \rangle_X \geqslant 0 \text{ and } \langle W_{t_1} \mathbf{x}, \mathbf{x} \rangle_X = 0 \text{ imply } \mathbf{x} = 0. \tag{3.3.19}$$

Hence our theorem follows. ■

COROLLARY 3.3.7 (Kobayashi (1978)). *The dynamical system* (3.2.1) *is approximately controllable in* $[0, t_1]$ *if and only if the operator* W_{t_1} *is invertible.*

Proof:

Invertibility of the operator W_{t_1} is equivalent to its positive definitness. Hence Corollary 3.3.7 follows from Theorem 3.3.4. ■

COROLLARY 3.3.8 (Megan and Hiris 1975)). *The dynamical system* (3.2.1) *is exactly controllable in* $[0, t_1]$ *if and only if the controllability operator* W_{t_1} *has a bounded nverse, i.e.,* $W_{t_1}^{-1} \in L(X, X)$.

Proof:

By Definition 3.2.3 the exact controllability in $[0, t_1]$ of the dynamical system (3.2.1) is equivalent to the statement that operator P_{t_1} has an image which covers the whole space X. Thus by (3.3.17) this is equivalent to the statement that the controllability operator W_{t_1} has a bounded inverse operator.

One easily verifies that some standard corollaries to the algebraic test for controllability in the finite-dimensional theory, can be extended to infinite-dimensional spaces.

3.4 Controllability of systems governed by parabolic partial differential equations

The general results concerning approximate controllability of dynamical systems defined in infinite-dimensional spaces can be used to analyse approximate controllability of dynamical systems described by partial differential equations of the parabolic type. Let us consider the dynamical system defined in a bounded domain $D \subset R^n$, with a sufficiently smooth boundary S, described by a time-invariant partial differential equation of the following form:

$$\frac{\partial w(\mathbf{x},t)}{\partial t} = \sum_{k=1}^{n} \frac{\partial^2 w(\mathbf{x},t)}{\partial x_k^2} + q(\mathbf{x})w(\mathbf{x},t), \quad \mathbf{x} \in D,\ t > 0 \tag{3.4.1}$$

with boundary condition

$$\frac{\partial w(\mathbf{x},t)}{\partial \mathbf{v}} + z(\mathbf{x})w(\mathbf{x},t) = \sum_{j=1}^{m} g_j(\mathbf{x})u_j(t), \quad \mathbf{x} \in S,\ t > 0, \tag{3.4.2}$$

where $\mathbf{v}$ is normal vector at S exterior to D, and with initial condition

$$w(\mathbf{x},0) = w_0(\mathbf{x}),\ \mathbf{x} \in D,\ w_0(\mathbf{x}) \in L^2(D). \tag{3.4.3}$$

Since in the above dynamical system the control is exercised through the boundary S of the domain D, it is called a *boundary control* (Fattorini (1975a), Fujii (1980), Russell (1971), Seidman (1978)).

Besides dynamical systems with boundary control, we shall also consider dynamical systems with so-called *distributed control*, described by the following equation:

$$\frac{\partial w(\mathbf{x},t)}{\partial t} = \sum_{k=1}^{n} \frac{\partial^2 w(\mathbf{x},t)}{\partial x_k^2} + q(\mathbf{x})w(\mathbf{x},t) + \sum_{j=1}^{m} b_j(\mathbf{x})u_j(t), \tag{3.4.4}$$

$$\mathbf{x} \in D,\ t > 0$$

with the boundary condition

$$\frac{\partial w(\mathbf{x},t)}{\partial \mathbf{v}} + z(\mathbf{x})w(\mathbf{x},t) = 0, \quad \mathbf{x} \in S,\ t > 0. \tag{3.4.5}$$

where $\mathbf{v}$ is a vector normal at S exterior to D, and with the initial condition

$$w(\mathbf{x},0) = w_0(\mathbf{x}), \quad \mathbf{x} \in D,\ w_0(\mathbf{x}) \in L^2(D). \tag{3.4.6}$$

It is assumed that the function $q(\mathbf{x})$ is continuous in the set $\bar{D} = D \cup S$, and functions $z(\mathbf{x})$ and $g_j(\mathbf{x})$, $j = 1, 2, \ldots, m$ are continuous in the set S.

Both these dynamical systems can be connected with a linear unbounded operator A: $D(A) \to L^2(D)$ defined as follows:

$$Aw(\mathbf{x}) = \sum_{k=1}^{n} \frac{\partial^2 w(\mathbf{x})}{\partial x_k^2} + q(\mathbf{x})w(\mathbf{x}), \quad w(\mathbf{x}) \in D(A), \tag{3.4.7}$$

$$D(A) = \Big\{ w(\mathbf{x}) \in L^2(D):\ Aw(\mathbf{x}) \in L^2(D),$$

$$\frac{\partial w(\mathbf{x})}{\partial \mathbf{v}} + z(\mathbf{x})w(\mathbf{x}) = 0,\ \mathbf{x} \in S \Big\}. \tag{3.4.8}$$

The domain $D(A)$ of the operator A is dense in the space $L^2(D)$, i.e., $\overline{D(A)} = L^2(D)$.

It is well known (see e.g. Fujii (1980), Seidman (1978), Triggiani (1976, 1978)) that the operator A satisfies all the hypotheses given in Section 3.3 before Theorem 3.3.2. It is enough to take as the space X the Hilbert space $L^2(D)$. However, it should be pointed out that the multiplicities $r(i)$, $i = 1, 2, \ldots$ of the eigenvalues s_i, $i = 1, 2, \ldots$ are finite for every $i < \infty$, but we do not always have $\sup_i r(i) < \infty$ (Fattorini and Russell (1971), Russell (1971), Triggiani (1976)). The number $r = \sup_i r(i)$ has an important meaning in investigating approximate controllability of dynamical systems described by partial differential equations.

Basing oneselves on formula (3.3.12) let us introduce the following notation:

$$\mathbf{B}_{Di} = \begin{bmatrix} \langle \mathbf{b}_1, \mathbf{x}_{i1}\rangle_{L^2(D)} & \langle \mathbf{b}_2, \mathbf{x}_{i2}\rangle_{L^2(D)} & \cdots & \langle \mathbf{b}_m, \mathbf{x}_{i1}\rangle_{L^2(D)} \\ \langle \mathbf{b}_1, \mathbf{x}_{i2}\rangle_{L^2(D)} & \langle \mathbf{b}_2, \mathbf{x}_{i2}\rangle_{L^2(D)} & \cdots & \langle \mathbf{b}_m, \mathbf{x}_{i2}\rangle_{L^2(D)} \\ \cdots & \cdots & \cdots & \cdots \\ \langle \mathbf{b}_1, \mathbf{x}_{ir(i)}\rangle_{L^2(D)} & \langle \mathbf{b}_2, \mathbf{x}_{ir(i)}\rangle_{L^2(D)} & \cdots & \langle \mathbf{b}_m, \mathbf{x}_{ir(i)}\rangle_{L^2(D)} \end{bmatrix}, \tag{3.4.9}$$

$i = 1, 2, 3, \ldots$

$$\mathbf{B}_{Si} = \begin{bmatrix} \langle \mathbf{b}_1, \mathbf{x}_{i1}\rangle_{L^2(S)} & \langle \mathbf{b}_2, \mathbf{x}_{i1}\rangle_{L^2(S)} & \cdots & \langle \mathbf{b}_m, \mathbf{x}_{i1}\rangle_{L^2(S)} \\ \langle \mathbf{b}_1, \mathbf{x}_{i2}\rangle_{L^2(S)} & \langle \mathbf{b}_2, x_{i2}\rangle_{L^2(S)} & \cdots & \langle \mathbf{b}_m, \mathbf{x}_{i2}\rangle_{L^2(S)} \\ \cdots & \cdots & \cdots & \cdots \\ \langle \mathbf{b}_1, \mathbf{x}_{ir(i)}\rangle_{L^2(S)} & \langle \mathbf{b}_2, \mathbf{x}_{ir(i)}\rangle_{L^2(S)} & \cdots & \langle \mathbf{b}_m, \mathbf{x}_{ir(i)}\rangle_{L^2(S)} \end{bmatrix}, \tag{3.4.10}$$

$i = 1, 2, 3, \ldots$

where

$$\langle \mathbf{b}_j, \mathbf{x}_{ik}\rangle_{L^2(D)} = \int_D \mathbf{b}_j(\mathbf{x})\mathbf{x}_{ik}(\mathbf{x})\,d\mathbf{x}, \quad j = 1, 2, \ldots, m, \ k = 1, 2, \ldots, r(i),$$

$i = 1, 2, \ldots,$

$$\langle \mathbf{b}_j, \mathbf{x}_{ik}\rangle_{L^2(S)} = \int_S \mathbf{b}_j(\mathbf{x})\mathbf{x}_{ik}(\mathbf{x})\,d\mathbf{x}, \quad j = 1, 2, \ldots, m, \ k = 1, 2, \ldots, r(i),$$

$i = 1, 2, 3, \ldots,$

$\mathbf{B}_{Di}$ and $\mathbf{B}_{Si}$, $i = 1, 2, \ldots$ are constant $(r(i) \times m)$-dimensional matrices.

THEOREM 3.4.1 (Fattorini and Russell (1971), Russell (1971)). *The dynamical system* (3.4.4), (3.4.5) *is approximately controllable if and only if*

$$\operatorname{rank} \mathbf{B}_{Di} = r(i) \quad \textit{for } i = 1, 2, 3, \ldots \tag{3.4.11}$$

Proof:

The dynamical system (3.4.4), (3.4.5) satisfies all the assumptions of Theorem 3.3.2. Moreover, the matrices $\mathbf{B}_{Di}$, $i = 1, 2, 3, \ldots$ correspond to the matrices $\mathbf{B}_i$ in (3.3.12). Hence Theorem 3.4.1 immediately follows from Theorem 3.3.2. ∎

THEOREM 3.4.2 (Fujii (1980), Seidman (1978)). *The dynamical system* (3.4.1), (3.4.2) *is approximately boundary controllable if and only if*

$$\operatorname{rank}\mathbf{B}_{Si} = r(i), \quad i = 1, 2, 3, \dots \tag{3.4.12}$$

Proof:

Using Green's formula and the properties of the operator A, we can express the dynamical system (3.4.1), (3.4.2) in the following equivalent form (Fujii 1980)):

$$\frac{d}{dt}\langle \mathbf{w}(\mathbf{x}, t), \mathbf{x}_{ik}\rangle_{L^2(D)} = s_i\langle \mathbf{w}(\mathbf{x}, t), \mathbf{x}_{ik}\rangle_{L^2(D)} + \sum_{j=1}^{m}\langle \mathbf{g}_j(\mathbf{x}), \mathbf{x}_{ik}\rangle_{L^2(S)} u_j(t),$$

$$i = 1, 2, 3, \dots, \quad k = 1, 2, 3, \dots, r(i). \tag{3.4.13}$$

Infinite systems of ordinary differential equations (3.4.13) can be rewritten in matrix form (Fujii (1980)):

$$\frac{d\mathbf{X}_i(t)}{dt} = \mathbf{A}_i\mathbf{X}_i(t) + \mathbf{B}_{Si}\mathbf{U}(t), \quad i = 1, 2, 3, \dots, \tag{3.4.14}$$

where

$$\mathbf{X}_i(t) = \begin{bmatrix} \langle w(\mathbf{x}, t), \mathbf{x}_{i1}\rangle_{L^2(D)} \\ \langle w(\mathbf{x}, t), \mathbf{x}_{i2}\rangle_{L^2(D)} \\ \cdots\cdots\cdots\cdots\cdots \\ \langle w(\mathbf{x}, t), \mathbf{x}_{ir(i)}\rangle_{L^2(D)} \end{bmatrix}, \quad i = 1, 2, 3, \dots,$$

$$\mathbf{A}_i = \operatorname{diag}[\underbrace{s_i \ s_i \ \dots \ s_i}_{r(i) \text{ times}}], \quad i = 1, 2, 3, \dots$$

From the results presented in Section 1.5 and a specially Theorem 1.5.1 we conclude that every system of the form (3.4.14) is controllable if and only if $\operatorname{rank}\mathbf{B}_{Si} = r(i)$. Since the eigenvectors $\mathbf{x}_{ik}$, $i = 1, 2, 3, \dots, k = 1, 2, 3, \dots, r(i)$ form a complete set in the Hilbert space $L^2(D)$, the dynamical system (3.4.1), (3.4.2) is approximately boundary controllable if and only if every finite-dimensional dynamical system (3.4.14) is controllable. Thus the dynamical system (3.4.1), (3.4.2) is approximately boundary controllable if and only if the condition (3.4.12) holds. ■

COROLLARY 3.4.1. *Suppose* $r(i) = 1$ *for* $i = 1, 2, 3, \dots$ *Then the dynamical system* (3.4.1), (3.4.2) *is approximately boundary controllable if and only if*

$$\sum_{j=1}^{m}\langle \mathbf{b}_j, \mathbf{x}_{ij}\rangle^2_{L^2(S)} \neq 0 \quad \textit{for } i = 1, 2, 3, \dots \tag{3.4.15}$$

Moreover, the dynamical system (3.4.4), (3.4.5) *is approximately controllable if and only if*

$$\sum_{j=1}^{m}\langle \mathbf{b}_j, \mathbf{x}_{i1}\rangle^2_{L^2(D)} \neq 0 \quad \textit{for } i = 1, 2, 3, \dots \tag{3.4.16}$$

Proof:

Corollary 3.4.1 follows from Theorem 3.4.2 and Corollary 3.3.5. ■

COROLLARY 3.4.2. *Suppose* $r(i) = 1$ *for* $i = 1, 2, 3, \ldots$ *and* $m = 1$. *Then the dynamical system* (3.4.1), (3.4.2) *is approximately boundary controllable if and only if*

$$\langle \mathbf{b}_1, \mathbf{x}_{i1} \rangle_{L^2(D)} \neq 0 \quad \text{for } i = 1, 2, 3, \ldots \tag{3.4.17}$$

Moreover, the dynamical system (3.4.4), (3.4.5) *is approximately controllable if and only if*

$$\langle \mathbf{b}_1, \mathbf{x}_{i1} \rangle_{L^2(D)} \neq 0 \quad \text{for } i = 1, 2, 3, \ldots \tag{3.4.18}$$

Proof:

Substituting $j = 1$ in inequalities (3.4.15) and (3.4.16), we directly obtain conditions (3.4.17) and (3.4.18). ■

The results obtained can easily be extended to cover the case of the dynamical systems described by partial differential equations of the parabolic type with coefficients depending on spatial variables x_l, $l = 1, 2, 3, \ldots, n$. Since such dynamical systems are still linear and time-invariant dynamical systems, we can use the preceding theorems and corollaries to investigate approximate controllability and approximate boundary controllability. However, it should be pointed out that the computations of the eigenfunctions x_{ik}, $i = 1, 2, 3, \ldots$, $k = 1, 2, 3, \ldots, r(i)$ is in general very tedious and difficult.

The same difficulties arise also in considering approximate controllability and approximate boundary controllability of dynamical systems (3.4.1), (3.4.2) and (3.4.4), (3.4.5). In practice, it is only in very special cases that we can effectively compute the eigenfunctions x_{ik}, where $i = 1, 2, 3, \ldots$ and $k = 1, 2, 3, \ldots, r(i)$.

A very important role in consireding the controllability of dynamical systems described by partial differential equations is played by the boundedness of the domain D. The bounded domain D implies that the operator A has only a pure discrete point spectrum consisting entirely of isolated real eigenvalues (Seidman (1978), Triggiani (1978)).

In the case where the domain D is unbounded operator A may have a so-called *continuous spectrum* (McGlothin (1974)). In this case the investigation of controllability is very difficult; some results connected with this topic are given in McGlothin (1978) for a very special dynamical system. However, it should be stressed that even in the case of an unbounded domain D there exist dynamical systems possessing only a pure point spectrum consisting of isolated eigenvalues (Fattorini (1974)). Approximate controllability of such dynamical systems is considered in Klamka (1982b).

The methods used in Klamka (1982b) are analogous to those described in Section 3.4.

Example 3.4.1 (Butkowski (1975), Fattorini (1975))
Let us consider the dynamical system

$$\frac{\partial w(x, t)}{\partial t} = \frac{\partial^2 w(x, t)}{\partial x^2} + b(x)u(t), \quad 0 \leqslant x \leqslant d, \ t > 0 \tag{3.4.19}$$

with the boundary conditions

$$w(0, t) = w(d, t) = 0, \ t > 0. \tag{3.4.20}$$

Hence the domain $D = [0, d]$ is a segment and the boundary $S = \{0, d\}$ is a set containing two points. Moreover, it is assumed that $b(\cdot) \in L^2([0, d], \boldsymbol{R})$.

It is well known (Fattorini (1975)) that the operator A corresponding to the dynamical system (3.4.19), (3.4.20) possesses only a pure point spectrum consisting entirely of isolated real eigenvalues $s_i = (i\pi/d)^2$, $i = 1, 2, 3, \ldots$ Moreover, the set of the eigenfunctions $x_i(x) = \sqrt{2/d}\sin(i\pi x/d)$, $i = 1, 2, 3, \ldots$, is complete in the Hilbert space $L^2([0, d], \boldsymbol{R})$. Since all the eigenvalues are singular, we have of course $r(i) = 1$ for all $i = 1, 2, 3, \ldots$ Thus we can use directly Corollary 3.4.2 in order to obtain criteria for aproximate controllability. Thus the dynamical system (3.4.19), (3.4.20) ia approximately controllable if and only if

$$\int_0^d b(x)x_i(x)dx = \int_0^d \sqrt{2/d}\sin(i\pi x/d)b(x)dx \neq 0 \quad \text{for } i = 1, 2, 3, \ldots \tag{3.4.21}$$

For instance, the function $b(x) = \text{const}$ does not ensure the approximate controllability of the dynamical system (3.4.19), (3.4.20).

Example 3.4.2 (Butkowski (1975), Fattorini (1975))
Let us consider the dynamical system

$$\frac{\partial w(x, t)}{\partial t} = \frac{\partial^2 w(x, t)}{\partial x^2} + qw(x, t) + b(x)u(t), \quad 0 \leqslant x \leqslant d, \ t > 0 \tag{3.4.22}$$

with boundary conditions (3.4.20), where $q \in \boldsymbol{R}$ is a constant coefficient.

Knowing the eigenvalues and eigenfunctions for the dynamical system (3.4.19), we can easily obtain the eigenvalues and eigenfunctions for the dynamical system (3.4.22). Namely, from the definitions of eigenvalues and eigenfunctions we obtain:

$$s_i = (i\pi/d)^2 + q, \quad i = 1, 2, 3, \ldots,$$
$$x_i(x) = \sqrt{2/d}\sin(i\pi x/d), \quad i = 1, 2, 3, \ldots$$

Hence the eigenvalues are shifted by the value q and the eigenfunctions are the same. Therefore the conditions of approximate controllability for the dynamical system (3.4.22) are the same as the conditions of approximate controllability for the dynamical system (3.4.19) and are given by inequalities (3.4.21).

Example 3.4.3 (Klamka (1982b))
Let us consider the dynamical system

$$\frac{\partial w(x,t)}{\partial t} = \frac{\partial^2 w(x,t)}{\partial x^2} + (k-x^2)w(x,t)+b(x)u(t),$$

$$-\infty \leqslant x \leqslant +\infty, \; t > 0 \tag{3.4.23}$$

with the initial condition $w(x,0) = w_0(x) \in L^2(R)$, where $b(x) \in L^2(R)$ and k is an integer.

This is an example of a dynamical system defined in an unbounded domain $D = (-\infty, +\infty)$ which, however, satisfies all the assumptions of Theorem 3.4.2 and the subsequent corollaries. This follows from the fact, that the operator A connected with the dynamical system (3.4.23) has only a pure point spectrum consisting entirely of isolated real eigenvalues $s_{ik} = -2i+k-1$ for $i = 0, 1, 2, \ldots$, each of multiplicity $r(i) = 1$. The corresponding eigenfunctions are

$$x_i(x) = (2^i i!)^{-0.5}(\pi)^{-0.25}H_i(x)\exp(-0.5x^2), \quad i = 0, 1, 2, \ldots,$$

where $H_i(x)$, $i = 0, 1, 2, \ldots$ are Hermite polynomials defined as follows:

$$H_i(x) = (-1)^i \exp(x^2)\frac{d^i}{dx^i}(\exp(-x^2)),$$

$$i = 0, 1, 2, \ldots, \quad -\infty \leqslant x \leqslant +\infty.$$

The Hermite polynomials $H_i(x)$, $i = 0, 1, 2, \ldots$ form a complete set in $L^2(B)$. By Corollary 3.4.2 the system (3.4.23) is approximately controllable if and only if

$$\int_{-\infty}^{+\infty} b(x)x_i(x)dx \neq 0 \quad \text{for } i = 0, 1, 2, \ldots \tag{3.4.24}$$

For instance, let $b(x) = \exp(x-0.5x^2) \in L^2(R)$. Thus by (3.4.24) we have

$$\int_{-\infty}^{+\infty} b(x)x_i(x)dx = (2^i i!)^{-0.5}(\pi)^{-0.25}\int_{-\infty}^{+\infty} \exp(x-0.5x^2)H_i(x) \times$$

$$\times \exp(-0.5x^2)dx = (2i!)^{-0.5}(\pi e)^{0.25} \neq 0 \quad \text{for } i = 0, 1, 2, \ldots$$

Therefore the dynamical system (3.4.23) is approximately controllable for the function $b(x) = \exp(x-0.5x^5)$ for an arbitrary integer k.

Since the eigenfunctions $x_i(x) \in L^2(R)$ are independent of the coefficient k for all $i = 0, 1, 2, 3, \ldots$, the approximate controllability of the dynamical system (3.4.23) is also independent of the coefficient k. The spectrum of the operator A does not contain a continuous part. This is an immediate consequence of the fact that in our case the function $q(x)$ has a very special form, i.e., $q(x) = k-x^2$.

Example 3.4.4
Let us consider the dynamical system

$$\frac{\partial w(x,t)}{\partial t} = \frac{\partial^2 w(x,t)}{\partial x^2} + (0.25x^{-2} - x^2 + k)w(x,t) + b(x)u(t), \qquad (3.4.25)$$

$$0 < x \leqslant +\infty, \quad t > 0$$

with the initial condition $w(x, 0) = w_0(x) \in L^2((0, \infty), R)$, where $b(x) \in L^2((0, \infty), R)$ and k is an integer.

This is an example of a dynamical system defined in an unbounded domain $D = (0, \infty)$ which, however, satisfies all the assumptions of Theorem 3.4.2 and the subsequent corollaries. This is an immediate consequence of the fact that the operator A connected with the dynamical system (3.4.25) has only a pure point spectrum, consisting entirely of real isolated eigenvalues $s_{ik} = -4i-2+k$, $i = 0, 1, 2, \ldots$, each of multiplicity $r(i) = 1$. The corresponding eigenfunctions are of the following form:

$$x_i(x) = \sqrt{x}\exp(-x^2/2)L_i(x^2), \quad i = 0, 1, 2, \ldots,$$

where $L_i(x)$ are Laguerre polynomials defined as

$$L_i(x) = e^x/i! \frac{d^i}{dx^i}(x^i e^{-x}), \quad i = 0, 1, 2, \ldots, \; x \in (0, \infty).$$

The Laguerre polynomials $L_i(x)$, $i = 0, 1, 2, \ldots$ form a complete set in the space $L^2((0, \infty), \boldsymbol{R})$. It should be pointed out that the eigenfunctions $x_i(x) \in L^2((0, \infty), \boldsymbol{R})$ do not depend on the parameter k.

By Corollary 3.4.2 the dynamical system (3.4.23) is approximately controllable if and only if

$$\int_0^{+\infty} b(x)x_i(x)\,dx \neq 0 \quad \text{for } i = 0, 1, 2, \ldots \qquad (3.4.26)$$

For instance, let $b(x) = \sqrt{x}\exp(-x^2/2) \in L^2((0, \infty), \boldsymbol{R})$. Then by (3.4.26) we immediately obtain

$$\int_0^{+\infty} b(x)x_i(x)\,dx = \int_0^{+\infty} \sqrt{x}\exp(-x^2/2)\sqrt{x}\exp(-x^2/2)L_i(x^2)\,dx$$

$$= 2^{-i} \neq 0 \quad \text{for } i = 0, 1, 2,$$

Thus the dynamical system (3.4.26) with the function $b(x) = \sqrt{x}\exp(-x^2/2)$ is approximately controllable for an arbitrary parameter k.

As in Example 3.4.3, the very special form of the function $q(x) = 0.25x^{-2} - -x^2 + k$ implies that the operator A corresponding to the dynamical system (3.4.26) does not contain the continuous part of the spectrum.

3.5 Controllability after the introduction of sampling

In this section we shall consider the controllability properties of the dynamical system (3.2.2) after the introduction of sampling with period T. We shall additionally assume that the dynamical system (3.2.2) satisfies the equalities (3.3.10) and (3.3.12).

We consider the case where the control $\mathbf{u}(t)$ is piecewise constant, i.e., the input $\mathbf{u}(t)$ changes value only at a discrete instant of time $t_k = kT$, $k = 0, 1, 2, \dots$, where $T > 0$ is called the *sampling period* for the dynamical system (3.2.2)

Hence the control $\mathbf{u}(t)$ is of the following form:

$$\mathbf{u}(t) = \mathbf{u}(k) \in \boldsymbol{R}^m \quad \text{for} \quad t \in [kT, (k+1)T) = [t_k, t_{k+1}), \quad k = 0, 1, 2, \dots \tag{3.5.1}$$

By (3.2.4) and (3.5.1) we derive

$$\mathbf{x}\big(t_k, \mathbf{x}(0), \mathbf{u}\big) = S(t_k)\mathbf{x}(0) + \int_0^{t_k} S(t_k - \tau) \sum_{j=1}^{m} \mathbf{b}_j \Big(\sum_{i=0}^{k-1} D_k(\tau)\mathbf{u}_j(k) \Big) d\tau, \tag{3.5.2}$$

where

$$D_k(\tau) = H(\tau - kT) - H\big(\tau - (k+1)T\big), \qquad \tau \in \boldsymbol{R},\ k = 0, 1, 2, \dots, \tag{3.5.3}$$

$H(\tau)$ is the so-called *Heaviside function.*

Without loss of generality, we can assume that the initial condition $\mathbf{x}(0)$ is equal to 0. Therefore by (3.5.2) we obtain the following formula for the value $\mathbf{x}(t_k, 0, \mathbf{u}) \in X$:

$$\begin{aligned} \mathbf{x}(t_k, 0, \mathbf{u}) &= \int_0^{t_k} S(t_k - \tau) \sum_{j=1}^{m} \mathbf{b}_j \sum_{i=0}^{k-1} D_k(\tau)\mathbf{u}_j(k)\big) d\tau \\ &= \sum_{i=0}^{k-1} \sum_{j=1}^{m} \int_0^{T} S\big((k-i)T - \tau\big)\mathbf{b}_j \mathbf{u}_j(k)\, d\tau. \end{aligned} \tag{3.5.4}$$

It can be proved (see e.g. Kobayashi (1980a, 1981b)) that $\mathbf{x}(t_k, 0, \mathbf{u})$ depends continuously on the sequence of controls $\mathbf{u}_j(k)$, $j = 1, 2, 3, \dots, m$, $k = 0, 1, 2, \dots$ for an arbitrary $k = 0, 1, 2, \dots$

Modifying Definitions 2.2.1 2.2.2, 3.2.1 and 3.2.2, we obtain the following definitions for approximate controllability and approximate controllability in N steps for the dynamical system (3.2.2) after the introduction of sampling (Kobayashi (1980a)).

DEFINITION 3.5.1. The dynamical system (3.5.2) is said to be *approximately controllable* if the linear subspace

$$L_\infty = \bigcup_{k=1}^{\infty} L_k = \bigcup_{k=1}^{\infty} \{x(t_k, 0, \mathbf{u}): \mathbf{u} \text{ is an arbitrary control of the form (3.5.1)}\}$$

is dense in the space X, i.e., $\overline{L_\infty} = X$.

DEFINITION 3.5.2. The dynamical system (3.5.2) is said to be *approximately controllable in N steps* if the linear subspace

$$L_N = \bigcup_{k=1}^{N} L_k$$

is dense in the space X, i.e., $\overline{L_N} = X$.

From Definitions 3.5.1 and 3.5.2 it follows that the approximate controllability in N steps implies the approximate controllability of the dynamical system (3.5.2). It should be stressed that the inverse implication does not hold. However, there exist special classes of discrete dynamical systems for which these two notions are equivalent (Fuhrmann (1972, 1974b)).

THEOREM 3.5.1 (Kobayashi (1980a)). *The dynamical system* (3.2.2) *after the introduction of sampling is approximately controllable if and only if*

$$\operatorname{rank} \mathbf{B}_i = r(i) \quad \text{for } i = 1, 2, 3, \ldots \tag{3.5.5}$$

where the $(r(i) \times m)$*-dimensional matrices* $\mathbf{B}_i$, $i = 1, 2, 3, \ldots$, *are given by formula* (3.3.12).

Proof:

From Definition 3.5.1 it immediately follows that the dynamical system (3.2.2) after the introduction of sampling is approximately controllable if and only if

$$\{\langle \mathbf{x}(t_k, 0, \mathbf{u}), \mathbf{z}\rangle_X = 0 \text{ for each control of the form (3.5.1) and } k = 1, 2, 3, \ldots\} \text{ implies } \mathbf{z} = 0. \tag{3.5.6}$$

The implication (3.5.6) is a consequence of the Hahn–Banach theorem used with respect to the linear subspace L_∞.

By (3.5.4) we obtain the following equality:

$$\langle \mathbf{x}(t_k, 0, \mathbf{u}), \mathbf{z}\rangle_X = \sum_{i=0}^{k-1} \sum_{j=1}^{m} \int_0^T \langle S((k-i)T-\tau)\mathbf{b}_j, \mathbf{z}\rangle_X \mathbf{u}_j(k)\, d\tau. \tag{3.5.7}$$

Since equality (3.5.7) must hold for each $\mathbf{u}_j(k)$, $j = 1, 2, 3, \ldots, m$, $k = 1, 2, 3, \ldots$, the dynamical system (3.2.2) is approximately controllable after the introduction of sampling if and only if

$$\left\{\int_0^T \langle S(t_k-\tau)\mathbf{b}_j, \mathbf{z}\rangle_X d\tau = 0 \text{ for } j = 1, 2, \ldots, m, k = 1, 2, 3, \ldots\right\} \text{ implies } \mathbf{z} = 0. \tag{3.5.8}$$

Now we shall utilize additional assumptions concerning the dynamical system (3.2.2). Namely we shall use formula (3.3.10), which allows us to present the semigroup $S(t)$, $t > 0$ in a special form. Thus, by (3.3.10), we can transform the left-hand side of implication (3.5.8) in the following manner:

$$\int_0^T \langle S(t_k-\tau)\mathbf{b}_j, \mathbf{z}\rangle_X d\tau = \int_0^T \Big\langle \sum_{i=1}^{\infty} \exp\big(s_i(t_k-\tau)\big) \sum_{l=1}^{r(i)} b_{il}^j \mathbf{x}_{il}, \sum_{i=1}^{\infty}\sum_{l=1}^{r(i)} z_{il}\mathbf{x}_{il}\Big\rangle_X d\tau$$

$$= \int_0^T \sum_{i=1}^{\infty} \exp\big(s_1(t_k-\tau)\big) \sum_{l=1}^{r(i)} \mathbf{b}_{il}^j \mathbf{z}_{il}\, d\tau$$

$$= \sum_{i=1}^{\infty} \exp\big(s_i(k-1)T\big)\big(1+\exp(s_i T)\big)/s_i \sum_{l=1}^{r(i)} \mathbf{b}_{il}^j \mathbf{z}_{il}, \tag{3.5.9}$$

where $b_{il}^j = \langle \mathbf{b}_j, \mathbf{x}_{il}\rangle_X$ and $z_{il} = \langle \mathbf{z}, \mathbf{x}_{il}\rangle_X$ for $i = 1, 2, 3, \ldots, l = 1, 2, \ldots, r(i)$, $j = 1, 2, \ldots, m$. Let us denote

$$a_i = \big(1+\exp(s_i T)\big)/s_i \neq 0 \quad \text{for } i = 1, 2, 3, \ldots$$

Then, by (3.5.9), we have

$$\int_0^T \langle S(t_k-\tau)\mathbf{b}_j, \mathbf{z}\rangle_X d\tau = \sum_{i=1}^{\infty} a_i \exp\big(s_i(k-1)T\big) \sum_{l=1}^{r(i)} b_{il}^j z_{il}. \tag{3.5.10}$$

Substituting (3.5.10) in the implication (3.5.8) and taking into account the formula (3.3.12) we obtain Theorem 3.5.1. ■

COROLLARY 3.5.1 (Kobayashi (1980a)). *The dynamical system* (3.2.2) *after the introduction of sampling is approximately controllable if and only if the dynamical system* (3.2.2) *without discretization is approximately controllable.*

Proof:

Comparing the conditions (3.3.13) in Theorem 3.3.2 and (3.5.5) in Theorem 3.5.1, we obtain Corollary 3.5.1. ■

COROLLARY 3.5.2 (Kobayashi (1980a)). *The dynamical system* (3.2.2) *after the introduction of sampling is not approximately controllable in N steps.*

Proof:

By Definition 3.5.2 and the Hahn–Banach theorem the dynamical system (3.2.2) after the introduction of sapling is approximately controllable in N steps if and only if

$$\{\langle \mathbf{x}(t_k, 0, \mathbf{u}), \mathbf{z}\rangle_X = 0 \text{ for each control of the form (3.5.2) and } k = 1, 2, 3, \ldots, N\} \text{ implies } \mathbf{z} = 0. \tag{3.5.11}$$

By (3.5.9) and (3.5.10) we have

$$\langle \mathbf{x}(t_k, 0, \mathbf{u}), \mathbf{z} \rangle_X = \sum_{i=1}^{\infty} a_i \exp(s_i kT) \sum_{l=1}^{r(i)} b_{il}^j z_i = 0, \tag{3.5.12}$$

$j = 1, 2, \ldots, m, \quad k = 0, 1, 2, \ldots, N-1.$

Formula (3.5.12) represents the system of mN linear, algebraic equations with unknowns z_{il}, $i = 1, 2, 3, \ldots$, $l = 1, 2, 3, \ldots, r(i)$. Therefore this system has infinitely many solutions and thus implication (3.5.11) does not hold. Hence the dynamical system (3.2.2) after the introduction of sampling is never approximately controllable in N steps (even if the dynamical system without discretization is approximately controllable). ■

3.6 Controllability of second order dynamical systems

In this section we shall consider approximate controllability of second order dynamical systems described by the following abstract differential equation:

$$\dot{\mathbf{x}}(t) = A\mathbf{x}(t) + \sum_{j=1}^{m} \mathbf{b}_j u_j(t), \quad t \geqslant 0, \tag{3.6.1}$$

where $\mathbf{x}(t) \in X$ is a separable Banach space, $u_j \in L^2_{\text{loc}}([0, \infty), \boldsymbol{R})$, $j = 1, 2, 3, \ldots$, are admissible controls, $A: X \supset D(A) \to X$ is the infinitesimal generator of a strongly continuous cosine function $C(t)$ of bounded linear operators in X, $-\infty \leqslant t \leqslant +\infty$.

These assumptions are necessary and sufficient for the homogeneous second order system to be uniformly well posed on the real line and, moreover, guarantee that the operator A is also the infinitesimal generator of strongly continuous semigroup $S(t)$: $X \to X$, $t \geqslant 0$, of bounded linear operators defined as follows (Triggiani (1978)):

$$S(t)\mathbf{x} = \frac{1}{\sqrt{\pi t}} \int_0^{\infty} \exp(-\tau^2/4t) C(\tau)\mathbf{x} d\tau, \quad t > 0, \ \mathbf{x} \in X. \tag{3.6.2}$$

Assuming zero initial conditions, i.e., $\mathbf{x}(0) = 0$ and $\dot{\mathbf{x}}(0) = 0$, the so-called mild solution of the dynamical system (3.6.1) is of the following form (Triggiani (1978)):

$$\mathbf{x}(t, 0, \mathbf{u}) = \int_0^t \left(\int_0^{t-\tau} C(s) \sum_{j=1}^{m} \mathbf{b}_j u_j(\tau) ds \right) d\tau, \quad t \geqslant 0, \tag{3.6.3}$$

$$\dot{\mathbf{x}}(t, 0, \mathbf{u}) = \int_0^t C(t-\tau) \sum_{j=1}^{m} \mathbf{b}_j u_j(\tau) d\tau, \quad t \geqslant 0. \tag{3.6.4}$$

The mild solution makes sense for any controls $u_j \in L^2_{\text{loc}}([0, \infty), \boldsymbol{R})$. In particular, if $u_j \in C^{(1)}([0, \infty), \boldsymbol{R})$, then the mild solution is indeed the strict solution which is twice strongly continuously differentiable.

Let K_t denote the attainable set at time t from the zero initial state for the dynamical system (3.6.1), i.e.,

$$K_t = \{(\mathbf{x}(t, 0, \mathbf{u}), \dot{\mathbf{x}}(t, 0, \mathbf{u})) \in X \times X\colon\ u_j \in L^2_{\mathrm{loc}}([0, \infty), \boldsymbol{R}),\ j = 1, 2, 3, \dots, m\}. \tag{3.6.5}$$

DEFINITION 3.6.1. The dynamical system (3.6.1) is said to be *approximately controllable in* $[0, t_1]$ if

$$\overline{K_{t_1}} = X \times X.$$

DEFINITION 3.6.2. The dynamical system (3.6.1) is said to be *approximately controllable in finite time* if

$$\overline{\bigcup_{0<t<\infty} K_t} = X \times X.$$

By Theorem 3.3.3 the dynamical system (3.6.1) is not exactly controllable in finite time, and thus of course it is not approximately controllable in $[0, t_1]$. This follows from the fact that the operator B is in this case a finite-dimensional operator and therefore it is a compact operator. Since the cosine abstract function $C(t)$, $-\infty \leqslant t \leqslant +\infty$, is not an analytic function, we cannot identify approximate controllability in finite time and the approximate controllability in $[0, t_1]$. However, the following inclusion holds

$$\overline{K_{t_1}} \subset \overline{\bigcup_{0<t<\infty} K_t}.$$

Hence for the dynamical system (3.6.1) the approximate controllability in $[0, t_1]$ implies approximate controllability in finite time. The inverse implication is not always true.

The relation between approximate controllability of the dynamical system (3.2.2) and approximate controllability in finite time of the dynamical system (3.6.1) is presented in the next theorem.

THEOREM 3.6.1 (Triggiani (1978)). *Suppose that the operator A satisfies all the assumptions given in Section* 3.3 *preceding Theorem* 3.3.2. *Then the dynamical system* (3.6.1) *is approximately controllable in finite time if and only if the dynamical system* (3.2.2) *is approximately controllable in an arbitrary time interval. Moreover, the dynamical system* (3.6.1) *is approximately controllable in an arbitrary time interval if the elements* $b_j \in X$, $j = 1, 2, \dots, m$ *can be expressed in the form*:

$$\mathbf{b}_j = S(\tau_j)\mathbf{g}_j = \sum_{i=1}^{\infty} \exp(s_i \tau_j) \sum_{k=1}^{r(i)} \langle \mathbf{d}_j, \mathbf{x}_{ik} \rangle_X \mathbf{x}_{ik}, \tag{3.6.6}$$

for some $\mathbf{d}_j \in X$ *and* $\tau_j > 0$, $j = 1, 2, \dots, m$.

Proof:

A detailed proof of the theorem is given in Triggiani (1978). Since it is based on some additional lemmas, it will be omitted here. ∎

COROLLARY 3.6.1 (Triggiani (1978)). *The dynamical system* (3.6.1) *is approximately controllable in finite time if and only if*

$$\operatorname{rank}\mathbf{B}_i = r(i) \quad \textit{for } i = 1, 2, 3, \ldots, \tag{3.6.7}$$

where the $(r(i)\times m)$*-dimensional matrices* $\mathbf{B}_i$, $i = 1, 2, 3, \ldots$ *are given by formula* (3.3.12). *Moreover, if equality* (3.6.6) *holds, then the dynamical system* (3.6.1) *is approximately controllable in an arbitrary time interval if and only if relation* (3.6.7) *holds.*

Proof:

By Theorem 3.3.2 condition (3.6.7) is the necessary and sufficient condition for approximate controllability in an arbitrary time interval for the dynamical system (3.2.2). Hence, by Theorem 3.6.1, this is also the necessary and sufficient condition for approximate controllability in finite time of the dynamical system (3.6.1). The second part of Corollary 3.6.1 immediately follows from Theorem 3.6.1. ∎

Further on in this section we shall investigate the approximate controllability of the dynamical system (3.6.1) after the introduction of sampling with period T. Assuming that the initial conditions are zero, and using the notation given in Section 3.5, we obtain by (3.6.3) the following formula defining $\mathbf{x}(t_k, 0, \mathbf{u}) \in X$ (Triggiani (1976, 1978)):

$$\mathbf{x}(t_k, 0, \mathbf{u}) = \int_0^{t_k} G(t_k-\tau)\sum_{j=1}^{m}\mathbf{b}_j\Big(\sum_{i=0}^{k-1} D_k(\tau)u_j(k)\Big)d\tau, \tag{3.6.8}$$

where

$$G(t-\tau)\mathbf{x} = \int_0^{t-\tau} C(s)\mathbf{x}\,ds, \qquad C(t)\mathbf{x} = \sum_{i=1}^{\infty}\cos\sqrt{-s_i}\,t\sum_{l=1}^{r(i)}\langle\mathbf{x}, \mathbf{x}_{il}\rangle_X\mathbf{x}_{il}.$$

Let

$$L_k = \{(\mathbf{x}(t_k, 0, \mathbf{u}), \dot{\mathbf{x}}(t_k, 0, \mathbf{u}))\in X\times X\colon \mathbf{u} \text{ is an arbitrary control of the form (3.5.1)}\}. \tag{3.6.9}$$

DEFINITION 3.6.3. The dynamical system (3.6.1) after the introduction of sampling is said to be *approximately controllable in an infinite number of steps* if

$$\overline{L_\infty} = \overline{\bigcup_{k=1}^{\infty} L_k} = X\times X.$$

DEFINITION 3.6.4. The dynamical system (3.6.1) after the introduction of sampling is said to be *approximately controllable in N steps* if

$$L_N = \overline{\bigcup_{k=1}^{N} L_k} = X \times X.$$

THEOREM 3.6.2 (Kobayashi (1980a)). *The dynamical system* (3.6.1) *after the introduction of sampling is approximately controllable in an infinite number of steps if and only if condition* (3.6.7) *holds and, moreover, the following inequalities are satisfied*:

$$\exp(\sqrt{s_i}T) \neq \exp(\sqrt{s_j}T) \quad \textit{for } i \neq j,\ i, j = 1, 2, 3, \ldots, \tag{3.6.10}$$

$$\exp(\sqrt{s_i}T) \neq \exp(-\sqrt{s_j}T) \quad \textit{for } i, j = 1, 2, 3, \ldots \tag{3.6.11}$$

Proof:

From Definition 3.6.3 it follows that the dynamical system (3.6.1) after the introduction of sampling is approximately controllable in an infinite number of steps if and only if

$$\left\{\int_0^T \langle G(t_k-\tau)\mathbf{b}_j, \mathbf{w}\rangle_X d\tau + \int_0^T \left\langle \frac{d}{dt} G(t_k-\tau)\mathbf{b}_j, \mathbf{z}\right\rangle_X d\tau = 0 \right.$$

$$\left.\text{for } j = 1, 2, 3, \ldots \text{ and } k = 1, 2, 3, \ldots\right\} \text{ implies } \mathbf{z} = 0 \text{ and } \mathbf{w} = 0. \tag{3.6.12}$$

By (3.3.10) and (3.6.2) we can represent the operator $G(t)$ in the following form:

$$G(t)\mathbf{x} = \sum_{i=1}^{\infty} \frac{1}{\sqrt{-s_i}} \sin(\sqrt{-s_i}t) \sum_{l=1}^{r(i)} \langle \mathbf{x}, \mathbf{x}_{il}\rangle_X \mathbf{x}_i. \tag{3.6.13}$$

Hence we obtain the following equalities:

$$\int_0^T \langle G(t_k-\tau)\mathbf{b}_j, \mathbf{w}\rangle_X d\tau + \int_0^T \left\langle \frac{d}{dt} G(t_k-\tau)\mathbf{b}_j, \mathbf{z}\right\rangle_X d\tau$$

$$= \int_0^T \left\langle \sum_{i=1}^{\infty} \sin\sqrt{s_i}(t_k-\tau)/\sqrt{s_i} \sum_{l=1}^{r(i)} b_{il}^j \mathbf{x}_{il}, \sum_{i=1}^{\infty}\sum_{l=1}^{r(i)} w_{il}\mathbf{x}_{il}/\sqrt{s_i}\right\rangle_X d\tau +$$

$$+ \int_0^T \left\langle \sum_{i=1}^{\infty} \cos\sqrt{s_i}(t_k-\tau) \sum_{l=1}^{r(i)} b_{il}^j \mathbf{x}_{il}, \sum_{i=1}^{\infty}\sum_{l=1}^{r(i)} z_{il}\mathbf{x}_{il}\right\rangle_X d\tau$$

$$= \int_0^T \sum_{i=1}^{\infty} \left(w_i^j \sin\sqrt{s_i}(t_k-\tau) + Z_i^j \cos\sqrt{s_i}(t_k-\tau)\right) d\tau$$

$$= \sum_{i=1}^{\infty} \frac{1}{\sqrt{s_i}} \big((\cos\sqrt{s_i}(k-1)T - \cos\sqrt{s_i}t_k) w_i^j -$$

$$- (\sin\sqrt{s_i}(k-1)T - \sin\sqrt{s_i}t_k) z_i^j \big), \tag{3.6.14}$$

where

$$w_{il} = \langle \mathbf{w}, \mathbf{x}_{il}/\sqrt{s_i} \rangle_X, \quad w_i^j = \sum_{l=1}^{r(i)} b_{il}^j w_{il}, \quad Z_i^j = \sum_{l=1}^{r(i)} b_{il}^j z_{il}.$$

Substituting (3.6.14) into the implication (3.6.12) and using the properties of the exponential function, we obtain Theorem 3.6.2. ∎

COROLLARY 3.6.2 (Kobayashi (1980a)). *The dynamical system* (3.6.1) *after the introduction of sampling is not approximately controllable in N steps.*

Proof:

The proof of Corollary 3.6.2 is the same as the proof of Corollary 3.5.2. ∎

COROLLARY 3.6.3 (Kobayashi (1980a)). *If the dynamical system* (3.6.1) *after the introduction of sampling is approximately controllable in an infinite number of steps, then the continuous dynamical system* (3.6.1) *is approximately controllable in finite time. The converse statement is not always true.*

Proof:

If the dynamical system (3.6.1) after the introduction of sampling is approximately controllable in an infinite number of steps, then by Theorem 3.6.2 we have

$$\operatorname{rank} \mathbf{B}_i = r(i) \quad \text{for } i = 1, 2, 3, \ldots$$

Thus, by Theorem 3.3.2, the dynamical system (3.6.1) is approximately controllable in finite time. If the continuous dynamical system (3.6.1) is approximately controllable in finite time, then by Corollary 3.6.1 condition (3.6.15) is fulfilled. However, it should be stressed that inequalities (3.6.10) and (3.6.11) are not always satisfied. Thus the approximate controllability in finite time of the continuous dynamical system (3.6.1) does not always imply the approximate controllability in infinite number of steps of the dynamical system (3.6.1) after the introduction of sampling. ∎

3.7 Topological properties of the set of controllable dynamical systems

In this section we shall consider the dynamical systems of the form (3.2.1) with a bounded operator A and separable Hilbert spaces X and U. Since the operator A is bounded, it is an infinitesimal generator of an analytic group $S(t)$, $-\infty < t < +\infty$, of bounded linear operators (Triggiani (1975a)). Thus Definitions 3.2.1 and 3.2.2 are equivalent and moreover so are Definitions 3.2.3 and 3.2.4. Hence in this section we shall investigate only two kinds of controllability of the dynamical system (3.2.1), namely, approximate controllability and exact controllability.

Under the above assumptions to each dynamical system (3.2.1) we can uniquely assign a selfadjoint controllability operator W of the following form (Megan and Hiris (1975)):

$$W = \int_0^1 S(-t)BB^*S^*(-t)dt = \int_0^1 \exp(-At)BB^*\exp(-A^*t)dt, \qquad (3.7.1)$$

where $\exp(-At)$ is a group of linear, bounded operators for $t \in (-\infty, +\infty)$. Moreover, since $A \in L(X, X)$, we have by Triggiani (1975a):

$$\exp(At) = \sum_{k=0}^{\infty} \frac{A^k t^k}{k!}, \qquad t \in (-\infty, +\infty). \qquad (3.7.2)$$

Let S denote the set of all dynamical systems of the form (3.2.1) with a bounded operator A, where $S_a \subset S$ and $S_d \subset S$ denote the subsets of approximately controllable systems and exactly controllable systems, respectively. Of course we have $S_d \subset S_a$ since the class of approximalety controllable dynamical systems is greater than the class of exactly controllable dynamical systems. The elements of the set S will be denoted by (A, B). The set S becomes a linear space after the introduction of algebraic operations defined as follows (Megan and Hiris (1975), Rastovic (1979)):

$$(A_1, B_1)+(A_2, B_2) = (A_1+A_2, B_1+B_2), \qquad (A_1, B_1) \in S,\ (A_2, B_2) \in S,$$

$$r(A, B) = (rA, rB), \qquad (A, B) \in S,\ r \in \mathbf{R}.$$

By the linearity of operators the above operations are well defined. In the linear space S we can define the norm in the following way:

$$\|(A, B)\| = \|A\|+\|B\|, \qquad (A, B) \in S.$$

Since the operators A and B are bounded, the norm in the space S is well defined. The set S with algebraic operations and the norm forms a linear normed space with a topology induced by the norm (Megan and Hiris (1975)). This space will also be denoted by S.

THEOREM 3.7.1 (Megan and Hiris (1975)). *If* $\dim X = \infty$, *then the set* S_a *is neither open nor closed in the linear space* S.

Proof:

Let $\{e_k\}_{k=1}^{k=\infty}$ be an orthonormal basis in the infinite-dimensional Hilbert space X and let B_n and B be operators defined as follows:

$$B_n e_k = \begin{cases} (1/\sqrt{k})e_k & \text{for } k \neq n, \\ 0 & \text{for } k = n, \end{cases} \qquad Be_k = (1/\sqrt{k})e_k.$$

Consider the dynamical systems $(0, B_n)$ and $(0, B)$. Since the operators B_n and B are selfadjoint, by (3.7.1) the appropriate controllability operators are of the forms B_n^2 and B^2, respectively. Since the operator B_n^2 is not invertible and the operator B^2 is invertible, the dynamical system $(0, B_n)$ is not approximately controllable, and thus $(0, B_n) \notin S_a$. However, the dynamical system $(0, B)$ is approximately controllable, i.e., $(0, B) \in S_a$. On the other hand,

$$\lim_{n\to\infty} \|(0, B_n)-(0, B)\| = \lim_{n\to\infty} \|B_n - B\| = \lim_{n\to\infty}(1/n) = 0.$$

Thus $(0, B_n) \underset{n\to\infty}{\to} (0, B)$, and hence the set S_a is not open in the space S. The set S_a is not closed in the space S either. In order to derive it, let us take the dynamical system $(0, I/n) \in S_d \subset S_a$ and observe that the sequence of the dynamical systems $(0, I/n)$, as $n \to \infty$, converges to the dynamical system $(0, 0)$, which is of course not approximately controllable. Therefore the set S_a is not closed in the space S. ■

THEOREM 3.7.2 (Megan and Hiris (1975)). *The set S_d is open in the topological space S.*

Proof:

If $(A, B) \in S_d$, then by the corollaries and lemmas given in Megan and Hiris (1975) we can choose numbers $c_1 > 0$ and $c_2 > 0$ such that

$$\|A - A_1\| < c_1 \text{ implies } (A_1, B) \in S_d$$

and

$$\|B - B_1\| < c_2 \text{ implies } (A, B_1) \in S_d.$$

Thus for $c = c_1 + c_2 > 0$ the following implication holds:

$$\|(A, B)-(A_1, B_1)\| < c \text{ implies } (A_1, B_1) \in S_d.$$

Thus the set S_d is open in the topological space S. ■

THEOREM 3.7.3 (Megan and Hiris (1975)). *If* $\dim X = \infty$ *and* $\dim U = 1$, *then* $\overline{S_a} \neq S$, *i.e., the set of approximately controllable dynamical systems is not dense in the space S.*

Proof:

The proof of Theorem 3.7.3 is based on the properties of the linear subspace $\overline{\text{span}}\{A^n b, n = 0, 1, 2, \ldots\}$ and is presented in detail in Megan and Hiris (1975). ■

3.8 Contrability with constrained controls

In this section the results given in Section 1.8 and concerning controllability with constrained controls of finite-dimensional dynamical systems will be extended to the case of infinite-dimensional dynamical systems. We shall consider the local and global approximate controllability, local and global exact controllability and various kinds of the so-called null-controllability (zero-controllability). We shall stress the differences between controllability without constraints on the controls and controllability with constrained controls. Moreover, we shall present the relations between constrained controllability in finite-dimensional and infinite-dimensional dynamical systems (Ahmed (1985), Peichl and Schappacher (1986), Carja (1988)).

In this section we shall consider only the dynamical systems of the form (3.2.1) with a bounded operator A. The operator A generates an analytic group of linear, bounded operators $S(t) = \exp(At)$, $t \in (-\infty, +\infty)$. Moreover, it is assumed that the admissible controls $\mathbf{u}(t)$, $t \in [0, \infty)$ are strongly measurable functions, almost everywhere bounded in the finite time interval $[0, t_1]$, and take their values from a given bounded subset of the Banach space U. The set of admissible controls in the time interval $[0, t_1]$ is denoted by $M_U(t_1)$ and is defined as follows:

$$M_U(t_1) = \{\mathbf{u}(\cdot) \in L_\infty([0, t_1], U)\}. \tag{3.8.1}$$

We additionally assume that the initial state $\mathbf{x}(0) = 0$ and the initial time $t_0 = 0$ Hence, to shorten the notation, the attainable set at time t_1 from the zero initial state and with admissible controls belonging to the set $M_U(t_1)$ will be denoted by $K_U(t_1)$. Since $S(t) = \exp(At)$ is a group, we can express the attainable set for the dynamical system (3.2.1) as follows:

$$K_U(t_1) = \left\{\mathbf{x} \in X\colon \mathbf{x} = -\int_0^{t_1} \exp(-At)B\mathbf{u}(t)\,dt\colon \mathbf{u}(\cdot) \in M_U(t_1)\right\}. \tag{3.8.2}$$

By (3.8.2) the attainable set at time t_1 from the zero initial state is the same as the set of initial states which can be steered to zero at time t_1. The set $K_U = \bigcup_{t_1 > 0} K_U(t_1)$ will be called the *attainable set in finite time* or the *null-controllable set in finite time.*

Now we shall present without proof two lemmas which will be useful in the subsequent considerations of this section.

LEMMA 3.8.1 (Korobov and Shon (1980a, 1980b, 1980c). *The set* $\overline{K_U(t_1)}$ *is convex. Moreover, if*

$$U \cap \operatorname{Ker} B \neq \emptyset, \tag{3.8.3}$$

i.e., if there exists a point $\mathbf{u}_0 \in U$ *such that* $B\mathbf{u}_0 = 0$, *then* $\overline{K_U}$ *is also a convex set*

LEMMA 3.8.2 (Korobov and Shon (1980a, 1980b, 1980c). *The following equality is valid:*

$$\overline{K_U(t_1)} = \overline{K_{\mathrm{CH}(U)}(t_1)}. \tag{3.8.4}$$

Lemma 3.8.2 is an extension to the infinite-dimensional case of the results) presented in Theorem 1.8.7 for the finite-dimensional case.

Adapting to the inifinite-dimensional case the definitions given in Section 1.8, we obtain the definitions of approximate local controllability and exact local controllability, respectively.

DEFINITION 3.8.1. The dynamical system (3.2.1) is said to be *approximately locally U-controllable at time* t_1 *(in finite time)* if $0 \in \operatorname{int}\overline{K_U(t_1)}$ $(0 \in \operatorname{int}\overline{K_U})$.

DEFINITION 3.8.2. The dynamical system (3.2.1) is said to be *exactly locally U-controllable at time* t_1 *(in finite time)*, if $0 \in \operatorname{int} K_U(t_1)$ $(0 \in \operatorname{int} K_U)$.

Definitions 3.8.1 and 3.8.2 immediately imply, that exact local U-controllability at time t_1 (in finite time) implies approximate local U-controllability at time t_1 (in finite time). The converse implication is not always true, which is illustrated by Example 3.8.1. However, in some special cases the approximate and the exact local U-controllabilities at time t_1 (in finite time) are equivalent. This situation will be illustrated in some detail in Theorem 3.8.1. In order to do so, let us introduce the notion of a *relatively internal set* $\operatorname{rint} U$, for a convex set U which contains zero as interior point, i.e., $0 \in \bar{U}$. A point $\mathbf{u}_0 \in U$ is said to be a *relatively internal point* of the set U if there exists an $\varepsilon > 0$ such that $(\mathbf{u}_0 + \varepsilon U_1) \cap \overline{\operatorname{sp}} U \subset U$, where U_1 is a unit ball centred at zero and $\operatorname{sp} U$ denotes the linear hull of the set U. The symbol $\operatorname{rint} U$ denotes the set of all relatively internal points of the set U, and $\overline{\operatorname{sp}} U$ is the closedness of the set $\operatorname{sp} U$. The set $\operatorname{rint} U$ is in fact the interior of the set $U \subset \overline{\operatorname{sp}} U$ with respect to the induced topology in the linear subspace $\overline{\operatorname{sp}} U$. It is well known (Korobov and Shon (1980b)), that in finite-dimensional spaces every convex set has a nonempty relative interior. However, this is no longer true in infinite-dimensional spaces.

Now we are in a position to formulate and prove some theorems and corollaries concerning different types of controllability of the dynamical system (3.2.1) with constrained controls.

THEOREM 3.8.1 (Korobov and Shon (1980b)). *Suppose the set* U *is bounded and* $0 \in \operatorname{rint} U$. *Then the dynamical system* (3.2.1) *is approximately locally U-controllable at time* t_1 *if and only if it is exactly locally U-controllable at time* t_1.

Proof:

It is obvious that exact local U-controllability at time t_1 implies the approximate local U-controllability at time t_1. Hence, now we shall prove the converse implication, i.e., we shall show that approximate local U-controllability at time t_1 implies exact local U-controllability at time t_1. First of all, let us observe that if the dynamical system (3.2.1) is approximately locally U-controllable at time t_1, then there exists a real number $r_0 > 0$, such that the following inclusion is valid:

$$r_0 X_1 \subset P_{t_1}(M_U(t_1)), \tag{3.8.5}$$

where X_1 is the unit neighbourhood of zero in the space X, and P_{t_1} is the operator defined by formula (3.3.16). Since the set U is bounded and $0 \in \operatorname{rint} U$, there exist real numbers $r_2 \geqslant r_1 > 0$ such that $r_1 \overline{\text{sp}}\, U_1 \subset r_2 \overline{\text{sp}}\, U_1$, where $\overline{\text{sp}}\, U_1$ denotes the unit neighbourhood of zero in the linear subspace $\overline{\text{sp}}\, U$. Let us denote

$$V_1 = \{\mathbf{v}(\cdot) \in I_\infty([0, t_1], \overline{\text{sp}}\, U) : \|\mathbf{v}(t)\| < 1 \text{ for } t \in [0, t_1]\}.$$

Hence $r_1 V_1 \subset M_U(t_1) \subset r_2 V_1$. Since the operator P_{t_1} is linear, then we have the inclusions:

$$r_1 P_{t_1}(V_1) \subset P_{t_1}(M_U(t_1)) \subset r_2 P_{t_1}(V_1).$$

By (3.8.5) and the second inclusion in formula (3.8.6) we have

$$(r_0/r_2)X_1 \subset \overline{P_{t_1}(V_1)}.$$

Therefore, by the completeness of the space X and the space $L_\infty([0, t_1], \overline{\text{sp}}\, U)$ and by the first inclusion in formula (3.8.6), we obtain the following relations:

$$(r_0 r_1/r_2)X_1 \subset P_{t_1}(M_U(t_1)) = K_U([0, t_1], 0) = K_U(t_1).$$

Thus $0 \in \operatorname{int} K_U(t_1)$ and hence the dynamical system (3.2.1) is exactly locally U-controllable at time t_1. ∎

COROLLARY 3.8.1 (Korobov and Shon (1980b)). *Suppose that condition* (3.8.3) *is satisfied and the dynamical system* (3.2.1) *is exactly locally U-controllable in finite time. Then it is also approximately locally U-controllable at time* $t_1 < \infty$. *Moreover, if the set U is bounded,* $0 \in \operatorname{rint} U$ *and the dynamical system* (3.2.1) *is exactly locally U-controllable in finite time, then it is also exactly locally U-controllable at time* t_1.

Proof:

The first part of Corollary 3.8.1 follows directly from the Hahn–Banach theorem and Baire's theorem on categories and from Definitions 3.8.1 and 3.8.2. The second part of Corollary 3.8.1 is a consequence of Theorem 3.8.1 and the first part of Corollary 3.8.1. ∎

THEOREM 3.8.2 (Korobov and Shon (1980c)). *If relation* (3.8.3) *holds, then the necessary condition, and if the set U is a cone with vertex at zero, also the sufficient condition for approximate local U-controllability at time* t_1 *(in finite time) of the dynamical system* (3.2.1) *is that for an arbitrary nonzero functional* $\mathbf{f} \in X^*$ *there should exist a point* $\mathbf{u}_\mathbf{f} \in U$ *and a time* $t_\mathbf{f} \in [0, t_1]$ *such that*

$$\langle \mathbf{f}, \exp(-At_\mathbf{f})B\mathbf{u}_\mathbf{f} \rangle < 0. \tag{3.8.7}$$

Proof:

A detailed proof will be presented only for the case of approximate local U-controllability in finite time. The generalization to approximate local controllability at time t_1 is then obvious.

Necessity. By contradiction. Suppose that there exists a nonzero functional $\mathbf{f}_0 \in X^*$ such that $\langle \mathbf{f}_0, \exp(-At)B\mathbf{u}\rangle \geqslant 0$ for an arbitrary $t \geqslant 0$ and an arbitrary $\mathbf{u} \in U$. Then for any point $\mathbf{x}_0 \in K_U(t_1)$ and an arbitrary $t_1 > 0$ we have

$$\langle \mathbf{f}_0, \mathbf{x}_0\rangle = -\int_0^{t_1} \langle \mathbf{f}_0, \exp(-At)B\mathbf{u}(t)\rangle\, dt < 0. \tag{3.8.8}$$

Relation (3.8.8) denotes that the set of null-controllability at time t_1 is contained in a closed subspace $H = \{\mathbf{x} \in X: \langle \mathbf{f}_0, \mathbf{x}\rangle \leqslant 0\}$. Thus $\overline{K_U} = \overline{\bigcup_{t_1>0} K_U(t_1)} \subset H$, whence $0 \notin \operatorname{int} \overline{K_U}$. Thus by Definition 3.8.1 we have a contradiction of the hypothesis of approximate local U-controllability in finite time of the dynamical system (3.2.1).

Sufficiency. By Lemma 3.8.1 the set $\overline{K_U}$ is convex. Since by assumption the set U is a cone with vertex at zero, also the set $\overline{K_U}$ is a cone with vertex at zero. The further part of the proof will be carried out by contradiction. Suppose that the dynamical system (3.2.1) is not approximately locally U-controllable in finite time, i.e., $0 \notin \operatorname{int} \overline{K_U}$. Thus $\overline{K_U} \neq X$ and there exists a nonzero functional $\mathbf{f}_0 \in X^*$ such that $\langle \mathbf{f}_0, \mathbf{x}_0\rangle \geqslant 0$ for all $x_0 \in \overline{K_U}$. Therefore for every $t_1 \geqslant 0$ and each admissible control $\mathbf{u}(t) \in M_U(t_1)$ we have

$$\left\langle \mathbf{f}_0, -\int_0^{t_1} \exp(-At)B\mathbf{u}(t)\,dt \right\rangle = \int_0^{t_1} \langle -\mathbf{f}_0, \exp(-At)B\mathbf{u}(t)\rangle dt \geqslant 0.$$

Therefore for $t \geqslant 0$ and $\mathbf{u} \in U$ we have $\langle -\mathbf{f}_0, \exp(-At)B\mathbf{u}\rangle \geqslant 0$, which contradicts condition (3.8.7). Hence our hypothesis that the dynamical system (3.2.1) is not approximately locally U-controllable in finite time is false. Thus Theorem 3.8.2 is proved. ■

In the next theorem we shall use the adjoint operator A^* for testing the exact local U-controllability of the dynamical system (3.2.1).

THEOREM 3.8.3 (Korobov and Shon (1980c)). *Suppose the set U is convex and* $\operatorname{rint}(U-\mathbf{u}_0) \neq \varnothing$ *for some* $\mathbf{u}_0 \in U \cap \operatorname{Ker} B$. *Then the dynamical system* (3.2.1) *is exactly locally U-controllable in finite time if and only if the following two conditions hold:*

(1) *the dynamical system without control constraints*

$$\dot{\mathbf{x}}(t) = A\mathbf{x}(t) + B\mathbf{v}(t), \qquad \mathbf{v}(t) \in V = \overline{\operatorname{sp}}(U-\mathbf{u}_0) \tag{3.8.9}$$

is exactly globally controllable in finite time,

(2) *there exists no eigenvector* $\mathbf{f}_0 \in X^*$ *of the operator* A^* *corresponding to the real eigenvalue* s_0 *such that* $\langle \mathbf{f}_0, B\mathbf{u}\rangle \geqslant 0$ *for every* $\mathbf{u} \in U$.

Proof:

Necessity. By contradiction. Suppose that the dynamical system (3.2.1) is exactly locally U-controllable in finite time, but condition (2) does not hold. Then there exist a real number s_0 and a nonzero functional $\mathbf{f}_0 \in X^*$ such that $A^*\mathbf{f}_0 = s_0\mathbf{f}_0$ and $\langle \mathbf{f}_0, Bu \rangle \geqslant 0$ for an arbitrary $\mathbf{u} \in U$. Therefore

$$\langle \mathbf{f}_0, \exp(-At)\mathbf{Bu} \rangle = \langle \exp(-At)\mathbf{f}_0, \mathbf{Bu} \rangle = \exp(-s_0 t)\langle \mathbf{f}_0, \mathbf{Bu} \rangle \geqslant 0$$

for each $\mathbf{u} \in U$ and every $t \geqslant 0$. Thus by Theorem 3.8.2 this contradicts the hypothesis of exact local U-controllability of the dynamical system (3.2.1).

Sufficiency. By contradiction. Suppose that conditions (1) and (2) are both satisfied, but the dynamical system (3.2.1) is not exactly locally U-controllable in finite time. Hence, using the notion of an attainable set and the Hanh–Banach theorem and the Baire theorem on categories, we can conclude in the same way as in Theorem 3.8.1 that there exists a real eigenvector $\mathbf{f}_0$ of the operator A^* such that $\langle \mathbf{f}_0, \mathbf{Bu} \rangle \geqslant 0$ for every $\mathbf{u} \in U$. Thus we have a contradiction of the condition (2). Hence our Theorem 3.8.3 follows. ∎

COROLLARY 3.8.2 (Korobov and Shon (1980c)). *Suppose that conditions* (1) *and* (2) *of Theorem* 3.8.3 *are satisfied and* rint $\mathrm{CH}(U - \mathbf{u}_0) \neq \emptyset$ *for a certain* $\mathbf{u}_0 \in U$ *such that* $\mathbf{Bu}_0 = 0$. *Then the dynamical system* (3.2.1) *is approximately locally U-controllable in finite time.*

Proof:

By contradiction. Suppose that all the assumptions of Corollary 3.8.2 are satisfied, but the dynamical system (3.2.1) is not approximately locally U-controllable in finite time. Thus we can show (the details are given in Korobov and Shon (1980a, 1980b)), that $\langle \mathbf{f}, \exp(-\mathbf{A}t)\mathbf{Bu} \rangle \leqslant 0$ for every $\mathbf{u} \in U$, each $t \geqslant 0$, and for an arbitrary element $\mathbf{f} \in \{r\mathbf{x} \in X\colon r \geqslant 0, \mathbf{x} \in \overline{K_U}\}$. Therefore by Korobov and Shon (1980c) we obtain a contradiction of conditions (1) and (2) in Theorem 3.8.3. Hence our Corollary 3.8.2 follows.

Example 3.8.1 (Korobov and Shon (1980b))

Let us consider a dynamical system described by a linear integro-differential equation of Volterra type:

$$\frac{\partial w(y,t)}{\partial t} = \int_0^y w(s,t)\,ds + v(y,t), \quad 0 \leqslant y \leqslant 1,\ t \geqslant 0, \tag{3.8.10}$$

where $w(y, t)$ is the state of the dynamical system and $v(y, t)$ is a control function.

Equation (3.8.10) can be rewritten as an abstract differential equation in the Banach space $L^1([0, 1], \boldsymbol{R})$, namely

$$\dot{x}(t) = Ax(t) + u(t), \tag{3.8.11}$$

where $x(t) = w(\cdot, t) \in L^1([0, 1], \boldsymbol{R})$, $u(t) = v(\cdot, t)$ and $A: L^1([0, 1], \boldsymbol{R}) \to L^1([0, 1], \boldsymbol{R})$ is a linear, bounded, integral operator of the Volterra type, given by the formula

$$(Af)(y) = \int_0^y f(s)\,ds.$$

It is assumed that the set of admissible controls is of the following form:

$$M_U(t_1) = \left\{v(y, t)\colon v(y, t) \in C([0, 1] \times [0, t_1], \boldsymbol{R})\colon \int_0^1 |v(y, t)|\,dy \leqslant 1\right\}.$$

By (3.8.2) the attainable set at time t_1 from the zero initial state is given by

$$K_U(t_1) = \left\{x \in L^1([0, 1], \boldsymbol{R})\colon x = -\int_0^{t_1} \exp(-At)v(y, t)\,dt;\right.$$

$$\left. v(y, t) \in M_U(t_1)\right\}.$$

Since $K_U(t_1) \subset C([0, 1], \boldsymbol{R})$, the interior of the attainable set $K_U(t_1)$ is empty in the state space $X = L^1([0, 1], \boldsymbol{R})$, i.e., $\operatorname{int} K\ (t_1) = \varnothing$. Thus the dynamical system (3.8.10) is not exactly locally U-controllable at time t_1.

However, we shall show that the dynamical system (3.8.10) is approximately locally U-controllable at time t_1. In order to do that, let us consider the neighbourhood of zero in the state space $X = L^1([0, 1], \boldsymbol{R})$ of the following form:

$$Q = \left\{x(y) \in L^1([0, 1], \boldsymbol{R})\colon \|x(\cdot)\| = \int_0^1 |x(y)|\,dy \leqslant \frac{t_1}{2\exp(\|A\|t_1)}\right\}.$$

Let $x_0(y) \in Q$ and $0 < \varepsilon < \dfrac{t_1}{2\exp(\|A\|t_1)}$. Since the set of continuous functions is dense in the space $L^1([0, 1], \boldsymbol{R})$, there exists $x_0(\cdot) \in C([0, 1], \boldsymbol{R})$ such that $\|x_0 - x\| < \varepsilon$. Choosing the control $\bar{v}_0(y, t) = \dfrac{1}{t_1}\exp(At)x_0(y)$, $t \geqslant 0$, we can easily verify that it is an admissible control, i.e., $\bar{v}_0(y, t) \in M_U(t_1)$ for every $t_1 > 0$. Moreover, this control steers the dynamical system (3.8.10) from the initial state $\bar{x}_0$ to zero at time t_1, i.e., $\bar{x}_0 \in K_U(t_1)$. Since for an arbitrary $x_0(\cdot) \in Q$ and any $\varepsilon > 0$ there exists an $\bar{x}_0 \in K_U(t_1)$ such that $\|\bar{x}_0 - x_0\| < \varepsilon$, the attainable set $K_U(t_1)$ is dense in the ball Q and hence the dynamical system (3.8.10) is approximately locally U-controllable at time t_1.

Example 3.8.1 shows that the notions of exact local U-controllability and approximate local U-controllability are essentially different and, moreover, the approximate local U-controllability of the dynamical system (3.2.1) does not always imply its exact local U-controllability. The lack of equivalence between exact and approximate local U-controllability at time t_1 for the dynamical system (3.8.10) may also be obtained by using Theorem 3.8.1. It is enough to observe that $0 \notin \operatorname{rint} U$. Thus the hypothesis of Theorem 3.8.1 is not satisfied and hence the notions of exact local U-controllability and approximate local U-controllability are not equivalent.

Example 3.8.2 (Korobov and Shon (1980b))
Let us consider the dynamical system defined in a Hilbert space $X = l_2$ and described by the following abstract differential equation:

$$\dot{\mathbf{x}}(t) = \mathbf{u}(t), \quad \mathbf{u}(t) \in U \subset l_2, \quad t \geqslant 0. \tag{3.8.12}$$

The set U is bounded and defined as follows:

$$U = \{\mathbf{u} \in l_2: \ |u^{(n)}| \leqslant 1/n, \ n = 1, 2, 3, \ldots\},$$

where $\mathbf{u}^{(n)}$ denotes the n-th coordinate of the vector $\mathbf{u} \in l_2$. It is fairly easy to verify that the attainable set at time t_1, $K_U(t_1) = t_1 U$ and therefore $\operatorname{int} \overline{K_U(t_1)} = \emptyset$. Thus the dynamical system (3.8.12) is not aprpoximately locally U-controllable at time t_1 for any $t_1 > 0$. However, we shall show that the dynamical system (3.8.12) is approximately globally U-controllable in finite time, i.e., the following inclusions hold:

$$\overline{K_U} = \overline{\bigcup_{t_1>0} K_U(t_1)} = \overline{\bigcup_{t_1>0} \overline{K_U(t_1)}} = \overline{\bigcup_{t_1>0} t_1 U} = l_2. \tag{3.8.13}$$

Indeed, let $\mathbf{x}_0 = \{x_0^{(1)}, x_0^{(2)}, \ldots\}$ be an arbitrary point in the space $l_2 = X$. Therefore for any $\varepsilon > 0$ and for sufficiently great integer numbers N we have $\|\mathbf{x}_0 - \mathbf{x}_N\| < \varepsilon$, where $\mathbf{x}_N = \{x_0^{(1)}, x_0^{(2)}, \ldots, x_0^{(N)}, 0, 0, \ldots\}$. On the other hand, the set U absorbs each finite vector in the space l_2, and hence there esists a T_1 such that $\mathbf{x}_N \in T_1 U = K_U(T_1) \subset K_U$. Thus the set K_U is everywhere dense in the space l_2, i.e., the relation (3.8.13) is true and therefore the dynamical system (3.8.12) is approximately globally U-controllable in finite time. The lack of local approximate U-controllability at time t_1 for the dynamical system (3.8.12) can also be explained in another way. Namely, let us observe that, although the set $K_U(t_1)$ is closed in the space l_2 for every $t_1 > 0$, it does not contain the interior points, i.e., $\operatorname{int} \overline{K_U(t_1)} = \emptyset$. Therefore the dynamical system (3.8.12) is an example of a system which is not approximately locally U-controllable at time t_1 for any $t_1 > 0$, but is approximately, globally U-controllable in finite time.

Example 3.8.3 (Korobov and Shon (1980c))
Let us consider a dynamical system defined in a Hilbert space $X = l_2$ and described by the following abstract differential equation:

$$\dot{\mathbf{x}}(t) = A\mathbf{x}(t) + \mathbf{u}(t), \quad t \geqslant 0, \tag{3.8.13}$$

where $A: l_2 \to l_2$ is a linear, bounded operator defined by the infinite-dimensional matrix

$$\mathbf{A} = \begin{bmatrix} 0 & -1 & 0 & \ldots \\ 0 & 0 & -1 & \ldots \\ 0 & 0 & 0 & \ldots \\ \vdots & \vdots & \vdots & \ddots \end{bmatrix}.$$

The control $\mathbf{u}(t) \in U = \{\mathbf{u} \in l_2 : u^{(n)} \geq 0, n = 1, 2, 3, \ldots\}$. The set U is convex, closed and $\operatorname{rint} \mathrm{CH}(U) = \emptyset$. Moreover, it is easily verified that the dynamical system (3.8.13) without control constraints is exactly, globally controllable in finite time. However, the dynamical system (3.8.13) is not approximately locally U-controllable in finite time. Indeed, since $\exp(-\mathbf{A}t)U \subset U$ for $t \geq 0$, then for vector $\mathbf{x}_0 = \{1, 0, 0, \ldots\} \in l_2$ we have $\langle \mathbf{x}_0, \exp(-\mathbf{A}t)\mathbf{u} \rangle \geq 0$ for $t \geq 0$ and any $\mathbf{u} \in U$. Thus, by Theorem 3.8.2, the dynamical system (3.8.13) is not approximately locally U-controllable in finite time (condition (3.3.7) does not hold). Therefore it is an example of a dynamical system which, although it satisfies conditions (1) and (2) of Theorem 3.8.3, and the set U is convex, is not approximately locally U-controllable in finite time. This is a consequence of the fact that $\operatorname{rint} \mathrm{CH}(U) = \emptyset$.

Example 3.8.4 (Korobov and Shon (1980c))
Let us consider a dynamical system defined in a Hilbert space $X = l_2$ and described by the following abstract differential equation:

$$\dot{\mathbf{x}}(t) = Ax(t) + \mathbf{b}u(t), \qquad t \geq 0, \tag{3.8.14}$$

where the operator A is the same as in Example 3.8.3 and vector $\mathbf{b} \in l_2$ is given by

$$\mathbf{b} = \begin{bmatrix} 1 \\ 1/2 \\ 1/3 \\ \vdots \end{bmatrix}.$$

Moreover, let $u(t) \in U = \{u \in R : 0 \leq u < \infty\}$. Since the sequence of vectors $\{\mathbf{b}, A\mathbf{b}, A^2\mathbf{b}, \ldots\}$ is everywhere dense in the space l_2, then by Corollary 3.3.3 the dynamical system (3.8.14) without control constraints is approximately globally controllable. However, the dynamical system (3.8.14) is not approximately locally U-controllable in finite time. Indeed, since for any $t \geq 0$ and each $u \in U$ and $\mathbf{x}_0 = \{1, 0, 0, \ldots\} \in l_2$ we have $\langle \mathbf{x}_0, \exp(-\mathbf{A}t)\mathbf{b}u \rangle \geq 0$, then by Theorem 3.8.2 the dynamical system (3.8.14) is not aproximately, locally U-controllable in finite time.

3.9 Remarks and comments

A valuable generalization of the dynamical systems of the form (3.2.1) is to consider the nonstationary (time-varying) dynamical systems with unbounded linear operator $A(t)$, whose domain is additionally variable in time. The controllability of such dynamical systems and their special cases is investigated in the publications of Chan and Li (1981, 1982), Hiris and Megan (1977a), Korobov and Rabach (1979), Megan (1978a), Megan and Hiris (1975), Triggiani (1975a). Adding to the state equation (3.2.1) the abstract operator-valued output equation, we can consider the approximate and exact output controllability of dynamical systems with an infinite-dimensional domain (Germani and Monaco (1981, 1983), Triggiani (1975a)). In paper of Chan and Lau (1982) stochastic controllability for infinite-dimensional dynamical systems has been investigated. The results of Section 3.7 can be extended to the case of unbounded linear operator A with bounded perturbations. This has been done in the papers of Frazho (1981) and O'Brien (1979).

Stabilizability problems for infinite-dimensional dynamical systems are more difficult than for finite-dimensional ones. It follows directly from the existence of several different definitions of stability, which are strongly dependent on the chosen topology (see, e.g. Benchimol (1978a, 1978b)). The relations between various kinds of stabilizability and approximate and exact controllability are considered and explained in papers of Benchimol (1978a, 1978b), Fuhrmann (1976), Gibson (1980), Kobayashi (1982), Lasiecka and Triggiani (1983a, 1983b), O'Brien (1978), Pandolfi (1978a, 1978b), Sakawa (1983, 1985), Slemrod (1974), Triggiani (1975b, 1975d), Wang (1972). The results obtained depend on the properties of the operator A, the definition of stability and the given state space X.

In the papers of Benchimol (1978a, 1978b), O'Brien (1978), Slemrod (1974) the so-called "weak stabilizability" of the dynamical systems (3.2.1) is considered and its relation to approximate controllability is analysed. In those papers it is additionally assumed that X is a Hilbert space and the operator A is the infinitesimal generator of the contraction semigroup $S(t)$, $t \geqslant 0$, ($\|S(t)\| \leqslant 1$, for $t \geqslant 0$) of linear bounded operators. This enables us to obtain the general form of stabilizing feedback gain. The stabilizability of nonstationary dynamical systems in Hilbert spaces has been considered in the papers of Megan (1974, 1976, 1978a, 1978b), Pandolfi (1978a, 1978b), Reghis and Megan (1977), Triggiani (1975d). The results given in those papers are based on the Riccati equation.

The attainable sets for dynamical systems (3.2.1) or (3.2.2) are investigated in the papers of Artstein (1973), Jurdjevic (1970), Mesarovic (1971), Seidman (1977). Moreover, papers of Boothby (1982), Ball, Marsden and Slemrod (1982) contain the conditions for approximate controllability of bilinear dynamical systems with distributed parameters, i.e., for systems described by partial differential equations.

CHAPTER 4

Controllability of Dynamical Systems with Delays

4.1 Introduction

Hereditary or time-delay dynamical systems have stimulated a great deal of interest among research scientists in recent years. This has been motivated, on the one hand, by the wide range of applications of hereditary models in various areas of science and enginnering and, on the other hand, by the interesting and difficult theoretical problems posed by such systems. The purpose of this chapter is to bring together various fundamental research results obtained in the field of controllability of hereditary systems within the last few years and to discuss their significance and consequences.

We shall discuss various kinds of controllability of linear time-invariant dynamical systems with delays. They include approximate controllability, exact controllability, Euclidean controllability and $\boldsymbol{R}^n(s)$-controllability of dynamical systems. We shall consider hereditary systems described by retarded functional differential equations with constant coefficients. In the case of such systems the true state is an element of some function space. Generally speaking, in the control theory of hereditary systems the use of a function space offers the possibilities of obtaining strong results in problems such as controllability and attainable sets. In most papers the space $C([-h, 0], \boldsymbol{R}^n)$ of continuous functions has been used as a state space. Recently, however, some other spaces have also been introduced, e.g., $W_2^{(1)}([-h, 0], \boldsymbol{R}^n)$ the Sobolev space of absolutely continuous functions with a square-integrable first derivative. The Sobolev space $W_2^{(1)}([-h, 0], \boldsymbol{R}^n)$ is a natural choice for dynamical systems described by linear time-invariant functional differential equations with controls belonging to the space $L^2([0, t], \boldsymbol{R}^m)$. Furthermore, some recent results show that the use of space $W_2^{(1)}([-h, 0], \mathbf{R}^n)$ yields an effective characterization of attainable sets and permits one to establish explicit conditions for the closedness of those sets, which is very important for the existence of optimal controls and the nontriviality of the necessary optimality conditions.

The basic results are given in the form of a mathematical reasoning with theorems, corollaries and proofs. At the end of the chapter several numerical illustrative examples are presented, indicating the practical, i.e., numerical methods of solving various controllability problems.

4.2 Mathematical models of hereditary systems and the basic definitions

From the mathematical point of view, hereditary systems can be described by ordinary or partial differential equations with a deviating argument, or, more generally, by functional differential equations. In this chapter we shall consider a simple case of a functional differential equations, namely the linear, time-invariant, neutral differential equation with lumped delay of the following form:

$$\dot{\mathbf{x}}(t) = \mathbf{A}_{-1}\dot{\mathbf{x}}(t-h)+\mathbf{A}_0\mathbf{x}(t)+\mathbf{A}_1\mathbf{x}(t-h)+\mathbf{B}\mathbf{u}(t), \quad t \geqslant 0 \tag{4.2.1}$$

with the initial condition $\mathbf{x}(t) = \mathbf{x}_0(t)$, $t \in [-h, 0]$, $\mathbf{x}_0 \in W_2^{(1)}([-h, 0], \mathbf{R}^n)$, where $h > 0$ is the constant delay or the retardation of the argument, $\mathbf{x}(t) \in \mathbf{R}^n$ is the so-called *instantaneous state* of the dynamical system, $\mathbf{u} \in L^2_{\mathrm{loc}}([0, \infty), \mathbf{R}^m)$ is the control function, $\mathbf{A}_{-1}, \mathbf{A}_0, \mathbf{A}_1$ are $(n \times n)$-dimensional matrices with real constant elements, $\mathbf{B}$ is an $(n \times m)$-dimensional matrix with real constant elements, $W_2^{(1)}([-h, 0], \mathbf{R}^n)$ is the Sobolev space of $\mathbf{R}^n$-valued functions, defined in $[-h, 0]$, with a square integrable first derivative.

For a given initial condition $\mathbf{x}_0 \in W_2^{(1)}([-h, 0], \mathbf{R}^n)$ and a control function $\mathbf{u} \in L^2_{\mathrm{loc}}([0, \infty), \mathbf{R}^m)$ there exists a unique, absolutely continuous solution $\mathbf{x}(t, \mathbf{x}_0, \mathbf{u})$ of equation (4.2.1) which is defined on the time interval $[-h, \infty)$ (Manitius (1976a), Rodas and Langenhop (1978)). This solution is explicitly given by the following formula:

$$\mathbf{x}(t, \mathbf{x}_0, \mathbf{u}) = \mathbf{x}(t, \mathbf{x}_0, 0)+\int_0^t \mathbf{F}(t-\tau)\mathbf{B}\mathbf{u}(\tau)d\tau, \quad t \geqslant 0, \tag{4.2.2}$$

where $\mathbf{F}(t)$ is an $(n \times n)$-dimensional, fundamental matrix for the dynamical system (4.2.1). The fundamental matrix $\mathbf{F}(t)$ satisfies the integral equation

$$\mathbf{F}(t) = \mathbf{I}+\mathbf{F}(t-h)\mathbf{A}_{-1}+\int_0^t \mathbf{F}(\tau)\mathbf{A}_0 d\tau+ \int_0^{t-h} \mathbf{F}(\tau)\mathbf{A}_1 d\tau, \quad t > 0, \tag{4.2.3}$$

with the initial conditions $\mathbf{F}(0) = \mathbf{I}$ and $\mathbf{F}(t) = 0$ for $t < 0$.

Moreover, $\mathbf{x}(t, \mathbf{x}_0, 0)$ is the solution of equation (4.2.1) with control $\mathbf{u}(t) = 0$ for $t \geqslant 0$, given by the following formula:

$$\mathbf{x}(t, \mathbf{x}_0, 0) = (\mathbf{F}(t) - \mathbf{F}(t-h)\mathbf{A}_{-1})\mathbf{x}_0(0) + \int_{-h}^{0} \mathbf{F}(t-\tau-h)(\mathbf{A}_1 \mathbf{x}_0(\tau) + \mathbf{A}_{-1}\dot{\mathbf{x}}_0(\tau))d\tau. \tag{4.2.4}$$

The fundamental matrix $\mathbf{F}(t)$ may also be computed as a solution of an appropriate differential equation (Manitius (1976d), Rodas and Langenhop (1978)). However, it should be stressed that in general the computation of the matrix $\mathbf{F}(t)$ is very difficult.

The solution of equation (4.2.1) may also be considered as a function with values in the Hilbert space $W_2^{(1)}([-h, 0], R^n)$.

In order to do that, let us define the function $\mathbf{x}_t(\cdot, \mathbf{x}_0, \mathbf{u})$. The function $\mathbf{x}_t(\cdot, \mathbf{x}_0, \mathbf{u})$ is called the *complete state of the dynamical system* (4.2.1), takes values in the Hilbert space $W_2^{(1)}([-h, 0], R^n)$ and is defined for $t \geqslant 0$ in the following manner:

$$\mathbf{x}_t(\cdot, \mathbf{x}_0, \mathbf{u}) = \mathbf{x}(t+\tau, \mathbf{x}_0, \mathbf{u}) \quad \text{for } \tau \in [-h, 0].$$

For every $t \geqslant 0$, let us define the linear, bounded operator $S(t)$: $W_2^{(1)}([-h, 0], R^n) \to W_2^{(1)}([-h, 0], R^n)$ as follows:

$$S(t)\mathbf{x}_0 = \mathbf{x}_t(\cdot, \mathbf{x}_0, 0), \quad \mathbf{x}_0 \in W_2^{(1)}([-h, 0], R^n).$$

It is well known (Jacobs and Langenhop (1976), O'Connor and Tarn (1983)), that the family of the operators $S(t)$. $t \geqslant 0$, forms a strongly continuous semigroup of linear, bounded operators in the Hilbert space $W_2^{(1)}([-h, 0], R^n)$ whose infinitesimal generator $\tilde{A}$: $D(\tilde{A}) \to W_2^{(1)}([-h, 0], R^n)$ is defined as follows:

$$\tilde{A}\mathbf{w} = \dot{\mathbf{w}} \quad \text{for } \mathbf{w} \in D(\tilde{A}),$$

$$D(\tilde{A}) = \{\mathbf{w} \in W_2^{(1)}([-h, 0], R^n): \dot{\mathbf{w}} \in W_2^{(1)}([-h, 0], R^n); \dot{\mathbf{w}}(0) = \mathbf{A}_{-1}\dot{\mathbf{w}}(-h) + \mathbf{A}_0\mathbf{w}(0) + \mathbf{A}_1\mathbf{w}(-h)\},$$

$$\overline{D(\tilde{A})} = W_2^{(1)}([-h, 0], R^n).$$

Moreover, the spectrum $\sigma(\tilde{A})$ of the operator $\tilde{A}$ is a countable, pure point spectrum, consisting of the roots of the characteristic equation of the operator $\tilde{A}$,

$$\det \mathbf{\Delta}(s) = 0,$$

where the $(n \times n)$-dimensional charactssistic matrix $\mathbf{\Delta}(s)$ is of the form

$$\mathbf{\Delta}(s) = s(\mathbf{I} - \mathbf{A}_{-1}e^{-sh}) - \mathbf{A}_0 - \mathbf{A}_1 e^{-sh}. \tag{4.2.5}$$

It is known (O'Connor and Tarn (1983)) that if $s \in \sigma(\tilde{A})$ then the k_s-dimensional eigenspace of the operator $\tilde{A}$ corresponding to the eigenvalue s is of the following form:

$$N_s = \bigcup_{r=1}^{\infty}(\tilde{A} - sI)^r = \operatorname{Ker}(sI - \tilde{A})^{k_s},$$

where $k_s < \infty$ is the multiplicity of the eigenvalue s.

Let $\mathbf{H}_s(\tau)$ be a $(k_s \times k_s)$-dimensional matrix whose columns form the base in the eigenspace N_s. Then the projection $\mathbf{x}_t(\cdot, \mathbf{x}_0, \mathbf{u})$ on the eigenspace N_s is the function $\mathbf{H}_s(\tau)\mathbf{c}_s(t)$, where $\mathbf{c}_s(t) \in \boldsymbol{R}^{k_s}$ satisfies the differential equation

$$\dot{\mathbf{c}}_s(t) = \mathbf{A}_s \mathbf{c}_s(t) + \mathbf{B}_s \mathbf{u}(t), \tag{4.2.6}$$

where $\mathbf{A}_s$ is a constant $(k_s \times k_s)$-dimensional matrix which has only the iegnvalue s of the multiplicity k_s, $\mathbf{B}_s = \mathbf{H}_s^{\mathrm{T}}(0)\mathbf{B}$ is a constant $(k_s \times m)$-dimensional matrix.

Moreover, for every $s \in \varrho(\tilde{A})$, the resolvent operator $R(s, \tilde{A}) = (sI - \tilde{A})^{-1}$ is linear and bounded in the space $W_2^{(1)}([-h, 0], \boldsymbol{R}^n)$ and has the form

$$((sI - \tilde{A})^{-1}\mathbf{g})(\tau) = e^{s\tau}\boldsymbol{\Delta}^{-1}(s)\mathbf{d} + \int_{\tau}^{0} e^{s(\tau - t)}\mathbf{g}(t)\,dt, \qquad \tau \in [-h, 0], \tag{4.2.7}$$

where $\mathbf{g} \in W_2^{(1)}([-h, 0], \boldsymbol{R}^n)$,

$$\boldsymbol{R}^n \ni \mathbf{d} = \mathbf{g}(0) + \mathbf{A}_1 \int_{-h}^{0} e^{s(-h-t)}\mathbf{g}(t)\,dt + s\mathbf{A}_{-1} \int_{-h}^{0} e^{s(\tau - t)}\mathbf{g}(t)\,dt. \tag{4.2.8}$$

Hence the solution $\mathbf{x}_t(\cdot, \mathbf{x}_0, \mathbf{u})$ is given by the formula (O'Connor and Tarn (1983))

$$\mathbf{x}_t(\cdot, \mathbf{x}_0, \mathbf{u}) = S(t)\mathbf{x}_0 + (sI - \tilde{A}) \int_{-h}^{0} S(t - \tau)\tilde{B}_s \mathbf{u}(\tau)\,d\tau, \tag{4.2.9}$$

where the linear, bounded operator $\tilde{B}_s\colon \boldsymbol{R}^m \to W_2^{(1)}([-h, 0], \boldsymbol{R}^n)$ is given by

$$(\tilde{B}_s \mathbf{u})(\tau) = e^{s\tau}\boldsymbol{\Delta}^{-1}(s)\mathbf{B}\mathbf{u}, \qquad \mathbf{u} \in \boldsymbol{R}^m,\ \tau \in [-h, 0],\ s \in \varrho(\tilde{A}). \tag{4.2.10}$$

Formula (4.2.9) called the *variation of constants formula*, presents the solution of equation (4.2.1) as a function with values in the space $W_2^{(1)}([-h, 0], \boldsymbol{R}^n)$.

Now, let us consider the special case of equation (4.2.1), where $\mathbf{A}_{-1} = 0$, i.e., a linear, time-invariant dynamical system with constant delay in the state variables, described by the differential equation with a deviating argument

$$\dot{\mathbf{x}}(t) = \mathbf{A}_0\mathbf{x}(t) + \mathbf{A}_1\mathbf{x}(t-h) + \mathbf{B}\mathbf{u}(t), \qquad t \geqslant 0, \tag{4.2.11}$$

where the initial condition is $\mathbf{x}_0 \in M^2([-h, 0], \boldsymbol{R}^n) = \boldsymbol{R}^n \times L^2([-h, 0], \boldsymbol{R}^n)$.

In this case the complete state $\mathbf{x}_t(\cdot, \mathbf{x}_0, \mathbf{u})$ of system (4.2.11) may be expressed as an element $\tilde{\mathbf{x}}(t) \in M^2([-h, 0], \boldsymbol{R}^n)$ in the following way (Manitius and Triggiani (1978c), Manitius and Olbrot (1978)):

$$\tilde{\mathbf{x}}(t) = \big(\mathbf{x}(t, \mathbf{x}_0, \mathbf{u}), \mathbf{x}_t(\cdot, \mathbf{x}_0, \mathbf{u})\big) \in M^2([-h, 0], \boldsymbol{R}^n).$$

Moreover, $\tilde{\mathbf{x}}(t)$ satisfies the variation of constants formula in the Hilbert space $M^2([-h, 0], \boldsymbol{R}^n)$,

$$\mathbf{x}(t) = \tilde{S}(t)\big(\mathbf{x}_0(0), \mathbf{x}_0\big) + \int_{0}^{t} \tilde{S}(t - \tau)\tilde{\mathbf{B}}\mathbf{u}(\tau)\,d\tau, \tag{4.2.12}$$

where $\tilde{S}(t)$, $t \geqslant 0$ is the extension of the semigroup $S(t)$, $t \geqslant 0$, from the space $W_2^{(1)}([-h, 0], \boldsymbol{R}^n)$ to the space $M^2([-h, 0], \boldsymbol{R}^n)$, the infinitesimal generator $\tilde{A}$ of the semigroup $\tilde{S}(t)$, $t \geqslant 0$ is defined as follows (Manitius and Triggiani (1978a), O'Connor and Tarn (1983), Rodas and Langenhop (1978)):

$$\tilde{A}(\mathbf{f}, \mathbf{g}) = (\mathbf{A}_0\mathbf{f}+\mathbf{A}_1\mathbf{g}(-h), \dot{\mathbf{g}}(\cdot)) \quad \text{for } (\mathbf{f}, \mathbf{g}) \in D(\tilde{A}),$$

$$D(\tilde{A}) = \{(\mathbf{f}, \mathbf{g}) \in M^2([-h, 0], \boldsymbol{R}^n)\colon \mathbf{g} \in L^2([-h, 0], \boldsymbol{R}^n), \mathbf{g}(0) = \mathbf{f}\},$$

$$\overline{D(\tilde{A})} = M^2([-h, 0], \boldsymbol{R}^n).$$

The linear, bounded operator $\tilde{B}\colon \boldsymbol{R}^m \to M^2([-h, 0], \boldsymbol{R}^n)$ is defined as follows:

$$\tilde{B}\mathbf{u} = (B\mathbf{u}, 0) \quad \text{for } \mathbf{u} \in \boldsymbol{R}^m.$$

Formula (4.2.12) may be considered as an extension of equality (4.2.9) from the space $W_2^{(1)}([-h, 0], \boldsymbol{R}^n)$ to the space $M^2([-h, 0], \boldsymbol{R}^n)$ (O'Connor and Tarn (1983)).

Moreover, $\tilde{\mathbf{x}}(t)$ defined by (4.2.12) is the solution of the following abstract differential equation in the Hilbert space $M^2([-h, 0], \boldsymbol{R}^n)$:

$$\dot{\tilde{\mathbf{x}}}(t) = \tilde{A}\tilde{\mathbf{x}}(t)+\tilde{B}\mathbf{u}(t), \quad t \geqslant 0. \tag{4.2.13}$$

An analogous differential equation for system (4.2.1) is of the form (O'Connor and Tarn (1983)):

$$\dot{\mathbf{y}}(t) = \tilde{A}\mathbf{y}(t)+\tilde{B}_s\mathbf{u}(t), \quad t > 0, \tag{4.2.14}$$

$$\mathbf{y}(0) = (sI-\tilde{A})^{-1}\mathbf{x}_0,$$

where $\mathbf{x}_t(\cdot, \mathbf{x}_0, \mathbf{u}) = (sI-\tilde{A})\mathbf{y}(t)$, or equivalently $\mathbf{y}(t) = (sI-\tilde{A})^{-1}\mathbf{x}_t(\cdot, \mathbf{x}_0, \mathbf{u})$.

The remaining part of this subsection will be devoted to several different types of controllability, for dynamical systems with delays.

DEFINITION 4.2.1. The dynamical system (4.2.1) is said to be *approximately controllable in* $[0, t_1]$ $(t_1 \geqslant h)$ if, for any $\mathbf{w}_0, \mathbf{w}_1 \in W_2^{(1)}([-h, 0], \boldsymbol{R}^n)$ and, for every real number $\varepsilon > 0$, there exists a control $\mathbf{u} \in L^2([0, t_1], \boldsymbol{R}^m)$ such that

$$\|\mathbf{x}_{t_1}(\cdot, \mathbf{w}_0, \mathbf{u})-\mathbf{w}_1\|_{W_2^{(1)}([-h, 0], \boldsymbol{R}^n)} \leqslant \varepsilon. \tag{4.2.15}$$

Definition 4.2.1 may also be formulated with the aid of the so-called attainable set K_t, $t > 0$, which is defined as follows:

$$K_t = \{\mathbf{x}_t(\cdot, 0, \mathbf{u}) \in W_2^{(1)}([-h, 0], \boldsymbol{R}^n)\colon \mathbf{u} \in L^2([0, t], \boldsymbol{R}^m)\}. \tag{4.2.16}$$

DEFINITION 4.2.2. The dynamical system (4.2.1) is said to be *approximately controllable in* $[0, t_1]$ if

$$\overline{K_{t_1}} = W_2^{(1)}([-h, 0], \boldsymbol{R}^n). \tag{4.2.17}$$

DEFINITION 4.2.3. The dynamical system (4.2.1) is said to be *approximately controllable in finite time* if

$$\overline{K_\infty} = \overline{\bigcup_{0<t<\infty} K_t} = W_2^{(1)}([-h, 0], \boldsymbol{R}^n). \tag{4.2.18}$$

DEFINITION 4.2.4. The dynamical system (4.2.1) is said to be *exactly controllable in* $[0, t_1]$ (*in finite time*) if

$$K_{t_1} = W_2^{(1)}([-h, 0], \boldsymbol{R}^n) \quad \Big(K_\infty = \bigcup_{0<t<\infty} K_t = W_2^{(1)}([-h, 0], \boldsymbol{R}^n)\Big). \tag{4.2.19}$$

It should be pointed out that approximate controllability and exact controllability are essentially different. Moreover, most dynamical systems with delays are only approximately controllable not exactly controllable. This problem will be studied more deeply in the next subsections.

Moreover, it should also be stressed that semigroups $S(t)$ and $\tilde{S}(t)$ are not analytic semigroups, whence the distinction between controllability in $[0, t_1]$ and controllability in finite time is essential.

Substituting in Definitions 4.2.1-4.2.4 the space $M^2([-h, 0], \boldsymbol{R}^n)$ for the space $W_2^{(1)}([-h, 0], \boldsymbol{R}^n)$, we immediately obtain the equivalent definitions for the dynamical system (4.2.11). In this case we speak about so-called M^2-approximate controllability. In a similar way we define L^2-approximate controllability of the dynamical system (4.2.11) (Manitius and Triggiani (1978a, 1978b)).

In the case where the function $\mathbf{w}_1 = 0$ we speak about *zero-approximate* or *zero-exact controllability in* $[0, t_1]$ or *in finite time*. Distinguishing zero-controllability is important, because of the existence of dynamical systems with delays, which are approximately (exactly) zero-controllable but not approximately (exactly) controllable. This phnomenon follows from the fact that the ranges of the semigroups $S(t)$ and $\tilde{S}(t)$ are not generally the whole spaces $W_2^{(1)}([-h,0], \boldsymbol{R}^n)$ or $M^2([-h,0], \boldsymbol{R}^n)$ (Jacobs and Langenhop (1976), O'Connor and Tarn (1983)). Hence it is "easier" to steer the dynamical system to the zero-function than to an arbitrary nonzero function.

In the case where we are interested only in controlling the dynamical system to some given point in the space $\boldsymbol{R}^n$ (and not to a given function in the space $W_2^{(1)}([-h, 0], \boldsymbol{R}^n)$) we speak about so-called *$\boldsymbol{R}^n$-controllability*, called also *Euclidean controllability* (Manitius and Triggiani (1978a), O'Conner and Tarn (1983)).

DEFINITION 4.2.5. The dynamical systems (4.2.1) or (4.2.11) are said to be *$\boldsymbol{R}^n$-controllable* in $[0, t_1]$ if, for any initial condition $\mathbf{x}_0$ or $(\mathbf{x}_0(0), \mathbf{x}_0)$ and for any vector $\mathbf{x}_1 \in \boldsymbol{R}^n$, there exists a control $\mathbf{u} \in L^2([0, t_1], \boldsymbol{R}^m)$ such that the corresponding trajectory of dynamical system (4.2.1) or (4.2.11) satisfies the final condition

$$\mathbf{x}(t_1, \mathbf{x}_0, \mathbf{u}) = \mathbf{x}_1. \tag{4.2.20}$$

Definition 4.2.5 can easily be extended to cover the case of ***R^n*-controllability** *in finite time* and ***R^n*-zero-controllability** *in* $[0, t_1]$ and *in finite time*. However there is no sense in distinguishing between approximate and exact R^n-controllability. This follows from the fact that the space R^n is finite-dimensional and all its linear subspaces are closed. Hence the equivalence between approximtae R^n-controllability and exact R^n-controllability is obvious.

The R^n-controllability of dynamical systems (4.2.1) and (4.2.11) is a weaker concept than function controllability, and the criteria for its investigation are quite similar to the criteria for the controllability of systems without delays formulated in Chapter 1.

A spectral analysis of dynamical systems with delays leads immediately to the definition of spectral controllability (Manitius and Triggiani (1978a), O'Connor and Tarn (1983)).

DEFINITION 4.2.6. The dynamical system (4.2.1) is said to be *spectrally controllable* if, for each $s \in \sigma(A)$, the linear, continuous, finite-dimensional dynamical system without delays 4.2.6 is controllable in the usual sense.

Spectral controllability is generally a weaker notion than approximate controllability (Bartosiewicz (1980), Manitius and Triggiani (1978a), O'Connor and Tarn (1983)). However, it should be stressed that these two notions are equivalent in the case where the dynamical system is spectrally complete (see the definition given below).

DEFINITION 4.2.7. The dynamical system (4.2.1) is said to be *spectrally complete in the space* $W_2^{(1)}([-h, 0], R^n)$ if

$$\overline{\text{sp}}\{N_s, s \in \sigma(\tilde{A})\} = W_2^{(1)}([-h, 0], R^n). \tag{4.2.21}$$

Spectral controllability implies that each element in the subspace $\text{sp}\{N_s, s \in \sigma(\tilde{A})\}$ can be reached. Hence spectral controllability and spectral completeness together imply that the attainable set K_∞ is dense in the space $W_2^{(1)}([-h, 0], R^n)$, whence the dynamical system (4.2.1) is approximately controllable in finite time (Bartosiewicz (1980), O'Connor and Tarn (1983)). The method of spectral completeness is also connected with the notion of so-called spectral completability (see Bartosiewicz (1980), O'Connor and Tarn (1983) for details).

DEFINITION 4.2.8. The dynamical system (4.2.1) is said to be *spectrally completable* if there exists a feedback containing delays and such that dynamical system (4.2.1) with that feedback is spectrally complete.

Definitions 4.2.6-4.2.8 apply also to the dynamical system (4.2.11) with the following modification: the space $W_2^{(1)}([-h, 0], R^n)$ must be replaced by the space $M^2([-h, 0], R^n)$ or $L^2([-h, 0], R^n)$. It should be pointed out however, that, whereas in the case of a neutral dynamical system (4.2.1) the spectrum of the operator $\tilde{A}$ is always countable, in the case of the dynamical system (4.2.11) the spectrum may be finite (Manitius and Triggiani (1978a), Manitius and Olbrot (1978), Marchenko (1981a)). Dynamical systems (4.2.11) with a finite spectrum are equivalent to linear, continuous, finite-dimensional dynamical systems without delays.

All the preceding definitions of controllability are based on the concepts taken from functional analysis and topology. In recent years the analysis of dynamical systems with delays has been making more and more use of algebraic methods (Kamen (1975a, 1976), Olbrot and Jakubczyk (1981), Sontag (1976a, 1976c)). In order to use such methods, let us recall some notions from abstract algebra.

$R[s]$ denotes a polynomial ring with coefficients from the field R. $R(s)$ denotes the field of rational fractions whose numerators and denominators are elements of $R[s]$. The symbol $\mathbf{R}^{m\times n}(s)$ denotes an $(n\times m)$-dimensional matrix with elements from the field $R(s)$. The rank of the matrix $\mathbf{A}\in\mathbf{R}^{m\times n}(s)$ is defined classically as the dimension of the highest nonzero subdeterminant.

Using the Laplace transform of equation (4.2.1) and assuming the initial conditions $\mathbf{x}_0 = 0$, we obtain the algebraic equation (Olbrot and Jakubczyk (1981))

$$\tilde{\mathbf{x}}(s) = (s\mathbf{I}-\mathbf{A}_0)^{-1}(s\mathbf{A}_{-1}+\mathbf{A}_1)s\tilde{\mathbf{x}}(s)+(s\mathbf{I}-\mathbf{A}_0)^{-1}\mathbf{B}\tilde{\mathbf{u}}(s) \tag{4.2.22}$$

where $\tilde{\mathbf{x}}(s)$ and $\tilde{\mathbf{u}}(s)$ are the Laplace transforms of the functions $\mathbf{x}(t)$ and $\mathbf{u}(t)$, respectively. For simplicity, let us introduce the following notation (Olbrot and Jakubczyk (1981)):

$$\mathbf{F}(s) = (s\mathbf{I}-\mathbf{A}_0)^{-1}(s\mathbf{A}_{-1}+\mathbf{A}_1)\in\mathbf{R}^{n\times n}(s), \tag{4.2.23}$$

$$\mathbf{G}(s) = (s\mathbf{I}-\mathbf{A}_0)^{-1}\mathbf{B}\in\mathbf{R}^{n\times m}(s). \tag{4.2.24}$$

Using the matrices $\mathbf{F}(s)$ and $\mathbf{G}(s)$, we can introduce a formal definition of $R^n(s)$-controllability for the dynamical system (4.2.1) (Olbrot and Jakubczyk (1981)).

Definition 4.2.9. The dynamical system (4.2.1) is said to be *$R^n(s)$-controllable* if the matrices $\mathbf{F}(s)\in\mathbf{R}^{n\times m}(s)$ and $\mathbf{G}(s)\in\mathbf{R}^{n\times m}(s)$ satisfy the condition:

$$\operatorname{rank}[\mathbf{G}\,\vdots\,\mathbf{F}\mathbf{G}\,\vdots\,\mathbf{F}^2\mathbf{G}\,\vdots\,\ldots\,\vdots\,\mathbf{F}^{n-1}\mathbf{G}] = n \text{ over the field } R(s). \tag{4.2.25}$$

Definition 4.2.9 is the special case of the "reachability" definition given in (Sontag (1976a)) for a wider class of dynamical systems. Owing to the special form of the matrices $\mathbf{F}(s)$ and $\mathbf{G}(s)$ we shall be able to derive more concrete and computable criteria for the investigation of $R^n(s)$-controllability, than those used in (Sontag (1976a)). Moreover, the concept of $R^n(s)$-controllability gives us a tool for constructing the canonical forms of the dynamical system (4.2.1).

4.3 Algebraic controllability criteria

In this subsection we shall formulate the algebraic controllability criteria for the dynamical systems (4.2.1) and (4.2.11). In order to do that, let us introduce some additional concepts and symbols. First of all, let us express the $(n \times n)$-dimensional matrix $\mathrm{adj}\boldsymbol{\Delta}(s)\mathbf{B}$ in the form

$$\mathrm{adj}\,\boldsymbol{\Delta}(s)\mathbf{B} = \sum_{k=0}^{n-1} \mathbf{P}_k(s)\mathbf{B}e^{-skh} = \mathbf{P}(s)\mathbf{V}(e^{-sh}),$$

where $\mathbf{P}(s)$ is an $(n \times nm)$-dimensional, polynomial matrix of the form

$$\mathbf{P}(s) = [\mathbf{P}_0(s)\mathbf{B} \vdots \mathbf{P}_1(s)\mathbf{B} \vdots \dots \vdots \mathbf{P}_k(s)\mathbf{B} \vdots \dots \vdots \mathbf{P}_{n-1}(s)\mathbf{B}], \tag{4.3.1}$$

where $\mathbf{P}_k(s)$, $k = 0, 1, 2, \dots, n-1$ are $(n \times n)$-dimensional polynomial matrices and $\mathbf{V}(e^{-sh})$ is an $(mn \times m)$-dimensional matrix of the form

$$\mathbf{V}(e^{-sh}) = [\mathbf{I}_m \vdots \mathbf{I}_m e^{-sh} \vdots \dots \vdots \mathbf{I}_m e^{-skh} \vdots \dots \vdots \mathbf{I}_m e^{-s(n-1)h}]^{\mathrm{T}},$$

where $\mathbf{I}_m$ is an $(m \times m)$-dimensional identity matrix.

The polynomial matrix $\mathbf{P}(s)$ plays an important role in investigating the controllability of dynamical systems with delays. Matrix $\mathbf{P}(s)$ depend sonly on the matrices $\mathbf{A}_{-1}$, $\mathbf{A}_0$, $\mathbf{A}_1$ and $\mathbf{B}$, and does not depend on the delay h.

The symbol $\mathrm{rank}\,\mathbf{P}(s)$ denotes the rank of the numerical matrix $\mathbf{P}(s)$ at a fixed point $s \in C$ (C is the field of complex numbers). Moreover, the equality $\mathrm{rank}_C\mathbf{P}(s) = n$ signifies that there exists a point $s_0 \in C$, such that $\mathrm{rank}_C\mathbf{P}(s_0) = n$. It is obvious that the equality $\mathrm{rank}_C\mathbf{P}(s) = n$ implies the equality $\mathrm{rank}\,\mathbf{P}(s) = n$ for almost every $s \in C$. This means that the matrix $\mathbf{P}(\cdot)$ has rank n over the polynomial ring.

We now introduce two lemmas, which will be useful in our further investigations. The proofs of these lemmas are given in the papers cited in the References

LEMMA 4.2.1 (Pandolfi (1986b)). *The dynamical system* (4.2.1) *is spectrally controllable if and only if, for every* $s \in \sigma(\tilde{A})$, *the following equality holds*:

$$\mathrm{rank}\,[\boldsymbol{\Delta}(s) \vdots \mathbf{B}] = n. \tag{4.3.2}$$

LEMMA 4.3.2 (Bartosiewicz (1980), O'Connor and Tarn (1983)). *The dynamical system* (4.2.1) *is spectrally complete if and only if*

$$\det[s\mathbf{A}_{-1} + \mathbf{A}_1] \neq 0. \tag{4.3.3}$$

Using Lemmas 4.3.1 and 4.3.2 and the remarks following Definition 4.2.7 we easily formulate a sufficient condition for approximate controllability in finite time for the dynamical system (4.2.1).

COROLLARY 4.3.1. *The dynamical system* (4.2.1) *is approximately controllable in finite time if simultaneously the following two conditions are satisfied*:

$$\det[s\mathbf{A}_{-1} + \mathbf{A}_1] \neq 0, \tag{4.3.4}$$

$$\mathrm{rank}[\boldsymbol{\Delta}(s) \vdots \mathbf{B}] = n \quad \textit{for every } s \in \sigma(\tilde{A}). \tag{4.3.5}$$

Condition (4.3.4) is equivalet to the spectral completeness of the dynamical system (4.2.1) (this follows immediately from Lemma 4.3.2) and can also be expressed in the form

$$\operatorname{rank}_C(s\mathbf{A}_{-1}+\mathbf{A}_1) = n. \tag{4.3.6}$$

From the Lemma 4.3.1 immediately implies that condition (4.3.5) is equivalent to the spectral controllability of the dynamical system (4.2.1).

Corollary 4.3.1 is only a sufficient condition for the approximate controllability in finite time of the dynamical system (4.2.1). In order to formulate the necessary and sufficient conditions for approximate controllability in finite time we shall use the concept of spectral completability given in Definition 4.2.8.

THEOREM 4.3.1 (Bartosiewicz (1980), O'Connor and Tarn (1983)). *The dynamical system* (4.2.1) *is approximately controllable in finite time if and only if*

$$\operatorname{rank}_C[s\mathbf{A}_{-1}+\mathbf{A}_1 \vdots \mathbf{B}] = n, \tag{4.3.7}$$

$$\operatorname{rank}[\boldsymbol{\Delta}(s) \vdots \mathbf{B}] = n \quad \textit{for every } s \in \sigma(\tilde{A}). \tag{4.3.8}$$

Proof:

Sufficient condition. We shall first prove, that there exists an $(m \times n)$-dimensional matrix $\mathbf{K}$, such that

$$\operatorname{rank}_C(s\mathbf{A}_{-1}+\mathbf{A}_1+\mathbf{BK}) = n. \tag{4.3.9}$$

Let us assume on the contrary, that such an $(m \times n)$-dimensional matrix $\mathbf{K}$ does not exist. Hence we can find an $(n \times 1)$-dimensional nonzero vector $\mathbf{p}(s)$ such that for all $s \in C$ and every $(m \times n)$-dimensional matrix $\mathbf{K}$ the following equality holds:

$$\mathbf{p}^{\mathrm{T}}(s)(s\mathbf{A}_{-1}+\mathbf{A}_1+\mathbf{BK}) = 0.$$

In particular, for $\mathbf{K} = 0$ we have

$$\mathbf{p}^{\mathrm{T}}(s)(s\mathbf{A}_{-1}+\mathbf{A}_1) = 0 \quad \text{for every } s \in C. \tag{4.3.10}$$

Hence for every $(m \times n)$-dimensional matrix $\mathbf{K}$ we have

$$\mathbf{p}^{\mathrm{T}}(s)\mathbf{BK} = 0 \quad \text{for every } s \in C.$$

Let $\mathbf{B} = [\mathbf{b}_1\ \mathbf{b}_2 \ldots \mathbf{b}_j \ldots \mathbf{b}_m]$ where $\mathbf{b}_j$, $j = 1, 2, \ldots, m$ are the j-th columns of the matrix $\mathbf{B}$. Moreover, let

$$\mathbf{K}_j = \begin{bmatrix} 0 & 0 & \ldots & 0 \\ 0 & 0 & \ldots & 0 \\ 1 & 1 & \ldots & 1 \\ 0 & 0 & & 0 \\ 0 & 0 & \ldots & 0 \end{bmatrix} \leftarrow j\text{-th row}.$$

Hence

$$\mathbf{p}^{\mathrm{T}}(s)\mathbf{BK}_j = 0 \quad \text{for} \quad j = 1, 2, \ldots, m \text{ and } s \in C. \tag{4.3.11}$$

Therefore

$$\mathbf{p}^{\mathrm{T}}(s)\mathbf{B} = 0 \quad \text{for every } s \in \boldsymbol{C}. \tag{4.3.12}$$

Conditions (4.3.10), (4.3.11) and (4.3.12) imply that $\operatorname{rank}_{\boldsymbol{C}}[s\mathbf{A}_{-1}+\mathbf{A}_1 \vdots \mathbf{B}] < n$. This inequality contradicts (4.3.7), whence there exists an $(m \times n)$-dimensional matrix $\mathbf{K}$ such that relation (4.3.9) is true.

In the second part of the proof of the sufficient condition, let us consider a dynamical system described by the following, neutral differential equation:

$$\dot{\mathbf{x}}(t) = \mathbf{A}_{-1}\dot{\mathbf{x}}(t-h)+\mathbf{A}_0\mathbf{x}(t)+(\mathbf{A}_1+\mathbf{BK})\mathbf{x}(t-h)+\mathbf{B}\mathbf{v}(t), \qquad t \geqslant 0. \tag{4.3.13}$$

From (4.2.9) it follows that the dynamical system (4.3.13) is spectrally complete and, moreover, if

$$\operatorname{rank}[\boldsymbol{\Delta}(s)-\mathbf{BK}e^{-sh} \vdots \mathbf{B}] = n \qquad \text{for every } s \in \sigma(\tilde{A}) \tag{4.3.14}$$

then it is also approximately controllable in finite time. However, we can easily verify that condition (4.3.14) is equivalent to the equality

$$\operatorname{rank}[\boldsymbol{\Delta}(s) \vdots \mathbf{B}] = n \qquad \text{for every } s \in \sigma(\tilde{A}).$$

Condition (4.3.8) and the above equality immediately imply approximate controllability in finite time of the dynamical system (4.2.1).

In the last part of the proof of the sufficient condition we shall prove, that approximate controllability in finite time of the dynamical system (4.3.13) imply approximate controllability in finite time of the dynamical system (4.2.1). Let $\mathbf{x}_t^1(\cdot, \mathbf{x}_0, \mathbf{u})$ and $\mathbf{x}_t^2(\cdot, \mathbf{x}_0, \mathbf{u})$ be, respectively, the trajectories of the dynamical systems (4.2.1) and (4.3.13). For any function $\mathbf{w}_1 \in W_2^{(1)}([-h, 0], \boldsymbol{R}^n)$ and every real number $\varepsilon > 0$, there exists a time t_1 and a control $\mathbf{v} \in L^2([0, t_1], \boldsymbol{R}^m)$, such that

$$\|\mathbf{x}_{t_1}^2(\cdot, 0, \mathbf{v})-\mathbf{w}_1\|_{W_2^{(1)}([-h, 0], \boldsymbol{R}^n)} \leqslant \varepsilon.$$

The above inequality follows from the approximate controllability in finite time of the dynamical system (4.3.13). Let us define the control $\mathbf{u} \in L^2([0, t_1], \boldsymbol{R}^m)$ as follows: $\mathbf{u}(t) = \mathbf{K}\mathbf{x}^2(t-h, 0, \mathbf{u})+\mathbf{v}(t)$, $t \geqslant 0$. Therefore we have

$$\mathbf{x}^1(t, 0, \mathbf{u}) = \mathbf{x}^2(t, 0, \mathbf{v}) \quad \text{and} \quad \|\mathbf{x}_{t_1}^1(\cdot, 0, \mathbf{u})-\mathbf{w}_1\|_{W_2^{(1)}([-h, 0], \boldsymbol{R}^n)} \leqslant \varepsilon.$$

Hence the dynamical system (4.2.1) is approximately controllable in finite time.

Necessary condition. Since approximate controllability in finite time implies spectral controllability, the necessity of condition (4.3.8) is obvious. Similarly, from Corollary 4.3.1 and Lemma 4.3.2 imply the necessity of condition (4.3.7). Hence our theorem follows. ∎

COROLLARY 4.3.2. *The dynamical system* (4.2.11) *is approximately controllable in finite time in the space* $L^2([-h, 0], \boldsymbol{R}^n)$ (*shortly*: *L^2-approximately controllable in finite time*) *if and only if*

$$\text{rank}[\mathbf{A}_1 \vdots \mathbf{B}] = n, \tag{4.3.15}$$

$$\text{rank}[\mathbf{\Delta}(s) \vdots \mathbf{B}] = n \quad \textit{for every } s \in \sigma(\tilde{A}). \tag{4.3.16}$$

Proof:

Since the space $W^{(1)}_2([-h, 0], R^n)$ is dense in the space $L^2([-h, 0], R^n)$, substituting $\mathbf{A}_1 = 0$ in equaity (4.3.7) we obtain

$$\text{rank}_C[\mathbf{A}_1 \vdots \mathbf{B}] = \text{rank}[\mathbf{A}_1 \vdots \mathbf{B}].$$

Hence, by Definition 4.2.3 our corollary follows. ∎

Since $\text{rank}\mathbf{\Delta}(s) = n$ for $s \notin \sigma(\tilde{A})$, the condition

$$\text{rank}[\mathbf{\Delta}(s) \vdots \mathbf{B}] = n \quad \text{for every } s \in \sigma(A)$$

is equivalent to the equality

$$\text{rank}[\mathbf{\Delta}(s) \vdots \mathbf{B}] = n \quad \text{for every } s \in C.$$

Using the above statements, we can modify conditions (4.3.5), (4.3.8) and (4.3.16).

As was mentioned before, the matrix $\mathbf{P}(s)$ plays an important role in formulating the controllability conditions for dynmaical systems (4.2.1) and (4.2.11). Using this matrix, we can obtain the next necessary condition for approximate controllability in finite time of the dynamical system (4.2.1).

LEMMA 4.3.3 (Manitius and Triggiani (1978a, 1978b), O'Connor and Tarn (1983)). *If the dynamical system* (4.2.1) *is approximately controllable in finite time, then*

$$\text{rank}_C\mathbf{P}(s) = n. \tag{4.3.17}$$

The complete, complicated proof of Lemma 4.3.3 is given in Manitius and Triggiani (1978) for the dynamical systems (4.2.11). An extension which covers the dynamical system (4.2.1) is presented in O'Connor and Tarn (1983). Paper by Manitius and Triggiani (1978b) contains only some comments and remarks concerning condition (4.3.17).

COROLLARY 4.3.3 (Manitius and Triggiani (1987a, 1978b)). *If the dynamical system* (4.2.11) *is L^2-approximately controllable in finite time, then equality* (4.3.17) *holds.*

Proof:

Corollary 4.3.3 immediately follows from Lemma 4.3.3. ∎

The next lemma characterizes the properties of the attainable set K_t for $t > nh$. It will be useful in considering the exact controllability in finite time of the dynamical system (4.2.1).

LEMMA 4.3.4 (Jacobs and Langenhop (1976)). *$K_{t_1} = K_{t_2}$ for all $t_1, t_2 > nh$, and if, moreover,*

$$\operatorname{rank}[\mathbf{B} \vdots \mathbf{A}_{-1}\mathbf{B} \vdots \mathbf{A}_{-1}^2\mathbf{B} \vdots \ldots \vdots \mathbf{A}_{-1}^{n-1}\mathbf{B}] = n,$$

then the attainable set K_t is closed in the space $W_2^{(1)}([-h,0], R^n)$ for $t > nh$, and its orthogonal complement $K_t^{\perp}$ is finite-dimensional.

THEOREM 4.3.2 (O'Connor and Tarn (1983)). *The dynamical system* (4.2.1) *is exactly controllable in finite time if and only if, simultaneously, the following two conditions are satisfied*:

(1) *the dynamical system* (4.2.1) *is approximately controllable in finite time,*

(2) $\operatorname{rank}[\mathbf{B} \vdots \mathbf{A}_{-1}\mathbf{B} \vdots \mathbf{A}_{-1}^2\mathbf{B} \vdots \ldots \vdots \mathbf{A}_{-1}^{n-1}\mathbf{B}] = n.$ (4.3.18)

Proof:

Sufficient condition. From condition (1) and Definition 4.2.3 it follows that the attainable set K_∞ is dense in the space $W_2^{(1)}([-h,0], R^n)$. Moreover, the condition (2) guarantees that the attainable set K_∞ is closed (by the Lemma 4.3.4). Hence from the density and closedness of the attainable set K_∞ follows the exact controllability in finite time of the dynamical system (4.2.1).

Necessary condition. By Definition 4.2.4 the exact controllability in finite time of the dynamical system (4.2.1) implies its approximate controllability in finite time and the closedness of the attainable set K_∞, which, by Lemma 4.3.4, implies equality (4.3.18). ■

COROLLARY 4.3.4 (Banks, Jacobs and Langenhop (1975)). *The dynamical system* (4.2.11) *is exactly controllable in finite time if and only if*

$$\operatorname{rank}\mathbf{B} = n. \tag{4.3.19}$$

Proof:

Sufficient condition. Since (4.3.19) immediately implies (4.3.7), (4.3.8) and (4.3.18), then by Theorems 4.3.1 and 4.3.2 we have the exact controllability in finite time of the dynamical system (4.2.11).

Necessary condition. Since for the dynamical system (4.2.1) we have $\mathbf{A}_{-1} = 0$, equality (4.3.19) is necessary for condition (4.3.18) to be satisfied. ■

From the analysis of Theorem 4.3.2 and Corollary 4.3.4 it follows that for the dynamical system (4.3.11) the criteria for exact controllability in finite time are very restrictive and hence very rarely satisfied in practice. However, for the dynamical systems (4.2.1) exact controllability in finite time may occur even for the scalar controls (in this case $\mathbf{B}$ is the vector).

Now we shall formulate some conditions for the $\mathbf{R}^n(s)$-controllability of the dynamical system (4.2.1). In order to do that, we pay special attention to Definition 4.2.9 and equality (4.2.25).

THEOREM 4.3.3 (Olbrot and Jakubczyk (1981)). *The dynamical system* (4.2.1) *is* $\mathbf{R}^n(s)$*-controllable if and only if at least one of the following conditions is satisfied:*

$$\operatorname{rank}[\mathbf{G}(s) \vdots \mathbf{F}(s)\mathbf{G}(s) \vdots \mathbf{F}^2(s)\mathbf{G}(s) \vdots \dots \vdots \mathbf{F}^{n-1}(s)\mathbf{G}(s)] = n \quad \text{over } R \text{ (or } C) \tag{4.3.20}$$

for at least one real (or complex) number s,

$$\operatorname{rank}[\mathbf{G}(s) \vdots \mathbf{F}(s)\mathbf{G}(s) \vdots \mathbf{F}^2(s)\mathbf{G}(s) \vdots \dots \vdots \mathbf{F}^{n-1}(s)\mathbf{G}(s)] = n \quad \text{over } R \text{ (or } C) \tag{4.3.21}$$

for almost all real (or complex) numbers s.

Proof:

Condition (4.2.25) is equivalent to the statement that at least one $(n \times n)$-dimensional subdeterminant of the matrix

$$[\mathbf{G}(s) \vdots \mathbf{F}(s)\mathbf{G}(s) \vdots \mathbf{F}^2(s)\mathbf{G}(s) \vdots \dots \vdots \mathbf{F}^{n-1}(s)\mathbf{G}(s)]$$

is a nonzero rational function in the complex indeterminate s. Hence it may be equal to zero only for a finite number of the values of the argument s. Hence immediately follow conditions (4.3.20) and (4.3.21) and their equivalence. ■

COROLLARY 4.3.5 (Olbrot and Jakubczyk (1981)). *The dynamical system* (4.2.1) *is* $\mathbf{R}^n(s)$*-controllable if and only if*

$$\operatorname{rank}[\tilde{\mathbf{G}} \vdots \tilde{\mathbf{F}}\tilde{\mathbf{G}} \vdots \tilde{\mathbf{F}}^2\tilde{\mathbf{G}} \vdots \dots \vdots \tilde{\mathbf{F}}^{n-1}\tilde{\mathbf{G}}] = n \quad \text{over the field } R(s), \tag{4.3.22}$$

where

$$\tilde{\mathbf{F}} = (s\mathbf{I}-\mathbf{A}_0)\mathbf{F}(s\mathbf{I}-\mathbf{A}_0)^{-1} = (s\mathbf{A}_{-1}+\mathbf{A}_1)(s\mathbf{I}-\mathbf{A}_0)^{-1}, \tag{4.3.23}$$

$$\tilde{\mathbf{G}} = (s\mathbf{I}-\mathbf{A}_0)\mathbf{G} = \mathbf{B}. \tag{4.3.24}$$

Proof:

From (4.2.23), (4.2.24), (4.3.23) and (4.3.24) we have

$$[\tilde{\mathbf{G}} \vdots \tilde{\mathbf{F}}\tilde{\mathbf{G}} \vdots \tilde{\mathbf{F}}^2\tilde{\mathbf{G}} \vdots \dots \vdots \tilde{\mathbf{F}}^{n-1}\tilde{\mathbf{G}}] = [s\mathbf{I}-\mathbf{A}_0][\mathbf{G} \vdots \mathbf{F}\mathbf{G} \vdots \mathbf{F}^2\mathbf{G} \vdots \dots \vdots \mathbf{F}^{n-1}\mathbf{G}].$$

Since the matrix $[s\mathbf{I}-\mathbf{A}_0]$ is nonsingular over the field $R(s)$, the equivalence of the conditions (4.2.25) and (4.3.22) is obvious. ■

COROLLARY 4.3.6 (Olbrot and Jakubczyk (1981)). *If the dynamical system* (4.2.1) *is* $\mathbf{R}^n(s)$*-controllable, then*

$$\operatorname{rank}[\mathbf{A}_{-1} \vdots \mathbf{A}_1 \vdots \mathbf{B}] = n. \tag{4.3.25}$$

Proof:

By contradiction. Let us assume, that condition (4.3.25) is not fulfilled. Therefore there exists a nonzero vector $\mathbf{q} \in R^n$ such that $\mathbf{q}^T\mathbf{A}_{-1} = \mathbf{q}^T\mathbf{A}_1 = 0$ and $\mathbf{q}^T\mathbf{B} = 0$. Hence $\mathbf{q}^T\tilde{\mathbf{G}} = \mathbf{q}^T\mathbf{B} = 0$ and $\mathbf{q}^T\tilde{\mathbf{F}} = \mathbf{q}^T(s\mathbf{A}_{-1}+\mathbf{A}_1)(s\mathbf{I}-\mathbf{A}_0)^{-1}$ is a zero rational function. Therefore Theorem 4.3.3 and Corollary 4.3.5 imply, that the $R^n(s)$-controllability of the dynamical system (4.2.1) is not satisfied. Hence (4.3.25) is the necessary condition for the $R^n(s)$-controllability of the dynamical system (4.2.1). ∎

COROLLARY 4.3.7 (Olbrot and Jakubczyk (1981)). *If*

$$\operatorname{rank}[\mathbf{B} \vdots \mathbf{A}_{-1}\mathbf{B} \vdots \mathbf{A}_{-1}^2\mathbf{B} \vdots \dots \vdots \mathbf{A}_{-1}^{n-1}\mathbf{B}] = n. \tag{4.3.26}$$

then the dynamical system (4.2.1) *is $R^n(s)$-controllable.*

Proof:

Substituting $s = 1/w$ in (4.2.23) and (4.3.23), we obtain,

$$\tilde{\mathbf{F}}(1/w) = (\mathbf{A}_{-1} + w\mathbf{A}_1)(\mathbf{I} - w\mathbf{A}_0)^{-1},$$

Furthermore, from (4.3.24) we have $\tilde{\mathbf{G}}(1/w) = \mathbf{B}$. Substituting $w = 0$ and using Theorem 4.3.3 and Corollary 4.3.5, we obtain Corollary 4.3.7. ∎

It is interesting to point out some connections between $R^n(s)$-controllability, approximate controllability, exact controllability and R^n-controllability. In order to do so, let us first formulate the lemma, which gives us the relation between attainable set K_t and the $R^n(s)$-controllability of the dynamical system (4.2.1).

LEMMA 4.3.5 (Olbrot and Jakubczyk (1981)). *The dynamical system* (4.2.1) *is $R^n(s)$ controllable if and only if the codimension of the closedness in the space $L^2([-h, 0], R^n$ of the attainable set $\overline{K_t}$, $t \geqslant nh$, is finite ($\overline{K_t}^{\perp}$ is finite-dimensional).*

Proof:

Necessary condition. By contradiction. Let us asume that $\overline{K_t}$, $t \geqslant nh$, is of finite dimension, but the dynamical system (4.2.1) is not $R^n(s)$-controllable, i.e., condition (4.2.25) does not hold. Therefore there exists a nonzero vector $\mathbf{q}(s) \in R^n(s)$ such that

$$\mathbf{q}^{\mathrm{T}}(s)\mathbf{F}^i(s)\mathbf{G}(s) = 0 \quad \text{(in the field } R(s)\text{) for } i = 0, 1, \dots, n-1. \tag{4.3.27}$$

Treating $\mathbf{q}^{\mathrm{T}}(s)$ as a transfer function of a linear, time-invariant dynamical system without delays with n inputs and a scalar output, we can construct a so-called *minimal realization* (Chen (1970), Chapter 6) for that system in the following form:

$$\dot{\mathbf{g}}(t) = \mathbf{A}\mathbf{g}(t) + \mathbf{B}\mathbf{y}(t),$$
$$\mathbf{w}(t) = \mathbf{C}\mathbf{g}(t).$$

Since a minimal realization is always controllable and observable, we can find a control $\mathbf{v}(t)$ such that the corresponding output $\mathbf{w}(t)$ is of the following form:

$$\mathbf{w}(t) = \begin{cases} \text{nonzero element from the space } W_2^{(1)}([0, h], R^n), \\ \text{function equal to zero in } (h, \infty). \end{cases} \tag{4.3.28}$$

Therefore the linear functional $f \in L^2([-h, 0], \boldsymbol{R})$ given by

$$f(z) = \int_{-h}^{0} \mathbf{w}^{\mathrm{T}}(-\tau)\mathbf{z}(\tau)d\tau, \qquad \mathbf{z} \in L^2([-h, 0], \boldsymbol{R}^n), \tag{4.3.29}$$

is nonzero, continuous and orthogonal to $\overline{K_t}$ for $t \geqslant nh$. A detailed proof of the last statement is given in (Olbrot and Jakubczyk (1981)) (the Laplace transform properties are used). Choosing a sequence of linear independent controls $\mathbf{v}_l(t)$ which generate outputs $\mathbf{w}_l(t)$ of the form (4.3.28) for $l = 0, 1, 2, \ldots$, we can construct a sequence of linear independent functionals $f_l \in L^2([-h, 0], \boldsymbol{R})$. All these functionals are orthogonal to the attainable set $\overline{K_t}$, $t \geqslant nh$. Therefore the dimension of $\overline{K_t^{\perp}}$, $t \geqslant nh$, is infinite. This is a contradiction, and thus the dynamical system (4.2.1) is $\boldsymbol{R}^n(s)$-controllable. Hence follows our necessary condition.

Sufficient condition. Let us assume that the dynamical system (4.2.1) is $\boldsymbol{R}^n(s)$-controllable, i.e., that condition (4.2.25) holds. We shall show that the linear subspace $\overline{K_t^{\perp}}$, $t \geqslant nh$ has finite dimension. In order to do that, let us consider an auxiliary dynamical system of the following form:

$$\dot{\mathbf{g}}_i(t) = \mathbf{A}_{-1}\dot{\mathbf{g}}_{i-1}(t) + \mathbf{A}_0\mathbf{g}_i(t) + \mathbf{A}_1\mathbf{g}_{i-1}(t) + \mathbf{B}\mathbf{v}_i(t), \qquad t \in [0, h], \tag{4.3.30}$$

where $\mathbf{g}_i(t) \in \boldsymbol{R}^n$, $\mathbf{v}_i \in L^1([0, h], \boldsymbol{R}^m)$, $i = 1, 2, \ldots, n$, and $\mathbf{g}_0(t) = 0$, $\mathbf{g}_1(0) = 0$. Using the Laplace transform and taking into account the initial condition, we have,

$$\mathbf{g}_n(s) = \mathbf{G}\mathbf{v}_n(s) + \mathbf{F}\mathbf{G}\mathbf{v}_{n-1}(s) + \ldots + \mathbf{F}^{n-1}\mathbf{G}\mathbf{v}_1(s). \tag{4.3.31}$$

Since from the assumption condition (4.2.25) holds, by choosing suitable $\mathbf{v}_i(s)$, $i = 1, 2, \ldots, n$, and using (4.3.31) we can obtain the vectors,

$$\mathbf{g}_n(s) = s^{-r}\mathbf{e}_j, \quad j = 1, 2, \ldots, n, \; r = 0, 1, 2, \ldots,$$

where $\mathbf{e}_j \in \boldsymbol{R}^n$ form the base in the space $\boldsymbol{R}^n$.

Using the inverse Laplace transform in the time domain, we derive the vectors $\mathbf{t}^r\mathbf{e}_j$, $r = 0, 1, 2, \ldots$, $j = 1, 2, \ldots, n$, $\mathbf{t} \in [0, h]$. The linear, convex hull of these vectors forms a dense set in the space $L^2([0, h], \boldsymbol{R}^n)$. Therefore the closedness of the linear subspace

$$L = \{\mathbf{g}_n \in L^2([0, h], \boldsymbol{R}^n): \mathbf{g}_n(t) \text{ satisfy } (4.3.30) \text{ for some } \mathbf{v}_i \in L^1([0, h], \boldsymbol{R}^m)\}$$

has finite codimension. Hence follows our sufficient condition. ∎

COROLLARY 4.3.8. (Olbrot and Jakubczyk (1981)). *If the dynamical system* (4.2.1) *is approximately controllable in finite time, then it is also* $\boldsymbol{R}^n(s)$*-controllable.*

Proof:

If the dynamical system (4.2.1) is approximately controllable in finite time, then, by Definition 4.2.3 and from the fact, that the attainable set is nondecreasing with respect to time, there exists a time $t \geqslant nh$, such that

$$\overline{K_t} = W_2^{(1)}([-h, 0], \boldsymbol{R}^n).$$

From the fact that the space $W_2^{(1)}([-h, 0], R^n)$ is dense in the topology of the space $L^2([-h, 0], R^n)$ it follows that $\overline{K_t^+}$, $t \geqslant nh$ has codimension equal to zero. Therefore by Lemma 4.3.5 the dynamical system (4.2.1) is $R^n(s)$-controllable. ∎

From the Corollary 4.3.8 it follows that the $R^n(s)$-controllability of the dynamical system (4.2.1) is the necessary condition for approximate controllability in finite time and hence also the necessary condition for exact controllability in finite time.

COROLLARY 4.3.9 (Bartosiewicz (1980), O'Connor and Tarn (1983), Olbrot and Jakubczyk (1981)). *If condition* (4.3.26) *is satisfied, then dynamical system* (4.2.1) *is $R^n(s)$-controllable and the attainable set K_t is closed for $t \geqslant nh$.*

Proof:

Corollary 4.3.9 follows immediately from Lemma 4.3.4 and 4.3.5. ∎

A detailed analysis of the properties of the attainable set K_t for $t \in [0, nh]$ is given in Olbrot and Jakubczyk (1981) which contains also the canonical forms of the dynamical systems (4.2.1) derived with the aid of $R^n(s)$-controllability. For example the Luenberger canonical form for system (4.2.1) is presented.

The last type of controllability considered in this subsection is the R^n-controllability (called also *Euclidean controllability*), defined in Definition 4.2.5. Using the general response formula (4.2.2) for the dynamical system (4.2.1) we can formulate the necessary and sufficient comdition for R^n-controllability in $[0, t_1]$.

LEMMA 4.3.6 (Jacobs and Langenhop (1975, 1976)). *The dynamical system* (4.2.1) *is R^n-controllable in* $[0, t_1]$ *if and only if*

$$\operatorname{rank}\mathbf{W}(t_1) = n, \tag{4.3.32}$$

where $\mathbf{W}(t_1)$ *is an* $(n \times n)$*-dimensional, symmetric matrix of the form*

$$\mathbf{W}(t_1) = \int_0^{t_1} \mathbf{F}(t-\tau)\mathbf{B}\mathbf{B}^{\mathrm{T}}\mathbf{F}^{\mathrm{T}}(t-\tau)\,d\tau = \int_0^{t_1} \mathbf{F}(t)\mathbf{B}\mathbf{B}^{\mathrm{T}}\mathbf{F}^{\mathrm{T}}(t)\,dt \tag{4.3.33}$$

and the fundamental matrix $\mathbf{F}(t)$ *is defined by formula* (4.2.3).

The main difficulty in applying Lemma 4.3.6 is the computation of the fundamental matrix $\mathbf{F}(t)$. In order to avoid this difficulty we shall use the following lemma.

LEMMA 4.3.7 (Gabasov and Kirilova (1971), Manitius (1976a)). *The dynamical system* (4.2.1) *is R^n-controllable in* $[0, t_1]$ *if and only if*

$$\mathbf{q}^{\mathrm{T}}\mathbf{F}(t)\mathbf{B} = 0 \quad \text{for } \mathbf{q} \in R^n,\ t \in [0, t_1] \Rightarrow \mathbf{q} = 0. \tag{4.3.34}$$

Proof:

Condition (4.3.32) is satisfied if and only if the rows of the matrix $\mathbf{F}(t)\mathbf{B}$ are linearly independent in $[0, t_1]$. This is equivalent to condition (4.3.34) and hence Lemma 4.3.7 follows from Lemma 4.3.6. ∎

Applying Lemma 4.3.7, we can formulate a necessary and sufficient condition for the R^n-controllability of the dynamical system (4.2.1) based on the matrices $\mathbf{A}_{-1}, \mathbf{A}_0, \mathbf{A}_1$ and $\mathbf{B}$ and not requiring the knowledge of the fundamental matrix $\mathbf{F}(t)$. In order to obtain this condition, let us first define the so-called *defining equation* (Gabasov and Kirilova (1971), Manitius (1976a)) for the dynamical system (4.2.1):

$$\dot{\mathbf{Q}}_k(t) = \mathbf{A}_{-1}\mathbf{Q}_k(t-h)+\mathbf{A}_0\mathbf{Q}_{k-1}(t)+\mathbf{A}_1\mathbf{Q}_{k-1}(t-h), \qquad (4.3.35)$$
$$k = 1, 2, \ldots, t \in [0, \infty)$$

with the initial conditions

$$\mathbf{Q}_0(t) = \begin{cases} \mathbf{B} & \text{for } t = 0, \\ 0 & \text{for } t > 0, \end{cases}$$
$$\mathbf{Q}_{-1}(t) = \mathbf{Q}_k(-h) = 0 \quad \text{for } k = 0, 1, 2, \ldots, t = 0, h, 2h, \ldots$$

For simplicity let us introduce the following notation:

$$\tilde{\mathbf{Q}}_n(t_1) = \{\mathbf{Q}_0(t), \mathbf{Q}_1(t), \mathbf{Q}_2(t), \ldots, \mathbf{Q}_{n-1}(t):\ t \in [0, t_1)\}. \qquad (4.3.36)$$

THEOREM 4.3.4 (Gabasov and Kirilova (1971), Manitius (1976a)). *The dynamical system* (4.2.1) *is R^n-controllable in* $[0, t_1]$ *if and only if*

$$\operatorname{rank}\tilde{\mathbf{Q}}_n(t_1) = n. \qquad (4.3.37)$$

Proof:

Sufficient condition. Let us consider the matrix function $\mathbf{G}(t) = \mathbf{F}(t)\mathbf{B}$ for $t \in [0, t_1]$. By (4.2.3) the function $\mathbf{G}(t)$ is piecewise l times differentiable for $l = 1, 2, 3, \ldots$ The only isolated points at which derivatives do not exist are the points, $t = \{t_1, t_1-h, t_1-2h, \ldots, t_1-jh\}$, where j is an integer satisfying the inequality $t_1-jh > 0$. Let

$$\Delta\mathbf{G}^{(k)}(t) = \mathbf{G}^{(k)}(t-0)-\mathbf{G}^{(k)}(t+0), \quad t \in [0, t_1], \qquad (4.3.38)$$

where $\mathbf{G}^{(k)}(t-0)$ and $\mathbf{G}^{(k)}(t+0)$ denote, respectively, the left-hand side and the right-hand side limits of the k-th derivatives of the matrix function $\mathbf{G}(t)$ at the points $t \in [0, t_1]$ for $k = 0, 1, 2, \ldots$ From (4.2.3) and (4.3.38) we obtain the formula

$$\Delta\mathbf{G}^0(t) = \begin{cases} 0 & \text{for } t \neq t_1-jh > 0, \\ \mathbf{A}^j\mathbf{B} & \text{for } t = t_1-jh > 0, \end{cases}$$

where the jump at the point t_1 follows from the noncontinuity of the fundamental matrix $\mathbf{F}(t)$ for $t = t_1-jh$. In a similar way, by induction, we derive the formulas

$$\Delta \mathbf{G}^{(k)}(t_1 - jh) = (-1)^k \mathbf{Q}_k(jh) \tag{4.3.39}$$
$$\text{for } j: t_1 - jh > 0 \text{ and } k = 1, 2, \ldots$$

The next part of the proof of the sufficient condition is given by contradiction. Let us assume that condition (4.3.37) is satisfied, but the dynamical system (4.2.1) is not $\mathbf{R}^n$-controllable in $[0, t_1]$. Hence by Lemma 4.3.7 there exists a non-zero vector $\mathbf{q} \in \mathbf{R}^n$ such that $\mathbf{q}^{\mathrm{T}}\mathbf{G}(t) = 0$ for $t \in [0, t_1]$. Therefore from (4.3.39) we obtain the equalities

$$0 = \mathbf{q}^{\mathrm{T}} \Delta \mathbf{G}^{(k)}(t_1 - jh) = (-1)^k \mathbf{q}^{\mathrm{T}} \mathbf{Q}_k(jh),$$
$$k = 0, 1, 2, \ldots, \quad j = 0, 1, 2, \ldots, \quad t_1 - jh > 0.$$

Hence vector $\mathbf{q}$ is orthogonal to each of the columns taken from the matrix set $\tilde{\mathbf{Q}}_n(t_1)$. This statement leads to a contradiction of the assumption of lack of $\mathbf{R}^n$-controllability of the dynamical system (4.2.1). Hence our sufficient condition follows.

Necessary condition. Proof by contradiction. Let us assume, that the dynamical system (4.2.1) is $\mathbf{R}^n$-controllable in $[0, t_1]$ but condition (4.3.37) does not hold. Therefore, from Lemma 4.3.7 it follows, that there exists a nonzero vcetor $\mathbf{q} \in \mathbf{R}^n$, such that

$$\mathbf{q}^{\mathrm{T}} \mathbf{Q}_i(t) = 0 \quad \text{for } i = 0, 1, 2, \ldots, \ t \in [0, t_1]. \tag{4.3.40}$$

Since the matrix function $\mathbf{G}(t)$ is piecewise analytic in $[0, t_1]$, (i.e. it is analytic except at a finite number of isolated points $t_1,\ t_1 - h,\ t_1 - 2h, \ldots, t_1 - jh > 0$), it follows from (4.3.38) and (4.3.39) that

$$\mathbf{q}^{\mathrm{T}} \mathbf{G}(t) = 0 \quad \text{for } t \in [0, t_1].$$

A detailed proof of the above statement requires, mathematical induction. Therefore, by Lemma 4.3.7, we obtain a contradiction, and hence our theorem follows. ∎

In considering $\mathbf{R}^n$-controllability for linear, time-invariant dynamical systems without delays the length of the time interval is inessential. However, for dynamical systems with delays (4.2.1) the situation is quite different. The length of the time interval $[0, t_1]$ plays an important role in investigating $\mathbf{R}^n$-controllability. This immediately follows from the expressions (4.3.36) and (4.3.37). For example for $t_1 \leqslant h$, from the defining equation (4.3.35) we have

$$\tilde{\mathbf{Q}}_n(t_1) = \{\mathbf{B}, \mathbf{A}_0\mathbf{B}, \mathbf{A}_0^2\mathbf{B}, \ldots, \mathbf{A}_0^{n-1}\mathbf{B}\}.$$

Hence, by condition (4.3.37) and Theorem 4.3.4, the $\mathbf{R}^n$-controllability of the dynamical system (4.2.1) is in this case equivalent to the controllability of a dynamical system without delays.

In increasing the length of the time interval $[0, t_1]$ the set of matrices $\tilde{\mathbf{Q}}_n(t_1)$ also increases. The additional elements of the set $\tilde{\mathbf{Q}}_n(t_1)$ depend on the matrices $\mathbf{A}_{-1}$ and $\mathbf{A}_1$.

It follows from these considerations that the dynamical system (4.2.1) which is not $\mathbf{R}^n$-controllable in a certain time interval $[0, t_1]$ may became $\mathbf{R}^n$-controllable in a certain longer time interval $[0, t_2]$, where $t_1 \leqslant t_2$. This property distinguishes linear, time-invariant dynamical systems with delays from linear, time-invariant dynamical systems without delays.

Now let us consider the special case of the dynamical system (4.2.11) where the matrix $\mathbf{A}_{-1} = 0$. In this case the defining equation (4.3.35) has the form

$$\mathbf{Q}_k(t) = \mathbf{A}_0 \mathbf{Q}_{k-1}(t) + \mathbf{A}_1 \mathbf{Q}_{k-1}(t-h), \qquad (4.3.41)$$
$$k = 1, 2, 3, \ldots, \quad t \in [0, \infty),$$

with the initial conditions

$$\mathbf{Q}_0(t) = \begin{cases} \mathbf{B} & \text{for } t = 0, \\ 0 & \text{for } t \neq 0. \end{cases}$$

The sequence of matrices $\mathbf{Q}_k(t)$ derived from equation (4.3.41) has the following form (Manitius (1976a)):

$$\mathbf{Q}_0(0) = \mathbf{B}, \quad \mathbf{Q}_0(jh) = 0 \quad \text{for } j = 1, 2, 3, \ldots$$
$$\mathbf{Q}_1(0) = \mathbf{A}_0\mathbf{B}, \quad \mathbf{Q}_1(h) = \mathbf{A}_1\mathbf{B}, \quad \mathbf{Q}_1(jh) = 0 \quad \text{for } j = 2, 3, 4, \ldots,$$
$$\mathbf{Q}_2(0) = \mathbf{A}_0^2\mathbf{B}, \quad \mathbf{Q}_2(h) = (\mathbf{A}_0\mathbf{A}_1 + \mathbf{A}_1\mathbf{A}_0)\mathbf{B}, \quad \mathbf{Q}_2(2h) = \mathbf{A}_2^1\mathbf{B},$$
$$\mathbf{Q}_2(jh) = 0 \quad \text{for } j = 3, 4, \ldots$$

The R^n-controllability in $[0, t_1]$ of the dynamical system (4.2.1) follows from exact controllability in time interval $[0, t_1+h]$. It is a consequence of Definitions 4.2.4 and 4.2.5. Definition 4.2.5 implies also that the R^n-controllability in $[0, t_1]$ of the dynamical system (4.2.1) implies its Euclidean controllability to zero. The inverse statement, however is not true. It is closely dependent on the so-called *degeneracy* of the dynamical system (see Manitius (1976a) for details).

4.4 Examples

Example 4.4.1

Consider the dynamical system (4.2.1) with the matrices $\mathbf{A}_{-1}, \mathbf{A}_0, \mathbf{A}_1, \mathbf{B}$ of the following form:

$$\mathbf{A}_{-1} = \begin{bmatrix} 0 & 1 \\ 0 & 0 \end{bmatrix}, \quad \mathbf{A}_0 = \begin{bmatrix} 0 & 0 \\ 0 & 0 \end{bmatrix}, \quad \mathbf{A}_1 = \begin{bmatrix} 1 & 0 \\ 0 & 0 \end{bmatrix}, \quad \mathbf{B} = \begin{bmatrix} 0 \\ 1 \end{bmatrix}. \qquad (4.4.1)$$

In order to investigate approximate controllability in finite time for the dynamical system (4.4.1), we first verify condition (4.3.7) in Theorem 4.3.1

$$\operatorname{rank}_C[s\mathbf{A}_{-1} + \mathbf{A}_1 \vdots \mathbf{B}] = \operatorname{rank}_C \left[\begin{array}{cc:c} 1 & s & 0 \\ 0 & 0 & 1 \end{array}\right] = 2 = n.$$

Next, we verify condition (4.3.8) for spectral controllability. In order to do that we first compute

$$[\boldsymbol{\Delta}(s) \vdots \mathbf{B}] = \left[\begin{array}{cc:c} s - e^{-sh} & -se^{-sh} & 0 \\ 0 & s & 1 \end{array}\right].$$

Considering all (2×2)-dimensional minors, we can easily check, that $\operatorname{rank}[\boldsymbol{\Delta}(s_0) \vdots \mathbf{B}] = 1 < 2 = n$, if and only if $s_0 \in \boldsymbol{C}$ satisfies simultaneously the following two equations:

$$s_0 - e^{-s_0 h} = 0,$$

$$s_0 e^{-s_0 h} = 0.$$

Since the above set of equations does not have a solution, we have for any $s \in \boldsymbol{C}$, $\operatorname{rank}[\boldsymbol{\Delta}(s) \vdots \mathbf{B}] = 2 = n$. Therefore Theorem 4.3.1 implies that the dynamical system (4.4.1) is approximately controllable in finite time and hence by Lemma 4.3.1 spectrally controllable and by Corollary 4.3.8 also $\boldsymbol{R}^2(s)$-controllable. Since

$$\operatorname{rank}[\mathbf{B} \vdots \mathbf{A}_1 \mathbf{B}] = \operatorname{rank}\begin{bmatrix} 0 & 1 \\ 1 & 0 \end{bmatrix} = 2 = n,$$

by Theorem 4.3.2 the dynamical system (4.4.1) is exactly controllable in finite time. However, the dynamical system (4.4.1) is not spectrally complete, because

$$\operatorname{rank}_{\boldsymbol{C}}[s\mathbf{A}_{-1} + \mathbf{A}_1] = \operatorname{rank}_{\boldsymbol{C}}\begin{bmatrix} 1 & s \\ 0 & 0 \end{bmatrix} = 1 < 2 = n.$$

Thus it is an example of a dynamical system which is exactly controllable in finite time, but simultaneously is not spectrally complete.

Example 4.4.2
Consider the dynamical system (4.2.1) with the matrices $\mathbf{A}_{-1}, \mathbf{A}_0, \mathbf{A}_1, \mathbf{B}$ of the following form:

$$\mathbf{A}_{-1} = \begin{bmatrix} 0 & 0 \\ 0 & 1 \end{bmatrix}, \quad \mathbf{A}_0 = \begin{bmatrix} 0 & 0 \\ 1 & 0 \end{bmatrix}, \quad \mathbf{A}_1 = \begin{bmatrix} 0 & 0 \\ -1 & 0 \end{bmatrix}, \quad \mathbf{B} = \begin{bmatrix} 1 \\ 0 \end{bmatrix}. \tag{4.4.2}$$

In order to investigate approximate controllability in finite time for the dynamical system (4.4.2), we first verify condition (4.3.7) in Theorem 4.3.1:

$$\operatorname{rank}_{\boldsymbol{C}}[s\mathbf{A}_{-1} + \mathbf{A}_1 \vdots \mathbf{B}] = \operatorname{rank}_{\boldsymbol{C}}\left[\begin{array}{cc:c} 0 & 0 & 1 \\ -1 & s & 0 \end{array}\right] = 2 = n.$$

Next we check condition (4.3.8) for spectral controllability. In order to do that we first compute

$$[\boldsymbol{\Delta}(s) \vdots \mathbf{B}] = \begin{bmatrix} s & 0 & 1 \\ e^{-sh} - 1 & s(1 - e^{-sh}) & 0 \end{bmatrix}.$$

Since for $s_0 = 0$ we have $\operatorname{rank}[\boldsymbol{\Delta}(s_0) \vdots \mathbf{B}] = \operatorname{rank}\left[\begin{array}{cc:c} 0 & 0 & 1 \\ 0 & 0 & 0 \end{array}\right] = 1 < 2 = n$, by Theorem 4.3.1 and Lemma 4.3.1 the dynamical system (4.4.2) is not spectrally controllable, and it is not approximately controllable in finite time either. In order to verify the

$R^2(s)$-controllability of the dynamical system (4.4.2) we determine, using formulas (4.2.23) and (4.2.24) the matrices $\mathbf{F}(s)$ and $\mathbf{G}(s)$

$$\mathbf{F}(s) = (s\mathbf{I}-\mathbf{A}_0)^{-1}(s\mathbf{A}_{-1}+\mathbf{A}_1) = \begin{bmatrix} s & 0 \\ -1 & s \end{bmatrix}^{-1} \begin{bmatrix} 0 & 0 \\ -1 & s \end{bmatrix}$$

$$= \begin{bmatrix} 1/s & 0 \\ 1/s^2 & 1/s \end{bmatrix} \begin{bmatrix} 0 & 0 \\ -1 & 1 \end{bmatrix} = \begin{bmatrix} 0 & 0 \\ -1/s & 1 \end{bmatrix},$$

$$\mathbf{G}(s) = (s\mathbf{I}-A_0)^{-1}\mathbf{B} = \begin{bmatrix} s & 0 \\ -1 & s \end{bmatrix}^{-1} \begin{bmatrix} 1 \\ 0 \end{bmatrix} = \begin{bmatrix} 1/s & 0 \\ 1/s^2 & 1/s \end{bmatrix} \begin{bmatrix} 1 \\ 0 \end{bmatrix} = \begin{bmatrix} 1/s \\ 1/s^2 \end{bmatrix}.$$

Hence

$$\text{rank}_{R(s(}[\mathbf{G}(s) \vdots \mathbf{F}(s)\mathbf{G}(s)] = \text{rank}_{R(s)} \begin{bmatrix} 1/s & 0 \\ 1/s^2 & 0 \end{bmatrix} = 1 < 2 = n.$$

Therefore, by Theorem 4.3.3 the dynamical system (4.4.2) is not $R^2(s)$-controllable.

It should also be stressed that the necessary condition for approximate controllability in finite time given in Lemma 4.3.3 is satisfied in this case, namely,

$$\text{adj}\, \mathbf{\Delta}(s) = \mathbf{P}_0(s)+e^{-sh}\mathbf{P}_1(s) = \begin{bmatrix} s & 0 \\ 1-e^{-sh} & s(e^{-th}-1) \end{bmatrix}$$

$$= \begin{bmatrix} s & 0 \\ 1 & -s \end{bmatrix} + e^{-sh} \begin{bmatrix} 0 & 0 \\ -1 & s \end{bmatrix}.$$

Hence

$$\mathbf{P}_0(s) = \begin{bmatrix} s & 0 \\ 1 & -s \end{bmatrix}, \quad \mathbf{P}_1(s) = \begin{bmatrix} 0 & 0 \\ -1 & s \end{bmatrix}.$$

By formula (4.3.1) the polynomial matrix $\mathbf{P}(s)$ is of the form

$$\mathbf{P}(s) = [\mathbf{P}_0(s)\mathbf{B} \vdots \mathbf{P}_1(s)\mathbf{B}] = \left[\begin{array}{c:c} s & 0 \\ 1 & -1 \end{array}\right].$$

Therefore $\text{rank}_C\mathbf{P}(s) = 2 = n$, and hence the necessary condition for the approximate controllability in finite time of the dynamical system (4.4.2) is satisfied, but this system is not approximately controllable in finite time. Moreover, it is an example of a dynamical system which is not $R^2(s)$-controllable either. However, this system is R^2-controllable in $[0, t_1]$ for any $t_1 > 0$. Indeed, for $t_1 \leqslant h$ the matrix $\tilde{\mathbf{Q}}_2(t_1)$ is by (4.3.36) of the form

$$\tilde{\mathbf{Q}}_2(t_1) = [\mathbf{B} \vdots \mathbf{A}_0\mathbf{B}] = \left[\begin{array}{c:c} 1 & 0 \\ 0 & 1 \end{array}\right].$$

Therefore $\text{rank}\, \tilde{\mathbf{Q}}_2(t_1) = 2 = n$, and by Theorem 4.3.4 dynamical system (4.4.2) is R^2-controllable in $[0, t_1]$ for any $t_1 > 0$. This example shows that even R^n-controllability in any interval does not imply approximate controllability in finite time.

Example 4.4.3

Consider the dynamical system (4.2.1) with the matrices $\mathbf{A}_{-1}, \mathbf{A}_0, \mathbf{A}_1, \mathbf{B}$ of the following form:

$$\mathbf{A}_{-1} = \begin{bmatrix} 0 & 0 \\ 0 & 1 \end{bmatrix}, \quad \mathbf{A}_0 = \begin{bmatrix} 0 & 0 \\ 1 & 0 \end{bmatrix}, \quad \mathbf{A}_1 = \begin{bmatrix} 0 & 0 \\ -1 & 0 \end{bmatrix}, \quad \mathbf{B} = \begin{bmatrix} 1 \\ 1 \end{bmatrix}. \qquad (4.4.3)$$

In order to investigate approximate controllability in finite time for the dynamical system (4.4.3), we first verify the condition (4.3.7) in Theorem 4.3.1

$$\operatorname{rank}_C[s\mathbf{A}_{-1} + \mathbf{A}_1 \vdots \mathbf{B}] = \operatorname{rank}_C \left[\begin{array}{cc:c} 0 & 0 & 1 \\ -1 & s & 1 \end{array}\right] = 2 = n.$$

Next we verify condition (4.3.8) for spectral controllability:

$$[\Delta(s) \vdots B] = \left[\begin{array}{cc:c} s & 0 & 1 \\ -1+e^{-sh} & s-se^{-sh} & 1 \end{array}\right] =$$

Since, for $s_0 = 0$, $\operatorname{rank}[\Delta(s_0) \vdots \mathbf{B}] = \operatorname{rank} \left[\begin{array}{cc:c} 0 & 0 & 1 \\ 0 & 0 & 1 \end{array}\right] = 1 < 2 = n$, by Lemma 4.3.1 and Theorem 4.3.1 the dynamical system (4.4.3) is not spectrally controllable and it is not approximately controllable in finite time either.

Next we check $R^2(s)$-controllability. Using formulas (4.2.23) and (4.2.24), we derive matrices $\mathbf{F}(s)$ and $\mathbf{G}(s)$:

$$\mathbf{F}(s) = (s\mathbf{I} - \mathbf{A}_0)^{-1}(s\mathbf{A}_{-1} + \mathbf{A}_1) = \begin{bmatrix} s & 0 \\ -1 & s \end{bmatrix}^{-1} \begin{bmatrix} 0 & 0 \\ -1 & s \end{bmatrix}$$

$$= \begin{bmatrix} 1/s & 0 \\ 1/s^2 & 1/s \end{bmatrix} \begin{bmatrix} 0 & 0 \\ -1 & s \end{bmatrix} = \begin{bmatrix} 0 & 0 \\ -1/s & 1 \end{bmatrix},$$

$$\mathbf{G}(s) = (s\mathbf{I} - \mathbf{A}_0)^{-1}\mathbf{B} = \begin{bmatrix} s & 0 \\ -1 & s \end{bmatrix}^{-1} \begin{bmatrix} 1 \\ 1 \end{bmatrix} = \begin{bmatrix} 1/s & 0 \\ 1/s^2 & 1/s \end{bmatrix} \begin{bmatrix} 1 \\ 1 \end{bmatrix} = \begin{bmatrix} 1/s \\ 1/s^2 + 1/s \end{bmatrix}.$$

Hence

$$\operatorname{rank}_{R(s)}[(\mathbf{G}(s) \vdots \mathbf{F}(s)\mathbf{G}(s)] = \operatorname{rank}_{R(s)} \left[\begin{array}{c:c} 1/s & 0 \\ 1/s^2 + 1/s & 1/s \end{array}\right] = 2 = n.$$

Therefore by Theorem 4.3.3, the dynamical system (4.4.3) is $R^2(s)$-controllable. Since the dynamical system (4.4.3) is not approximately controllable in finite time, it is not exactly controllable in finite time either.

In order to check R^2-controllability, we first compute the set of matrices $\tilde{\mathbf{Q}}_2(t_1)$. Assuming $t_1 \leqslant h$, we obtain from (4.3.36) the following expression:

$$\tilde{\mathbf{Q}}_2(t_1) = [\mathbf{B} \vdots \mathbf{A}_0\mathbf{B}] = \left[\begin{array}{c:c} 1 & 0 \\ 1 & 1 \end{array}\right].$$

Therefore, $\operatorname{rank}\tilde{\mathbf{Q}}_2(t_1) = 2 = n$, and hence by Theorem 4.3.4, the dynamical system (4.4.3) is R^2-controllable in $[0, t_1]$ for any $t_1 > 0$.

Hence it is an example of dynamical system which is $\boldsymbol{R}^2$-controllable in every time interval $[0, t_1]$, $t_1 > 0$, and moreover, $\boldsymbol{R}^2(s)$-controllable, but is not approximately controllable in finite time.

Example 4.4.4

Consider the dynamical system (4.2.1) with the matrices $\mathbf{A}_{-1}, \mathbf{A}_0, \mathbf{A}_1, \mathbf{B}$ of the following form:

$$\mathbf{A}_{-1} = \begin{bmatrix} 0 & 0 \\ 1 & 0 \end{bmatrix}, \quad \mathbf{A}_0 = \begin{bmatrix} 0 & 1 \\ 0 & 0 \end{bmatrix}, \quad \mathbf{A}_1 = \begin{bmatrix} 1 & 0 \\ 0 & 0 \end{bmatrix}, \quad \mathbf{B} = \begin{bmatrix} 0 \\ 1 \end{bmatrix}. \tag{4.4.4}$$

In order to investigate approximate controllability in finite time for the dynamical system (4.4.4), we first verify condition (4.3.7) in Theorem 4.3.1:

$$\operatorname{rank}_C[s\mathbf{A}_{-1} + \mathbf{A}_1 \vdots \mathbf{B}] = \operatorname{rank}\begin{bmatrix} 1 & 0 & \vdots & 0 \\ s & 0 & \vdots & 1 \end{bmatrix} = 2 = n.$$

Next we check condition (4.3.8) for spectral controllability

$$[\boldsymbol{\Delta}(s) \vdots \mathbf{B}] = \begin{bmatrix} s - e^{-sh} & -1 & \vdots & 0 \\ -se^{-sh} & s & \vdots & 1 \end{bmatrix}.$$

Taking into account the last two columns of the above matrix, we can easily verify, that $\operatorname{rank}[\boldsymbol{\Delta}(s) \vdots \mathbf{B}] = 2 = n$ for every $s \in \boldsymbol{C}$. Hence by Theorem 4.3.1, the dynamical system (4.4.4) is approixmately controllable in finite time, and so, by Lemma 4.3.1, it is also spectrally controllable. Furthermore, by Corollary 4.3.8, it is $\boldsymbol{R}^2(s)$-controllable. Since

$$\operatorname{rank}[\mathbf{B} \vdots \mathbf{A}_{-1}\mathbf{B}] = \operatorname{rank}\begin{bmatrix} 0 & \vdots & 0 \\ 1 & \vdots & 0 \end{bmatrix} = 1 < 2 = n,$$

hence by Theorem 4.3.2 the dynamical system (4.4.4) is not exactly controllable in finite time. Moreover, since

$$\operatorname{rank}_C[s\mathbf{A}_{-1} + \mathbf{A}_1] = \operatorname{rank}_C\begin{bmatrix} 1 & 0 \\ s & 0 \end{bmatrix} = 1 < 2 = n,$$

the dynamical system (4.4.4) is not spectrally complete.

In order to verify $\boldsymbol{R}^2$-controllability, we first compute the set of matrices $\tilde{\mathbf{Q}}_2(t_1)$. Assuming $t_1 \leqslant h$, we obtain by formula (4.3.36) the following equalities:

$$\tilde{\mathbf{Q}}_2(t_1) = [\mathbf{B} \vdots \mathbf{A}_0\mathbf{B}] = \begin{bmatrix} 0 & \vdots & 1 \\ 1 & \vdots & 0 \end{bmatrix}.$$

Therefore $\operatorname{rank}\tilde{\mathbf{Q}}_2(t_1) = 2 = n$, and so, by Theorem 4.3.4 the dynamical system (4.4.4) is $\boldsymbol{R}^2$-controllable in $[0, t_1]$ for any $t_1 > 0$.

This is an example of a dynamical system which is approximately controllable in finite time, is not spectrally complete, is not exactly controllable in finite time, but is $\boldsymbol{R}^2$-controllable in $[0, t_1]$, for any $t_1 > 0$.

This example shows that Euclidean controllability of a dynamical system in an arbitrary time interval does not imply its exact controllability in finite time.

Example 4.4.5
Consider the dynamical system (4.2.1) with matrices $\mathbf{A}_{-1}, \mathbf{A}_0, \mathbf{A}_1, \mathbf{B}$ of the following form:

$$\mathbf{A}_{-1} = \begin{bmatrix} 1 & 0 \\ 0 & 1 \end{bmatrix}, \quad \mathbf{A}_0 = \begin{bmatrix} 0 & 1 \\ 0 & 0 \end{bmatrix}, \quad \mathbf{A}_1 = \begin{bmatrix} 0 & 0 \\ 0 & 0 \end{bmatrix}, \quad \mathbf{B} = \begin{bmatrix} 0 \\ 1 \end{bmatrix}. \tag{4.4.5}$$

In order to investigate approximate controllability in finite time for the dynamical system (4.4.5), we first verify condition (4.3.7) in Theorem 4.3.1

$$\text{rank}_{\mathbf{C}}[s\mathbf{A}_{-1} + \mathbf{A}_1 \vdots \mathbf{B}] = \text{rank}_{\mathbf{C}} \left[\begin{array}{cc:c} s & 0 & 0 \\ 0 & s & 1 \end{array}\right] = 2 = n.$$

Next we check spectral controllability by using condition (4.3.8)

$$[\boldsymbol{\Delta}(s) \vdots \mathbf{B}] = \left[\begin{array}{cc:c} s - se^{-sh} & -1 & 0 \\ 0 & s - se^{-sh} & 1 \end{array}\right].$$

Taking into account the last two columns of the above matrix, we can easily check that $\text{rank}[\boldsymbol{\Delta}(s) \vdots \mathbf{B}] = 2 = n$ for any $s \in \mathbf{C}$. Therefore, by Theorem 4.3.1, the dynamical system (4.4.5) is approximately controllable in finite time and so by Lemma 4.3.1, also spectrally controllable and, by Corollary 4.3.8, $\boldsymbol{R}^2(s)$-controllable. Since

$$\text{rank}[\mathbf{B} \vdots \mathbf{A}_{-1}\mathbf{B}] = \text{rank}\left[\begin{array}{c:c} 0 & 0 \\ 1 & 1 \end{array}\right] = 1 < 2 = n,$$

by Theorem 4.3.2 the dynamical system (4.4.5) is not exactly controllable in finite time. However, the dynamical system (4.4.5) is spectrally complete, which immediately follows from the Lemma 4.3.2 and the equalities

$$\text{rank}_{\mathbf{C}}[s\mathbf{A}_{-1} + \mathbf{A}_1] = \text{rank}_{\mathbf{C}} \begin{bmatrix} s & 0 \\ 0 & s \end{bmatrix} = 2 = n.$$

In order to establish $\boldsymbol{R}^2$-controllability for the dynamical system (4.4.5), it is necessary first to compute $\mathbf{Q}_2(t_1)$. Assuming $t_1 \leqslant h$, we have by (4.3.36),

$$\tilde{\mathbf{Q}}_2(t_1) = [\mathbf{B} \vdots \mathbf{A}_0\mathbf{B}] = \left[\begin{array}{c:c} 0 & 1 \\ 1 & 0 \end{array}\right].$$

Hence $\text{rank}\tilde{\mathbf{Q}}_2(t_1) = 2 = n$, and so, by Theorem 4.3.4, the dynamical system (4.4.5) is $\boldsymbol{R}^2$-controllable in $[0, t_1]$ for any $t_1 > 0$.

This is an example of a dynamical system which is spectrally controllable, $\boldsymbol{R}^2(s)$-controllable, approximately controllable in finite time and $\boldsymbol{R}^2$-controllable in any time interval, but is not exactly controllable in finite time. It should also be pointed out that the dynamical system (4.4.5) is spectrally complete.

This example shows that exact controllability of a dynamical system is the "strongest" of all types of controllability.

Example 4.4.6
Consider a dynamical system with matrices $\mathbf{A}_{-1}, \mathbf{A}_0, \mathbf{A}_1, \mathbf{B}$ of the following form:

$$\mathbf{A}_{-1} = \begin{bmatrix} 0 & 1 \\ 0 & 0 \end{bmatrix}, \quad \mathbf{A}_0 = \begin{bmatrix} 0 & 0 \\ -1 & 0 \end{bmatrix}, \quad \mathbf{A}_1 = \begin{bmatrix} 1 & 0 \\ 0 & 0 \end{bmatrix}, \quad \mathbf{B} = \begin{bmatrix} 0 \\ 1 \end{bmatrix}. \quad (4.4.6)$$

In order to investigate approximate controllability for the dynamical system (4.4.6) we first verify the condition (4.3.7) in Theorem 4.3.1:

$$\operatorname{rank}_C[s\mathbf{A}_{-1} + \mathbf{A}_1 \vdots \mathbf{B}] = \operatorname{rank}_C \left[\begin{array}{cc:c} 1 & s & 0 \\ 0 & 0 & 1 \end{array}\right] = 2 = n.$$

Next we check condition (4.3.8) for spectral controllability

$$[\boldsymbol{\Delta}(s) \vdots \mathbf{B}] = \left[\begin{array}{cc:c} s - e^{-sh} & -se^{-sh} & 0 \\ 1 & s & 1 \end{array}\right].$$

For $s = 0$ the first and third columns are linearly independent, and, similarly, for $s \neq 0$, $s \in C$, the second and third columns are linearly independent. Hence for any $s \in C$ we have $\operatorname{rank}[\boldsymbol{\Delta}(s)|\mathbf{B}] = 2 = n$, and so the dynamical system (4.4.6) is by Theorem 4.3.1 approximately controllable in finite time. Moreover, by Lemma 4.3.1, it is also spectrally controllable and, by Corollary 4.3.8, R^2-controllable. Since

$$\operatorname{rank}[\mathbf{B} \vdots \mathbf{A}_{-1}\mathbf{B}] = \operatorname{rank} \left[\begin{array}{c:c} 0 & 1 \\ 1 & 0 \end{array}\right] = 2 = n,$$

by Theorem 4.3.2 the dynamical system (4.4.6) is exactly controllable in finite time. However, dynamical system (4.4.6) is not spectrally complete. This follows immediately from the inequality

$$\operatorname{rank}_C[s\mathbf{A}_{-1} + \mathbf{A}_1] = \operatorname{rank}_C \begin{bmatrix} 1 & s \\ 0 & 0 \end{bmatrix} = 1 < 2 = n.$$

In order to check R^2-controllability in $[0, t_1]$, we compute $\tilde{\mathbf{Q}}_2(t_1)$. In the case where $t_1 \leqslant h$, we have by (4.3.36)

$$\tilde{\mathbf{Q}}_2(t_1) = [\mathbf{B} \vdots \mathbf{A}_0\mathbf{B}] = \left[\begin{array}{c:c} 0 & 0 \\ 1 & 0 \end{array}\right].$$

Therefore $\operatorname{rank} \tilde{\mathbf{Q}}_2(t_1) = 1 < 2 = n$, and so dynamical system (4.4.6) is not R^2-controllable for $t_1 \leqslant h$. Now let us consider the time interval $[0, t_1]$ for $t_1 \in (h, 2h]$. By (4.3.36) and the defining equation (4.3.35) we have the following equalities:

$$\begin{aligned} \operatorname{rank} \tilde{\mathbf{Q}}_2(t_1) &= \operatorname{rank}[\mathbf{B} \vdots \mathbf{A}_0\mathbf{B} \vdots \mathbf{A}_{-1}\mathbf{B} \vdots (\mathbf{A}_1 + \mathbf{A}_{-1}\mathbf{A}_0 + \mathbf{A}_0\mathbf{A}_{-1})\mathbf{B}] \\ &= \operatorname{rank} \left[\begin{array}{c:c:c:c} 0 & 0 & 1 & 0 \\ 1 & 0 & 0 & -1 \end{array}\right] = 2 = n. \end{aligned}$$

Therefore, by Theorem 4.3.4, the dynamical system (4.4.6) is R^2-controllable in $[0, t_1]$ for any $t_1 > h$.

This is an example of a dynamical system which is exactly controllable in finite time but is not spectrally complete and, moreover, is $\boldsymbol{R}^2$-controllable in $[0, t_1]$ for any $t_1 > h$.

Example 4.4.7

Consider the dynamical system (4.2.1) with the matrices $\mathbf{A}_{-1}, \mathbf{A}_0, \mathbf{A}_1, \mathbf{B}$ of the following form:

$$\mathbf{A}_{-1} = \begin{bmatrix} 0 & 1 & 0 \\ 0 & 0 & 1 \\ 0 & 0 & 0 \end{bmatrix}, \quad \mathbf{A}_0 = \begin{bmatrix} 1 & 0 & 0 \\ 0 & 1 & 0 \\ 0 & 0 & 1 \end{bmatrix}, \quad \mathbf{A}_1 = \begin{bmatrix} -1 & 1 & 0 \\ 1 & 0 & 1 \\ 0 & 1 & 1 \end{bmatrix},$$

$$\mathbf{B} = \begin{bmatrix} 0 & 0 \\ 1 & 0 \\ 0 & 1 \end{bmatrix}. \tag{4.4.7}$$

In order to verify approximate controllability in finite time for the dynamical system (4.4.7), we first check condition (4.3.7) in Theorem 4.3.1:

$$\mathrm{rank}_C[s\mathbf{A}_{-1}+\mathbf{A}_1\mathbf{B}] = \mathrm{rank}_C\left[\begin{array}{ccc|cc} -1 & s+1 & 0 & 0 & 0 \\ 1 & 0 & s+1 & 1 & 0 \\ 0 & 1 & 1 & 0 & 1 \end{array}\right] = 3 = n.$$

Next we verify condition (4.3.8) for spectral controllability

$$[\boldsymbol{\Delta}(s)\vdots\mathbf{B}] = \left[\begin{array}{ccc|cc} s+e^{-sh}-1 & -(s+1)e^{-sh} & 0 & 0 & 0 \\ -e^{-sh} & s-1 & -(s+1)e^{-sh} & 1 & 0 \\ 0 & -e^{-sh} & s-e^{-sh}-1 & 0 & 1 \end{array}\right].$$

Considering all (3×3)-dimensional minors, we can easily verify that $\mathrm{rank}[\boldsymbol{\Delta}(s_0)\vdots\mathbf{B}$ $= 3 = n$ if and only if $s_0 \in \boldsymbol{C}$ satisfies simultaneously the following two equations:

$$s_0+e^{-s_0h}-1 = 0,$$
$$s_0+1 = 0.$$

The above set of equations has a solution only for $h = \ln 2$. Therefore, for $h \neq \ln 2$, the dynamical system (4.4.7) is, by Theorem 4.3.1, approximately controllable in finite time and so, by Lemma 4.3.1, it is also spectrally controllable and, by Corollary 4.3.8, $\boldsymbol{R}^3(s)$-controllable. However, for $h = \ln 2$ the dynamical system (4.4.7) is not spectrally controllable and thus it is not approximately controllable in finite time either. Since

$$\mathrm{rank}_C[s\mathbf{A}_{-1}+\mathbf{A}_1] = \mathrm{rank}_C\begin{bmatrix} -1 & s+1 & 0 \\ 1 & 0 & s+1 \\ 0 & 1 & 1 \end{bmatrix} = 3 = n,$$

the dynamical system (4.4.7) is, by Lemma 4.3.2, spectrally complete. Moreover, for $h \neq \ln 2$ it is exactly controllable in finite time. This statement follows from Theorem 4.3.2 and the equalities

$$\operatorname{rank}[\mathbf{B} \vdots \mathbf{A}_{-1}\mathbf{B} \vdots \mathbf{A}_{-1}^2\mathbf{B}] = \operatorname{rank}\begin{bmatrix} 0 & 0 & 1 & 0 & 0 & 1 \\ 1 & 0 & 0 & 1 & 0 & 0 \\ 0 & 1 & 0 & 0 & 0 & 0 \end{bmatrix} = 3 = n.$$

Since for $t_1 \leqslant h$ we have

$$\operatorname{rank}\tilde{\mathbf{Q}}_3(t_1) = \operatorname{rank}[\mathbf{B} \vdots \mathbf{A}_0\mathbf{B} \vdots \mathbf{A}_0^2\mathbf{B}] = \operatorname{rank}\begin{bmatrix} 0 & 0 & 0 & 0 & 0 & 0 \\ 1 & 0 & 1 & 0 & 1 & 0 \\ 0 & 1 & 0 & 1 & 0 & 1 \end{bmatrix}$$
$$= 2 < 3 = n,$$

by Theorem 4.3.4 the dynamical system (4.4.7) is not $\boldsymbol{R}^2$-controllable in $[0, t_1]$ for $t_1 \leqslant h$. It is easy to verify, that the dynamical system (4.4.7) is $\boldsymbol{R}^3$-controllable in $[0, t_1]$ for any $t_1 > h$ and an arbitrary delay $h > 0$.

It is thus an example of a dynamical system for which approximate controllability and exact controllability in finite time are essentially dependent on the delay h.

4.5 Topological properties of the set of controllable systems

In practical applications it is important to know how the properties of a dynamical system change with the change of its parameters. In this subsection we shall consider the above problem in relation to various kinds of controllability of the dynamical system (4.2.1). Some special cases will also be dealt with.

First of all, let us map the set of dynamical systems of the form (4.2.1) into the space of parameters, i.e., the Euclidean space $\boldsymbol{R}^{3n^2+mn}$. For a fixed delay $h > 0$, every dynamical system (4.2.1) is completely characterized by a quadruple of real matrices $(\mathbf{A}_{-1}, \mathbf{A}_0, \mathbf{A}_1, \mathbf{B}) \in \boldsymbol{R}^{3n^2+mn}$, and represents one concrete point in the space $\boldsymbol{R}^{3n^2+mn}$. In this space the norm is defined as follows (Jacobs and Langenhop (1976)):

$$||L||_{mn} = ||\mathbf{A}_{-1}|| + ||\mathbf{A}_0|| + ||\mathbf{A}_1|| + ||\mathbf{B}||, \tag{4.5.1}$$

where the symbol L denotes a dynamical system of the form (4.2.1) with the matrices $\mathbf{A}_{-1}, \mathbf{A}_0, \mathbf{A}_1, \mathbf{B}$ and $||\cdot||$ denotes the common norm of the matrix. The space $\boldsymbol{R}^{3n^2+mn}$ with the norm given by (4.5.1) will be denoted by S_{mn} and identified with the space of all dynamical systems (4.2.1) (where the dimensions n and m are assumed to be fixed).

THEOREM 4.5.1 (Jacobs and Langenhop (1976)). *The set of dynamical systems* (4.2.1) *which are exactly controllable in* $[0, t_1]$, $t_1 > nh$, *is open and dense in the space* S_{mn}.

Proof:

By Lemma 4.3.4 and Theorem 4.3.2 the dynamical system (4.2.1) which is exactly controllable in finite time (or equivalently in any time interval $[0, t_1]$, $t_1 > nh$), must satisfy simultaneously the following conditions (O'Connor and Tarn (1983)):

$$\text{rank}[\mathbf{B} \vdots \mathbf{A}_{-1}\mathbf{B} \vdots \mathbf{A}_1\mathbf{B} \vdots \ldots \vdots \mathbf{A}_{-1}^{n-1}\mathbf{B}] = n, \tag{4.5.2}$$

$$\text{rank}[\mathbf{\Delta}(s) \vdots \mathbf{B}] = n \quad \text{for every } s \in \mathbf{C}. \tag{4.5.3}$$

By Theorem 1.4.1 and Corollary 1.6.1 the set of dynamical systems of the form (4.2.1) satisfying condition (4.5.2) is open and dense in the space S_{mn}. Since the dependence (4.5.3) has an analytic character in relation to the parameters of the system the set of dynamical systems satisfying condition (4.5.3) is also open and dense in the space S_{mn}. Hence, since the topological product of open sets is also an open set and, moreover, the topological product of dense sets is also a dense set, our theorem follows. ■

Remark 4.5.1

A property of the points of a topological space will be said to be *generic* if the set of points having the property is a dense open subset. Hence, by Theorem 4.5.1, exact controllability in $[0, t_1]$, $t_1 > nh$, is a generic property of the dynamical systems (4.2.1).

In the case of the dynamical system (4.2.11) exact controllability in finite time (or equivalently in $[0, t_1]$, $t_1 > nh$) may occur only when $m = n$ (see Jacobs and Langenhop (1976) for details). Let $\tilde{S}_n$ be a space $\mathbf{R}^{3n^2}$ with the norm

$$||\tilde{L}||_n = ||\mathbf{A}_0|| + ||\mathbf{A}_1|| + ||\mathbf{B}||, \tag{4.5.4}$$

where the symbol $\tilde{L}$ denotes a dynamical system of the form (4.2.11) and matrix $\mathbf{B}$ is $(n \times n)$-dimensional.

COROLLARY 4.5.1. *If $m = n$, then the set of dynamical systems* (4.2.11) *which are exactly controllable in finite time is open and dense in the space $\tilde{S}_n$.*

Proof:

By Corollary 4.3.4, the necessary and sufficient condition for the exact controllability in finite time of the dynamical system (4.2.11) is the nonsingularity of the matrix $\mathbf{B}$. Since the set of nonsingular $(n \times n)$-dimensional matrices is open and dense in the set of all $(n \times n)$-dimensional matrices, and, moreover, exact controllability in finite time does not depend on matrices $\mathbf{A}_1$ and $\mathbf{A}_0$, the set of dynamical systems (4.2.11) which are exactly controllable in finite time is open and dense in the space $\tilde{S}_n$. ■

Remark 4.5.2

Corollary 4.5.1 can also be proved directly from the theorem 4.5.1.

Remark 4.5.3

Exact controllability in finite time is a generic property of the dynamical system (4.2.11).

In the very special case of a dynamical system without delays, from Theorem 4.5.1 or Corollary 4.5.1 we obtain Corollary 1.6.3, concerning dynamical systems of the form (1.2.8).

Remark 4.5.4
Since the exact controllability in finite time of the dynamical system (4.2.1) implies its approximate controllability in finite time, then Theorem 4.5.1 and Corollary 4.5.1 are also true for approximate controllability in finite time.

Similar considerations may be applied to derive results about other types of controllability of dynamical systems (4.2.1) or (4.2.11) (e.g., Euclidean controllability, $\boldsymbol{R}^n(s)$-controllability, or spectral controllability).

An interesting open problem is to construct controllability measures for dynamical systems (4.2.1) and (4.2.11), modelled on the measures obtained for dynamical systems without delays.

4.6 Dynamical systems with delays in control

In this section we shall consider the dynamical systems with time-variable, lumped delays in control. We shall restrict ourselves to linear, nonstationary dynamical systems, described by the following ordinary differential equation:

$$\dot{\mathbf{x}}(t) = \mathbf{A}(t)\mathbf{x}(t) + \sum_{i=0}^{M} \mathbf{B}_i(t)\mathbf{u}(v_i(t)), \quad t \geqslant t_0, \tag{4.6.1}$$

where $\mathbf{x}(t) \in \boldsymbol{R}^n$ is the instantaneous state vector, $\mathbf{u} \in L^2_{\text{loc}}([t_0, \infty), \boldsymbol{R}^m)$ is the control, $\mathbf{A}(t)$ is an $(n \times n)$-dimensional matrix with elements $a_{kj} \in L^2_{\text{loc}}([t_0, \infty), \boldsymbol{R})$ for $k, j = 1, 2, \ldots, n$, $\mathbf{B}_i(t)$, $i = 0, 1, 2, \ldots, M$ are $(n \times m)$-dimensional matrices with elements $b_{ikj} \in L^2_{\text{loc}}([t_0, \infty), \boldsymbol{R})$, for $j = 1, 2, \ldots, m$ and $k = 1, 2, \ldots, n$.

It is assumed that the functions v_i: $[t_0, \infty) \to \boldsymbol{R}$, $i = 0, 1, 2, \ldots, M$ are absolutely continuous and strictly increasing ($\dot{v}_i(t) > 0$ almost everywhere in $[t_0, \infty)$) and satisfy the following inequalities:

$$v_M(t) < v_{M-1}(t) < \ldots < v_k(t) < \ldots < v_1(t) < v_0(t) = t \quad \text{for } t \in [t_0, \infty). \tag{4.6.2}$$

The functions $v_i(t)$, $i = 0, 1, 2, \ldots, M$ can also be expressed in the following form:

$$v_i(t) = t - h_i(t), \quad i = 0, 1, 2, \ldots, M \text{ for } t \in [t_0, \infty), \tag{4.6.3}$$

where $h_i(t) \geqslant 0$, $i = 0, 1, 2, \ldots, M$ are time-variable control delays.

Dynamical systems (4.6.1) can be transformed to linear, nonstationary dynamical systems without delays of the form given by (1.2.1). The problems of reducing dynamical systems (4.6.1) to the form (1.2.1) are extensively investigated in the literature (see, e.g. Artstein (1982b), Artstein and Tadmor (1982), Klamka (1976a, 1976c, 1976d, 1976f, 1977a, 1977b, 1978b, 1980a, 1980b, 1981b), Lewis (1979), Manitius and Olbrot (1972), Olbrot (1972)). More general dynamical systems with distributed delays in control have been considered in Klamka (1976d, 1977a). The dynamical systems defined in infinite-dimensional spaces and with delays in control are investigated in Klamka (1978b). Moreover, in Manitius (1972) the controllability of a linear, stationary dynamical system with distributed delays in control and state variables has been considered.

In order to reduce the dynamical system (4.6.1) to the form (1.2.1), let us fix the time $t_1 > t_0$ and the time interval $[t_0, t_1]$ in which the system (4.6.1) is defined. It should be pointed out that without loss of generality it can be assumed that for some natural $k \leqslant M$ we have $v_k(t_1) = t_0$.

To simplify the notation, let us introduce the so-called *lead function*, which is the inverse function for $v_i(t)$. The leading functions r_i: $[v_i(t_0), v_i(t_1)] \to [t_0, t_1]$ satisfy the following relation:

$$r_i(v_i(t)) = t \quad \text{for } i = 0, 1, 2, \dots, M, \; t \in [t_0, t_1]. \tag{4.6.4}$$

For the dynamical systems of the form (4.6.1), besides the *instantaneous state* $\mathbf{x}(t) \in \boldsymbol{R}^n$, we introduce also the notion of the so-called *complete state* at time t, $\mathbf{z}(t) = \{\mathbf{x}(t), \mathbf{u}_t(s)\}$, where $\mathbf{u}_t(s) = \mathbf{u}(s)$ for $s \in [v_M(t), t)$. Therefore we distinguish two basic notions of controllability for dynamical systems (4.6.1), namely: relative controllability and absolute controllability.

DEFINITION 4.6.1. The dynamical system (4.6.1) is said to be *relatively controllable in* $[t_0, t_1]$, if for any initial complete state $\mathbf{z}(t_0)$ and any vector $\mathbf{x}_1 \in \boldsymbol{R}^n$, there exists a control $\mathbf{u} \in L^2([t_0, t_1], \boldsymbol{R}^m)$ such that the corresponding trajectory $\mathbf{x}(t, \mathbf{z}(t_0), \mathbf{u})$ of the dynamical system (4.6.1) satisfies the following condition:

$$\mathbf{x}(t_1, \mathbf{z}(t_0), \mathbf{u}) = \mathbf{x}_1, \tag{4.6.5}$$

DEFINITION 4.6.2. The dynamical system (4.6.1) is said to be *absolutely controllable in* $[t_0, t_1]$ if for any initial complete state $\mathbf{z}(t_0)$, any vector $\mathbf{x}_1 \in \boldsymbol{R}^n$, and an arbitrary function $\mathbf{w} \in L^2([0, t_1 - v_M(t_1)], \boldsymbol{R}^m)$ there exists a control $\mathbf{u} \in L^2([t_0, t_1 - v_M(t_1)], \boldsymbol{R}^m)$ such that the complete state at time t_1 of the dynamical system (4.6.1) satisfies the following condition:

$$\mathbf{z}(t_1) = \{\mathbf{x}_1, \mathbf{w}\}. \tag{4.6.6}$$

The complete state at time t for the dynamical system (4.6.1) is an element of the Hilbert space $\boldsymbol{R}^n \times L^2([0, t - v_M(t)], \boldsymbol{R}^m)$, which is an infinite-dimensional space. From Definitions 4.6.1 and 4.6.2 it follows that the absolute controllability in $[t_0, t_1]$ of the dynamical system (4.6.1) implies relative controllability in the same time interval. The converse implication is not always true. Examples of dynamical systems which are relatively controllable in $[t_0, t_1]$ and not absolutely controllable in the same time interval are given in Klamka (1976a), Olbrot (1972), Sebakhy and Bayoumi (1973, 1976). Definition 4.6.2 immediately implies that the absolute controllability of a dynamical system (4.6.1) has sense only for a sufficiently great time interval, i.e., when

$$t_0 < v_M(t_1). \tag{4.6.7}$$

It should be also stressed, that relative controllability of a dynamical system (4.6.1) is defined for an arbitrary time interval $[t_0, t_1]$, $t_0 < t_1$.

Now let us consider a special case of the dynamical system (4.6.1), i.e., a linear system with multiple, lumped time-invariant delays in control of the form:

$$\dot{\mathbf{x}}(t) = \mathbf{A}\mathbf{x}(t) + \sum_{i=0}^{M} \mathbf{B}_i(t)\mathbf{u}(t-h_1), \quad t \geqslant t_0. \tag{4.6.8}$$

The dynamical system (4.6.8) can be obtained from system (4.6.1) by substituting

$$v_i(t) = t - h_i, \quad i = 0, 1, 2, \ldots, M, \; t \in [t_0, \infty), \tag{4.6.9}$$

where

$$0 = h_0 < h_1 < \ldots < h_i < \ldots < h_{M-1} < h_M$$

are constant delays.

In this case the lead functions $r_i(t)$ are of the following form:

$$r_i(t) = t + h_i, \quad i = 0, 1, 2, \ldots, M, \; t \in [t_0, \infty). \tag{4.6.10}$$

The simplest case of the dynamical system with delays in control is the linear stationary dynamical system, described by the following differential equation:

$$\dot{\mathbf{x}}(t) = \mathbf{A}\mathbf{x}(t) + \sum_{i=0}^{M} \mathbf{B}_i \mathbf{u}(t-h_i), \quad t \geqslant 0. \tag{4.6.11}$$

The dynamical system (4.6.1) has a unique, absolute continuous solution $\mathbf{x}(t, \mathbf{z}(t_0), \mathbf{u})$, given by the following formula (Klamka (1976f, 1976e, 1977a):

$$\mathbf{x}(t, \mathbf{z}(t_0), \mathbf{u}) = \mathbf{F}(t, t_0)\mathbf{x}(t_0) + \int_{t_0}^{t} \mathbf{F}(t, \tau) \sum_{i=0}^{M} \mathbf{B}_i(\tau)\mathbf{u}(v_i(\tau))d\tau, \quad t \geqslant t_0, \tag{4.6.12}$$

where $\mathbf{F}(t, \tau)$ is an $(n \times n)$-dimensional transition matrix for the autonomous system. Using the assumption of the absolute continuity of the functions v_i and the notion of the leading function r_i, we can express the formula (4.6.12) in a more convenient form:

$$\mathbf{x}(t, \mathbf{z}(t_0), \mathbf{u}) = \mathbf{F}(t, t_0)\mathbf{x}(t_0) + \sum_{i=0}^{M} \int_{v_i(t_0)}^{v_i(t_1)} \mathbf{F}(t, r_i(\tau))\mathbf{B}_i(r_i(\tau))\dot{r}_i(\tau)\mathbf{u}(\tau)d\tau. \tag{4.6.13}$$

Without loss of generality we can assume that $t_0 = v(t_1)$. Therefore for $t = t_1$ formula (4.6.13) can be expressed in the following form:

$$\begin{aligned}\mathbf{x}(t_1, \mathbf{z}(t_0), \mathbf{u}) = {} & \mathbf{F}(t_1, t_0)\mathbf{x}(t_0) + \\ & + \sum_{i=0}^{k} \int_{v_i(t_0)}^{t_0} \mathbf{F}(t_1, r_i(\tau))\mathbf{B}_i(r_i(\tau))\dot{r}_i(\tau)\mathbf{u}_{t_0}(\tau)\,d\tau + \\ & + \sum_{i=k+1}^{M} \int_{v_i(t_0)}^{v_i(t_1)} \mathbf{F}(t_1, r_i(\tau))\mathbf{B}_i(r_i(\tau))\dot{r}_i(\tau)\mathbf{u}_{t_0}(\tau)\,d\tau + \\ & + \sum_{i=0}^{k} \int_{t_0}^{v_i(t_1)} \mathbf{F}(t_1, r_i(\tau))\mathbf{B}_i(r_i(\tau))\dot{r}_i(\tau)\mathbf{u}(\tau)\,d\tau. \end{aligned} \tag{4.6.14}$$

The first three terms on the right-hand side of formula (4.6.14) depend only on the complete state at time t_0, $\mathbf{z}(t_0)$, and do not depend on the control $\mathbf{u}$. Using the properties of the transition matrix $\mathbf{F}(t, \tau)$, we can express these three terms in a shorter form as follows:

$$\begin{aligned}\mathbf{q}(\mathbf{z}(t_0)) = {} & \mathbf{F}(t_1, t_0)\Big(\mathbf{x}(t_0) + \sum_{i=1}^{k} \int_{v_i(t_0)}^{t_0} \mathbf{F}(t_0, r_i(\tau))\mathbf{B}_i(r_i(\tau))\dot{r}_i(\tau)\mathbf{u}_{t_0}(\tau)\,d\tau + \\ & + \sum_{i=k+1}^{M} \int_{v_i(t_0)}^{v_i(t_1)} \mathbf{F}(t_0, r_i(\tau))\mathbf{B}_i(r_i(\tau))\dot{r}_i(\tau)\mathbf{u}_{t_0}(\tau)\,d\tau\Big). \end{aligned} \tag{4.6.15}$$

To simplify the notation, let us denote:

$$\mathbf{B}_{t_1}(t) = \sum_{j=0}^{i} \mathbf{F}(t_0, r_j(t))\mathbf{B}_j(r_j(t))\dot{r}_j(t), \tag{4.6.16}$$

$$t \in [v_{i+1}(t_1), v_i(t_1)),\ i = 0, 1, 2, \ldots, k-1.$$

$\mathbf{B}_{t_1}(t)$ is an $(n \times m)$-dimensional matrix defined in the time interval $[t_0, t_1]$ (since $t_0 = v_k(t_1)$. The elements of the matrix $\mathbf{B}_{t_1}$ are square integrable functions.

LEMMA 4.6.1 (Artstein (1982), Artstein and Tadmor (1982), Klamka (1976a, 1976b, 1976c)). *Let*

$$\dot{\mathbf{y}}(t) = \mathbf{A}(t)\mathbf{y}(t) + \mathbf{B}_{t_1}(t)\mathbf{u}(t), \quad t \in [t_0, t_1] \tag{4.6.17}$$

be the linear, nonstationary dynamical system without delays in control. Then

$$\mathbf{x}(t, \mathbf{z}(t_0), \mathbf{u}) = \mathbf{y}(t, \mathbf{q}(\mathbf{z}(t_0)), \mathbf{u}) \quad \textit{for } t \in [t_0, t_1]. \tag{4.6.18}$$

Proof:

Equality (4.6.18) follows immediately from the formulas (1.2.2), (4.6.14), (4.6.15) and (4.6.16) and the basic properties of the transition matrix $\mathbf{F}(t, \tau)$. ∎

Lemma 4.6.1 allows us to reduce dynamical system with delays in control (4.6.1) to the equivalent dynamical system without delays in control (4.6.17). This simplifies the investigation of controllability of the dynamical system (4.6.1). The relative controllability in $[t_0, t_1]$ of the dynamical system (4.6.1) is equivalent to the controllability in $[t_0, t_1]$ of the dynamical system (4.6.17). Since the dynamical system (4.6.17) is a system without delays, then in order to check its controllability in $[t_0, t_1]$ it is enough to use the necessary and sufficient conditions for controllability in $[t_0, t_1]$ formulated in Chapter 1. These conditions will be extensively used in the subsequent part of this section. The equivalence between the dynamical system with delays in control and the dynamical system without delays can also be used in another problems, e.g., in considering the minimum energy control and the time optimal control.

In order to use the results presented in Chapter 1, let us introduce the *relative controllability matrix* $\mathbf{W}_k(t_0, t_1)$, defined in $[t_0, t_1]$. $\mathbf{W}_k(t_0, t_1)$ is an $(n\times n)$-dimensional symmetric matrix given by the following integral formula (Klamka (1976f)):

$$\begin{aligned}\mathbf{W}_k(t_0, t_1) &= \int_{t_0}^{t_1} \mathbf{F}(t_1, t)\mathbf{B}_{t_1}(t)\mathbf{B}_{t_1}^{\mathrm{T}}(t)\mathbf{F}^{\mathrm{T}}(t_1, t)\,dt \\ &= \sum_{i=0}^{k-1} \int_{v_{i+1}(t_1)}^{v_i(t_1)} \Big[\sum_{j=0}^{i} \mathbf{F}(t_1, r_j(t))\mathbf{B}_j(r_j(t))\dot{r}_j(t)\Big] \times \\ &\quad \times \Big[\sum_{t=0}^{i} [\mathbf{F}(t_1, r_j(t))\mathbf{B}_j(r_j(t))\dot{r}_j(t)\Big]^{\mathrm{T}} dt. \end{aligned} \tag{4.6.19}$$

THEOREM 4.6.1 (Klamka (1976f)). *The dynamical system* (4.6.1) *is relatively controllable in* $[t_0, t_1]$ *if and only if*

$$\operatorname{rank}\mathbf{W}_k(t_0, t_1) = n. \tag{4.6.20}$$

Proof:

From (4.6.19) it follows that the relative controllability matrix $\mathbf{W}_k(t_0, t_1)$ in $[t_0, t_1]$ for the dynamical system (4.6.1) is simultaneously the controllability matrix in $[t_0, t_1]$ for the dynamical system without delays (1.6.17). Hence by Lemma 4.6.1 and Theorem 1.3.1 the dynamical system (4.6.1) is relatively controllable in $[t_0, t_1]$ if and only if condition (4.6.10) is satisfied. ∎

COROLLARY 4.6.1 (Klamka (1976f)). *The dynamical system* (4.6.8) *is relatively controllable in* $[t_0, t_1]$ *if and only if*

$$\begin{aligned}\operatorname{rank} &\sum_{i=0}^{k-1} \int_{t_1-h_{i+1}}^{t_1-h_i} \Big[\sum_{j=0}^{i} \mathbf{F}(t_1, t+h_j)\mathbf{B}_j(t+h_j)\Big] \times \\ &\times \Big[\sum_{j=0}^{i} \mathbf{F}(t_1, t+h_j)\mathbf{B}_j(t+h_j)\Big]^{\mathrm{T}} dt = n. \end{aligned} \tag{4.6.21}$$

Proof:

Substituting (4.6.10) in formula (4.6.19) and taking into account that

$$\dot{r}_i(t) = 1 \quad \text{for} \quad i = 0, 1, 2, \dots, M, \quad t \in [t_0, \infty) \tag{4.6.22}$$

by Theorem 4.6.1 we obtain Corollary 4.6.1. ∎

COROLLARY 4.6.2 (Klamka (1976f)). *The dynamical system* (4.6.11) *is relatively controllable in* $[0, t_1]$ *if and only if*

$$\operatorname{rank} \sum_{i=0}^{k-1} \int_{t_1-h_{i+1}}^{t_1-h_i} \Big[\sum_{j=0}^{i} \exp(\mathbf{A}(t_1-h_j-t))\mathbf{B}_j\Big] \times$$

$$\times \Big[\sum_{j=0}^{i} [\exp(\mathbf{A}(t_1-h_j-t))\mathbf{B}_j\Big]^{\mathrm{T}} dt = n. \tag{4.6.23}$$

Proof:

Corollary 4.6.2 immediately follows from Corollary 4.6.1 and the formula (1.2.10). ∎

In order to use in practice Theorem 4.6.1 and Corollary 4.6.1 it is necessary to compute the transition matrix $\mathbf{F}(t_1, t)$, which, in the general case, is very difficult. Hence, it is important to derive the algebraic conditions for relative controllability, which are entirely based on the knowledge of the matrices $\mathbf{A}(t)$, $\mathbf{B}_i(t)$ and the functions $v_i(t)$, $i = 0, 1, 2, \dots, M$. Since the matrix $\mathbf{B}_{t_1}(t)$ is a noncontinuous function in $[t_0, t_1]$, we cannot directly use Theorems 1.3.2 or 1.3.3 presented in Chapter 1. The matrix function $\mathbf{B}_{t_1}(t)$ has discontinuous points for $t = v_i(t_1)$, $i = 0, 1, 2, \dots, k-1$.

In order to solve this problem, let us introduce some symbols and additional assumptions concerning the matrices $\mathbf{A}(t)$, $\mathbf{B}_i(t)$, and the function $v_i(t)$, $i = 0, 1, 2, \dots, M$. Namely, let us assume that the matrix $\mathbf{A}(t)$ is $p-2$ times continuously differentiable on $[t_0, t_1]$, the matrices $\mathbf{B}_i(t)$ are $p-1$ times continuously differentiable on $[r_i(t_0), t_1]$ for $i = 0, 1, 2, \dots, M$, and the functions $v_i(t)$ are p times continuously differentiable on $[t_0, t_1]$ for $i = 0, 1, 2, \dots, M$. Let

$$\mathbf{Q}(t) = [L^0\mathbf{B}_0(t) \; \dots \; L^0\mathbf{B}_{k-1}(t) \; L^1\mathbf{B}_0(t) \; \dots \; L^1\mathbf{B}_{k-1}(t) \; \dots \; L^jB_0(t)$$

$$\dots \; L^j\mathbf{B}_{k-1}(t) \; \dots \; L^p\mathbf{B}_0(t) \; \dots \; L^p\mathbf{B}_{k-1}(t)] \tag{4.6.24}$$

denote an $(n \times mkp)$-dimensional matrix, defined for $t \in [r_{k-1}(t_0), t_1]$, where the operator L is defined as follows:

$$L^0\mathbf{B}_i(t) = \mathbf{B}_i(t)\dot{r}_i(v_i(t)), \quad i = 0, 1, 2, \dots, k-1, \quad t \in [r_i(t_0), t_1],$$

$$L^j\mathbf{B}_i(t) = -\mathbf{A}(t)\dot{r}_i(v_i(t))L^{j-1}\mathbf{B}_i(t) + \frac{d}{dt}L^{j-1}\mathbf{B}_i(t),$$

$$j = 0, 1, 2, \dots, p, \quad i = 0, 1, 2, \dots, k-1, \quad t \in [r_i(t_0), t_1], \tag{4.6.25}$$

Using the adjoint equation for the transition matrix $\mathbf{F}(t, t_0)$, we can derive the j-th derivatives of the matrix function $\mathbf{B}_{t_1}(t)$ at the regular points in the interval $[t_0, t_1]$ (Klamka (1967f)), namely:

$$\frac{d^j}{dt^j}\mathbf{B}_{t_1}(t) = \sum_{i=0}^{k} \mathbf{F}(t_0, r_i(t)) L^j \mathbf{B}_i(r_i(t)) = \mathbf{B}_{t_1}^{(j)},$$
$$j = 0, 1, 2, \ldots, p-1. \tag{4.6.26}$$

The jumps of the j-th derivatives $\mathbf{B}_{t_1}^{(j)}$ at the nonregular points $t = v_i(t_1), j = 0, 1, 2, \ldots$ $\ldots, p-1$, $i = 0, 1, 2, \ldots, k-1$ can be computed by using the operator L; namely, by (4.6.25) and (4.6.26) we have:

$$\Delta\mathbf{B}_{t_1}^{(j)}(v_i(t_1)) = L^j\mathbf{B}_i(t_1),$$
$$j = 0, 1, 2, \ldots, p-1, \; i = 0, 1, 2, \ldots, k-1. \tag{4.6.27}$$

Making use of the previously introduced symbols, we can formulate the algebraic conditions for the relative controllability in $[t_0, t_1]$ of the dynamical system (4.6.1). These conditions generalize to the case of delays the well-known controllability conditions for dynamical systems without delays presented in Section 1.3. The assumptions and the results are quite similar to those given in Section 1.3.

THEOREM 4.6.2 (Klamka (1976)). *Suppose that the matrix $A(t)$ is $n-2$ times continuously differentiable in $[t_0, t_1]$, the matrices $\mathbf{B}_i(t)$, $i = 0, 1, 2, \ldots, k-1$ are $n-1$ times continuously differentiable in $[r_i(t_0), t_1]$ and the functions $v_i(t)$, $i = 0, 1, 2, \ldots$ $\ldots, k-1$ are n limes continuously differentiable in $[t_0, t_1]$. Then, if*

$$\operatorname{rank}\mathbf{Q}_n(t_1) = n, \tag{4.6.28}$$

the dynamical system (4.6.1) *is relatively controllable in* $[t_0, t_1]$.

Proof:

By contradiction. Let us assume that condition (4.6.28) is satisfied, but the dynamical system (4.6.1) is not relatively controllable in $[t_0, t_1]$. Thus, by Lemma 1.3.2 and Theorem 1.3.1 and the properties of the transition matrix (Klamka (1976f)), there exists a nonzero vector $\mathbf{y} \in \boldsymbol{R}^n$ such that $\mathbf{y}^{\mathrm{T}}\mathbf{B}_{t_1}(t) = 0$ almost everywhere in $[t_0, t_1]$. Since the matrix function $\mathbf{B}_{t_1}(t)$ is piecewise of a class C^{n-1} in $[t_0, t_1]$, we have for all $t \in [t_0, t_1]$ except the points $t = v(t_1)$, $i = 0, 1, 2, \ldots, k-1$

$$\mathbf{y}^{\mathrm{T}}\mathbf{B}_{t_1}(t) = 0. \tag{4.6.29}$$

Differentiating with respect to t, equality (4.6.29) at the points $t \neq v_i(t_1)$, $i = 0, 1, 2, \ldots, k-1$ and computing $\mathbf{y}^{\mathrm{T}}\Delta\mathbf{B}_{t_1}^{(j)}(v_i(t_1)), j = 0, 1, 2, \ldots, n-1$ we obtain by (4.6.27) the following relation:

$$\mathbf{y}^{\mathrm{T}}L^j\mathbf{B}_i(t_1) = 0 \quad \text{for } i = 0, 1, 2, \ldots, k-1. \tag{4.6.30}$$

Since $\mathbf{y} \neq 0$, then (4.6.30) implies the inequality $\operatorname{rank} \mathbf{Q}_n(t_1) < n$, which leads to a contradiction. Thus the dynamical system (4.6.1) is relatively controllable in the time interval $[t_0, t_1]$. ■

Theorem 4.6.2 is only a sufficient condition for the relative controllability in $[t_0, t_1]$ of the dynamical system (4.6.1). A necessary condition may be obtained under some additional assumptions concerning the analyticity of the matrices $\mathbf{A}(t)$, $\mathbf{B}_i(t)$ and the functions $v_i(t)$.

THEOREM 4.6.3 (Klamka (1976f)). *Let us assume that the matrix* $\mathbf{A}(t)$ *is analytic on* $[t_0, t_1]$, *the matrices* $\mathbf{B}_i(t)$, $i = 0, 1, 2, \ldots, k-1$ *are analytic, respectively, on* $[r_i(t_0), t_1]$ and *the functions* $v_i(t)$, $i = 0, 1, 2, \ldots, k-1$ *are analytic on* $[t_0, t_1]$. *Then the dynamical system* (4.6.1) *is relatively controllable in* $[t_0, t_1]$ *if and only if*

$$\operatorname{rank} \mathbf{Q}_\infty(t_1) = n. \tag{4.6.31}$$

Proof:

Sufficiency. Since the assumptions of Theorem 4.6.2 are all satisfied, then the proof of the sufficient condition is analogous to the proof of Theorem 4.6.2.

Necessity. By contradiction. Let us assume that the dynamical system (4.6.1) is relatively controllable in $[t_0, t_1]$, but condition (4.6.31) is not satisfied, i.e., $\operatorname{rank} \mathbf{Q}_\infty(t_1) < n$. Therefore the subspace spanned by the columns of the matrix $\mathbf{Q}_\infty(t_1)$ is the proper subspace of the space $\boldsymbol{R}^n$ and thus there exists a nonzero vector $\mathbf{y} \in \boldsymbol{R}^n$ such that for all $i = 0, 1, 2, \ldots, k-1$ and all $j = 0, 1, 2, \ldots$, we have

$$\mathbf{y}^{\mathrm{T}} L^j \mathbf{B}_i(t_1) = 0. \tag{4.6.32}$$

Therefore by (4.6.27) we have

$$\mathbf{y}^{\mathrm{T}} \Delta \mathbf{B}_{t_1}^{(j)}(v_i(t)_1) = 0 \quad \text{for} \quad i = 0, 1, 2, \ldots, k-1, \; j = 0, 1, 2, \ldots \tag{4.6.33}$$

Since the matrix function $\mathbf{B}_{t_1}(t) = 0$ for $t > t_1$, we have

$$\mathbf{y}^{\mathrm{T}} \mathbf{B}_{t_1}^{(j)}(t_1 + 0) = 0 \quad \text{for } j = 0, 1, 2, \ldots \tag{4.6.34}$$

By (4.6.33) and (4.6.34) we obtain the following equalities:

$$\mathbf{y}^{\mathrm{T}} \mathbf{B}_{t_1}^{(j)}(t_1 - 0) = 0 \quad \text{for } j = 0, 1, 2, \ldots \tag{4.6.35}$$

Since the matrix function $\mathbf{B}_{t_1}(t)$ is piecewise analytic in $[t_0, t_1]$, then by (4.6.35) we have $\mathbf{y}^{\mathrm{T}} \mathbf{B}_{t_1}(t) = 0$ in $[v_1(t_1), t_1]$. In a similar way we can prove that $\mathbf{y}^{\mathrm{T}} \mathbf{B}_{t_1}(t) = 0$ in $[v_i(t_1), v_{i-1}(t_1)]$ for $i = 2, 3, \ldots, k$. Therefore $\mathbf{y}^{\mathrm{T}} \mathbf{B}_{t_1}(t) = 0$ in $[t_0, t_1]$, which by Lemma 4.6.1 contradicts the assumption of the relative controllability of the dynamical system (4.6.1) in $[t_0, t_1]$. Hence our theorem follows. ■

For the case of constant delays, i.e., if we consider the dynamical system (4.6.8), by Theorem 4.6.3 we can easily obtain the results presented in (Olbrot (1972)) and even extend those results to cover the case of time-variable delays in control.

THEOREM 4.6.4 (Klamka (1976f)). *Let the assumptions of Theorem* 4.6.3 *be all satisfied. Then the dynamical system* (4.6.1) *is relatively controllable in* $[t_0, t_1]$ *if and only if the dynamical system without delays in control, of the form*

$$\dot{\mathbf{x}}(t) = \mathbf{A}(t)\mathbf{x}(t)+\tilde{\mathbf{B}}(t)\mathbf{w}(t), \quad t \in [t_0, t_1], \tag{4.6.36}$$

where

$$\tilde{\mathbf{B}}(t) = [\mathbf{B}_0(t) \vdots \mathbf{B}_1(t) \vdots \dots \vdots \mathbf{B}_{k-1}(t)], \quad t \in [t_0, t_1], \ \mathbf{w} \in R^{km}, \tag{4.6.37}$$

is controllable in $[r_{k-1}(t_0), t_1]$.

Proof:

By (4.6.16) the matrix function $\mathbf{B}_{t_1}(t)$ can be expressed as follows:

$$\mathbf{B}_{t_1}(t) = \sum_{i=0}^{j} \mathbf{F}(t_0, r_i(t))\mathbf{B}_i(r_i(t))\dot{r}_i(t), \tag{4.6.38}$$
$$t \in (v_{j+1}(t_1), v_j(t_1)], \ j = 0, 1, 2, \dots, k-1.$$

By Theorem 4.6.1 and formula (4.6.38) the dynamical system (4.6.1) is relatively controllable in $[t_0, t_1]$ if and only if, for $\mathbf{y} \in R^n$, the relation

$$\mathbf{y}^{\mathrm{T}}\Big(\sum_{i=0}^{j} \mathbf{F}(t_0, r_i(t))\mathbf{B}_i(r_i(t))\dot{r}_i(t)\Big) = 0, \tag{4.6.39}$$
$$t \in [v_{j+1}(t_1), v_j(t_1)], \ j = 0, 1, 2, \dots, k-1$$

implies that $\mathbf{y} = 0$. By the analyticity assumptions about matrices $\mathbf{A}(t)$, $\mathbf{B}_i(t)$ and the functions $v_i(t)$, the preceding statement is equivalent to the statement that

$$\mathbf{y}^{\mathrm{T}}\mathbf{F}(t_0, r_j(t))\mathbf{B}_j(r_j(t))\dot{r}_j(t) = 0 \tag{4.6.40}$$
$$\text{for } t \in (v_{j+1}(t_1), v_j(t_1)], \ j = 0, 1, 2, \dots, k-1$$

implies $\mathbf{y} = 0$. Making use of the analyticity assumptions once again, we can express the formula (4.6.40) in a more convenienet form:

$$\mathbf{y}^{\mathrm{T}}\mathbf{F}(t_0, r_j(t))\mathbf{B}_j(r_j(t))\dot{r}_j(t) = 0$$
$$\text{for } t \in [t_0, v_j(t_1)], \ j = 0, 1, 2, \dots, k-1,$$

or equivalently

$$\mathbf{y}^{\mathrm{T}}\mathbf{F}(t_0, t)\mathbf{B}_j(t) \quad \text{for } t \in [r_j(t_0), t_1], \ j = 0, 1, 2, \dots, k-1. \tag{4.6.41}$$

By (4.6.41) the following statement is true: the dynamical system (4.6.1) is relatively controllable in $[t_0, t_1]$ if and only if, for $\mathbf{y} \in R^n$, the relation

$$\mathbf{y}^{\mathrm{T}}[\mathbf{F}(t_0, t)\mathbf{B}_0(t) \vdots \mathbf{F}(t_0, t)\mathbf{B}_1(t) \vdots \dots \vdots \mathbf{F}(t_0, t)\mathbf{B}_{k-1}(t)] = 0$$
$$\text{for } t \in [r_{k-1}(t_0), t_1]$$

implies $\mathbf{y} = 0$. On the other hand, the above implication is a necessary and sufficient condition of controllability in $[r_{k-1}(t_0), t_1]$ for the dynamical system without delays in control of the form (4.6.36). Hence our theorem follows. ∎

COROLLARY 4.6.3 (Klamka (1976f)). *If the matrix* $\mathbf{A}(t)$ *is analytic in* $[t_0, t_1]$ *and the matrices* $\mathbf{B}_i(t)$, $i = 0, 1, 2, \ldots, k-1$ *are analytic, respectively, in* $[t_0+h_i, t_1]$, *then the dynamical system* (4.6.8) *is relatively controllable in* $[t_0, t_1]$ *if and only if the dynamical system without delays* (4.6.36) *is controllable in* $[t_0+h_{k-1}, t_1]$.

Proof:

Substituting (4.6.9) and (4.6.10) in the appropriate relations of Theorem 4.6.4, we directly obtain Corollary 4.6.3. ∎

A result analogous to Corollary 4.6.3 has been derived in the paper of Olbrot (1972), by means of a slighty different method. Similar results are also obtained in the papers of Sebakhy and Bayoumi (1973, 1976).

In the general case, without the analyticity assumption about matrices $\mathbf{A}(t)$ and $\mathbf{B}_0(t)$ in (t_0, t_1), controllability in $[t_0, t_1]$ of the dynamical system without delays

$$\dot{\mathbf{x}}(t) = \mathbf{A}(t)\mathbf{x}(t)+\mathbf{B}_0(t)\mathbf{u}(t) \tag{4.6.42}$$

does not imply relative controllability in $[t_0, t_1]$ for the dynamical systems with delays in control. This situation is shown in Example 4.6.1.

Example 4.6.1

Let us consider a dynamical system of the form (4.6.8) with $t_0 < t_1-h$, Let $\mathbf{A}(t)$ be any $(n\times n)$-dimensional matrix with the corresponding transition matrix $\mathbf{F}(t, \tau)$. Moreover, let $\mathbf{B}_0(t)$ and $\mathbf{B}_1(t)$ be the $(n\times n)$-dimensional matrices given by

$$\mathbf{B}_0(t) = \begin{cases} \mathbf{I} & \text{for } t\in[t_0, t_1-h], \\ 0 & \text{for } t\in(t_1-h, t_1], \end{cases}$$

$$\mathbf{B}_1(t) = \begin{cases} 0 & \text{for } t\in[t_0, t_0+h], \\ -\mathbf{F}(t, t-h) & \text{for } t\in(t_0+h, t_1]. \end{cases}$$

Matrices $\mathbf{B}_0(t)$ and $\mathbf{B}_1(t)$ are of course nonanalytic in $[t_0, t_1]$.

Using the formula (4.6.19), we can compute the relative controllability matrix

$$\mathbf{W}_1(t_0, t_1) = \int_{t_0}^{t_1-h} (\mathbf{F}(t_0, \tau)-\mathbf{F}(t_0, \tau+h)\mathbf{F}(\tau+h, \tau))(\mathbf{F}(t_0, \tau)-$$
$$-\mathbf{F}(t_0, \tau+h)\mathbf{F}(\tau+h, \tau))^{\mathrm{T}}d\tau = 0.$$

Hence, by Theorem 4.6.1 or Corollary 4.6.1, the dynamical system under consideration is not relatively controllable in $[t_0, t_1]$.

In the next part of Example 4.6.1 we shall consider the dynamical system without delays in control (4.6.42) with the same matrices $\mathbf{A}(t)$, $\mathbf{B}_0(t)$ as before. Using the properties of the transition matrix and relation (1.3.8), we have

$$\operatorname{rank} \int_{t_0}^{t_1} \mathbf{F}(t_0, \tau)\mathbf{B}_0(\tau)\mathbf{B}_0^{\mathrm{T}}(\tau)\mathbf{F}^{\mathrm{T}}(t_0, \tau)d\tau$$

$$= \operatorname{rank} \int_{t_0}^{t_1-h} \mathbf{F}(t_0, \tau)\mathbf{F}^{\mathrm{T}}(t_0, \tau)d\tau = n.$$

Hence by Theorem 1.3.1 the dynamical system under consideration and without delays in control, is controllable in $[t_0, t_1]$. Therefore in view of the lack of analyticity assumption for the appropriate matrices, the controllability in $[t_0, t_1]$ of the dynamical system without delays in control does not imply the relative controllability in $[t_0, t_1]$ of the dynamical system with delays in control.

However, if the matrices $\mathbf{A}(t)$ and $\mathbf{B}_0(t)$ are analytic in (t_0, t_1), then controllability in $[t_0, t_1]$ of the dynamical system without delays in control (4.6.42) implies the relative controllability in $[t_0, t_1]$ of the dynamical system with delays in control. This immediately follows from Theorem 4.6.4.

The dynamical system without delays in control of the form (4.6.42) need not be controllable in $[t_0, t_1]$, but the dynamical system with delays in control and with the same matrices $\mathbf{A}(t), \mathbf{B}_0(t)$ may be relatively controllable in $[t_0, t_1]$. This property follows directly from Theorem 4.6.1 and specially from the form of the relative controllability matrix $\mathbf{W}_k(t_0, t_1)$ given by formula (4.6.19). This situation with respect to stationary dynamical system is shown in Example 4.6.2, and with respect to nonstationary dynamical system in Example 4.6.3. The nature of this phenomenon depends strongly on the properties of analytic functions. However, it shoud be stressed that this situation is rather exceptional and occurs only for specially chosen matrices $\mathbf{A}(t)$ and $\mathbf{B}_0(t)$.

Example 4.6.2

Let us consider a stationary dynamical system of the form (4.6.11) with $M = 1$ and matrices $\mathbf{A}, \mathbf{B}_0, \mathbf{B}_1$ given by

$$\mathbf{A} = \begin{bmatrix} 1 & 2 \\ 2 & 1 \end{bmatrix}, \quad \mathbf{B}_0 = \begin{bmatrix} 1 \\ 1 \end{bmatrix}, \quad \mathbf{B}_1 = \begin{bmatrix} 0 \\ 1 \end{bmatrix}, \quad h = 1.$$

Hence we obtain the following relations:

$$\operatorname{rank}[\mathbf{B}_0 \vdots \mathbf{A}\mathbf{B}_0] = \operatorname{rank}\begin{bmatrix} 1 & 3 \\ 1 & 3 \end{bmatrix} = 1 < 2 = n$$

and

$$\operatorname{rank}[\mathbf{B}_0 \vdots \mathbf{B}_1 \vdots \mathbf{A}\mathbf{B}_0 \vdots \mathbf{A}\mathbf{B}_1] = \operatorname{rank}\begin{bmatrix} 1 & 0 & 3 & 2 \\ 1 & 1 & 3 & 1 \end{bmatrix} = 2 = n.$$

Therefore, by Corollary 1.4.1 the dynamical system in question is not controllable in [0, 1]. However, by Theorem 4.6.4, this dynamical system is relatively controllable in $[0, t_1]$ for any $t_1 > 1$.

Example 4.6.3

Let us consider a nonstationary dynamical system of the form (4.6.1) defined in the time interval [1, 2], with $M = 3$ and matrices $\mathbf{A}(t)$, $\mathbf{B}_0(t)$, $\mathbf{B}_1(t)$, $\mathbf{B}_2(t)$ and $\mathbf{B}_3(t)$ given by

$$\mathbf{A}(t) = \mathbf{A} = \begin{bmatrix} -0.5 & 0 \\ 0 & -1 \end{bmatrix},$$

$$\mathbf{B}_0(t) = \begin{cases} \begin{bmatrix} 0 & 0 \\ 0 & 0 \end{bmatrix} & \text{for } t \in [1, 1.5), \\ \begin{bmatrix} 1 & \exp(0,5t) \\ 0 & 0 \end{bmatrix} & \text{for } t \in [1.5, 2], \end{cases}$$

$$\mathbf{B}_1(t) = \begin{cases} \begin{bmatrix} t & 0 \\ \exp t & t^2 \end{bmatrix} & \text{for } t \in [1, 4/3), \\ \begin{bmatrix} 0 & 0 \\ 2 & 1 \end{bmatrix} & \text{for } t \in [4/3, 2], \end{cases}$$

$$\mathbf{B}_2(t) = \begin{bmatrix} \exp t & t^2 \\ t & 0 \end{bmatrix} \quad \text{for } t \in [1, 2],$$

$$\mathbf{B}_3(t) = \begin{bmatrix} t^2 & \exp t \\ 0 & t \end{bmatrix} \quad \text{for } t \in [1, 2].$$

Moreover, $v_0(t) = t$, $v_1(t) = 0.75t$, $v_2(t) = 0.5t$, $v_3(t) = 0.25t$ for $t \in [1, 2]$.

Since the matrix $\mathbf{A}(t) = \mathbf{A}$ is time independent, it is easy to compute the transition matrix $\mathbf{F}(t, \tau)$ for the dynamical system in question, namely, for $t_0 = 1$ we have:

$$\mathbf{F}(t, t_0) = \exp(\mathbf{A}(t-t_0)) = \begin{bmatrix} \exp(0.5-0.5t) & 0 \\ 0 & \exp(1-t) \end{bmatrix}.$$

Moreover, the leading function are as follows: $r_0(t) = t$, $\dot{r}_0(t) = 1$, $r_1(t) = 4/3t$, $\dot{r}_1(t) = 4/3$, $r_2(t) = 2t$, $\dot{r}_2(t) = 2$, $r_3(t) = 4t$, $\dot{r}_3(t) = 4$. Moreover, for $t_1 = 2$ we have: $v_0(2) = 2$, $v_1(2) = 1.5$, $v_2(2) = 1$, $v_3(2) = 0.5$. Hence in formula (4.6.19) we have index $k = 1$.

Using the above equalities and by (4.6.19), we obtain the formula for the relative controllability matrix $\mathbf{W}_3(t_0, t_1)$, namely:

$$\mathbf{W}_3(t_0, t_1) = \mathbf{W}_3(1, 2) = \int_{1,5}^{2} \exp(\mathbf{A}(1-\tau))\mathbf{B}_0(\tau)\mathbf{B}_0^{\mathrm{T}}(\tau)\exp(\mathbf{A}^{\mathrm{T}}(1+\tau))d\tau +$$

$$+ \int_{1}^{1.5} \Big(\exp(\mathbf{A}(1-\tau))\mathbf{B}_0(\tau) +$$

$$+ \exp(\mathbf{A}(1+4/3\tau)\mathbf{B}_1(4/3\tau)4/3\big)\big(\exp(\mathbf{A}(1-\tau))\mathbf{B}_0(\tau) +$$

$$+ \exp(\mathbf{A}(1-4/3\tau)\mathbf{B}_1(4/3\tau)4/3\Big)^{\mathrm{T}} d\tau.$$

Taking into account the form of the matrices $\mathbf{A}(t)$, $\mathbf{B}_0(t)$ and $\mathbf{B}_1(t)$ and the formula for the computation of the exponent matrix function, we have

$$\mathbf{W}_3(1,2) = \begin{bmatrix} \exp 3-\exp 2+\exp 1-\exp 0.5 & 0 \\ 0 & 10/3(\exp 2-\exp 2/3) \end{bmatrix}.$$

Hence rank $\mathbf{W}_3(1, 2) = 2 = n$.

Thus, by Theorem 4.6.1, the dynamical system in question is relatively controllable in [1, 2].

In the next part of the example we shall consider the dynamical system without delays in control of the form (4.6.42) and with the same matrices $\mathbf{A}(t)$ and $\mathbf{B}_0(t)$. By formula (1.3.8) the controllability matrix in the time interval [1, 2] is of the following simple form:

$$\mathbf{W}(t_0, t_1) = \mathbf{W}(1,2) = \int_1^2 \exp\big(\mathbf{A}(1-\tau)\big)\mathbf{B}_0(\tau)\mathbf{B}_0^T(\tau)\exp\big(\mathbf{A}^T(1-\tau)\big)d\tau.$$

Since the matrix $\exp\big(\mathbf{A}(t-\tau)\big)$ is diagonal, and the matrix $\mathbf{B}_0(t)$ contains the zero row, it is easy to deduce, that rank$\mathbf{W}(1, 2) = 1 < 2 = n$. Therefore, by Theorem 1.3.1, the examined dynamical system in question is not controllable in [1, 2].

Examples 4.6.1–4.6.3 illustrate the interrelations between various kinds of the controllability in $[t_0, t_1]$ of the dynamical system without delays in control and the relative controllability in $[t_0, t_1]$ of the dynamical system with delays in control given by the state equation (4.6.1). These three examples explain also the nature of the mutual implications between various kinds of dynamical systems and their controllability properties. Finally, it should be pointed out that analyticity of the system matrices and delayed functions play an important role in the investigation of controllability.

In order to consider absolute controllability in $[t_0, t_1]$ for the dynamical system (4.6.1) (Definition 4.6.2), it is necessary to assume that

$$t_0 < v_M(t_1). \tag{4.6.43}$$

This means that the time interval $[t_0, t_1]$ should be appropriately long. This requirement follows directly from Definition 4.6.2.

By inequality (4.6.43) formulas (4.6.14) and (4.6.15) can be expressed in the following, more convenient form:

$$\begin{aligned} \mathbf{x}\big(t_1, \mathbf{z}(t_0), \mathbf{u}\big) &= \mathbf{F}(t_1, t_0)\mathbf{x}(t_0) + \\ &+ \sum_{i=0}^{M} \int_{v_i(t_0)}^{t_0} \mathbf{F}\big(t_1, r_i(\tau)\big)\mathbf{B}_i\big(r_i(\tau)\big)\dot{r}_i(\tau)\mathbf{u}_{t_0}(\tau)d\tau + \\ &+ \sum_{i=0}^{M} \int_{t_0}^{v_i(t_1)} \mathbf{F}\big(t_1, r_i(\tau)\big)\mathbf{B}_i\big(r_i(\tau)\big)\dot{r}_i(\tau)\mathbf{u}(\tau)d\tau, (\tau)d\tau, \end{aligned} \tag{4.6.44}$$

$$\mathbf{q}\big(\mathbf{z}(t_0)\big) = \mathbf{F}(t_1, t_0)\Big(\mathbf{x}(t_0) + \sum_{i=0}^{M} \int_{v_i(t_0)}^{t_0} \mathbf{F}\big(t_0, r_i(\tau)\big)\mathbf{B}_i\big(r_i(\tau)\big)\dot{r}_i(\tau)\mathbf{u}_{t_0}(\tau)d\tau\Big). \tag{4.6.45}$$

Using a similar method to that ones in the case of relative controllability in $[t_0, t_1]$, we can formulate and prove the necessary and sufficient conditions for absolute controllability in $[t_0, t_1]$ of the dynamical system (4.6.1).

THEOREM 4.6.5 (Klamka (1977c)). *The dynamical system* (4.6.1) *is absolutely controllable in* $[t_0, t_1]$ *if and only if the dynamical system without delays in control of the form*

$$\dot{\mathbf{x}}(t) = \mathbf{A}(t)\mathbf{x}(t)+\hat{\mathbf{B}}(t)\mathbf{u}(t), \tag{4.6.46}$$

where

$$\hat{\mathbf{B}}(t) = \sum_{i=0}^{M} \mathbf{F}(t, r_i(t))\mathbf{B}_i(r_i(t))\dot{r}_i(t) \tag{4.6.47}$$

is controllable in $[t_0, v_M(t_1)]$.

Proof:

Taking into account inequality (4.6.43) and formula (4.6.45), we can express relation (4.6.44) in the following form:

$$\begin{aligned}
&\mathbf{x}(t_1, \mathbf{z}(t_0), \mathbf{u}) \\
&= \mathbf{F}(t_1, t_0)\mathbf{x}(t_0)+\sum_{i=0}^{M} \int_{v_i(t_0)}^{t_0} \mathbf{F}(t_1, r_i(\tau))\mathbf{B}_i(r_i(\tau))\dot{r}_i(\tau)\mathbf{u}_{t_0}(\tau)\, d\tau+ \\
&\quad+\sum_{i=0}^{M} \int_{t_0}^{v_M(t_1)} \mathbf{F}(t_1, r_i(\tau))\mathbf{B}_i(r_i(\tau))\dot{r}_i(\tau)\mathbf{u}(\tau)\, d\tau+ \\
&\quad+\sum_{i=0}^{M} \int_{v_M(t_1)}^{v_i(t_1)} \mathbf{F}(t_1, r_i(\tau))\mathbf{B}_i(r_i(\tau))\dot{r}_i(\tau)\mathbf{u}(\tau)\, d\tau \\
&= \mathbf{q}(\mathbf{z}(t_0))+\sum_{i=0}^{M} \int_{v_M(t_1)}^{v_i(t_1)} \mathbf{F}(t_1, r_i(\tau))\mathbf{B}_i(r_i(\tau))\dot{r}_i(\tau)\mathbf{u}_{t_1}(\tau)\, d\tau+ \\
&\quad+\sum_{i=0}^{M} \int_{t_0}^{v_M(t_1)} \mathbf{F}(t_1, r_i(\tau))\mathbf{B}_i(r_i(\tau))\dot{r}_i(\tau)\mathbf{u}(\tau)\, d\tau.
\end{aligned} \tag{4.6.48}$$

According to Definition 4.6.2 the complete state at time t_1 of the dynamical system (4.6.1), $\mathbf{z}(t_1) = \{\mathbf{x}(t_1, \mathbf{z}(t_0), \mathbf{u}), \mathbf{u}_{t_1}\} = \{\mathbf{x}_1, \mathbf{w}\}$, is arbitrary. Hence the dynamical system (4.6.1) is absolutely controllable in $[t_0, t_1]$ if and only if in (4.6.48) the term

$$\sum_{i=0}^{M} \int_{t_0}^{v_M(t_1)} \mathbf{F}(t_1, r_i(\tau))\mathbf{B}_i(r_i(\tau))\dot{r}_i(\tau)\mathbf{u}(\tau)\, d\tau,$$

which, in fact, depends only on the control values in the time interval $[t_0, v_M(t_1)]$, can take arbitrary values in the space $\boldsymbol{R}^n$. Using the properties of the transition matrix $\mathbf{F}(t, t_0)$ and formula (5.6.47), we can express this term in the following form:

$$\sum_{i=0}^{M} \int_{t_0}^{v_M(t_1)} \mathbf{F}(t_1, r_i(\tau))\mathbf{B}_i(r_i(\tau))\dot{r}_i(\tau)\mathbf{u}(\tau)\, d\tau = \int_{t_0}^{v_M(t_1)} \mathbf{F}(t_1, \tau)\mathbf{B}(\tau)\mathbf{u}(\tau)\, d\tau. \tag{4.6.49}$$

By comparing the right-hand side of relation (4.6.49) and formula (1.2.3) it is easy to deduce that the dynamical system (4.6.1) is absolutely controllable in $[t_0, t_1]$ if and only if the dynamical system without delays in control (4.6.46) is controllable in $[t_0, v_M(t_1)]$. ■

COROLLARY 4.6.4 (Klamka (1977c)). *The dynamical system* (4.6.1) *is absolutely controllable in* $[t_0, t_1]$ *if and only if*

$$\operatorname{rank} \int_{t_0}^{v_M(t_1)} \mathbf{F}(t_1, t)\hat{\mathbf{B}}(t)\hat{\mathbf{B}}^{\mathrm{T}}(t)\mathbf{F}^{\mathrm{T}}(t_1, t)\,dt = n.$$

Proof:
Corollary 4.6.4 immediately follows from Theorems 4.6.5 and 1.3.1, ■

In the case of stationary dynamical systems (4.6.11) by using Theorem 4.6.5 it is relatively easy to deduce the results presented in Banks, Jacobs and Latina (1977), Olbrot (1972).

COROLLARY 4.6.5 (Banks, Jacobs and Latina (1977), Klamka (1977c), Olbrot (1972)). *The dynamical system* (4.6.11) *is absolutely controllable in* $[t_0, t_1]$ *if and only if*

$$\operatorname{rank}[\mathbf{K} \,\vdots\, \mathbf{AK} \,\vdots\, \mathbf{A}^2\mathbf{K} \,\vdots\, \ldots \,\vdots\, \mathbf{A}^{n-1}\mathbf{K}] = n, \tag{4.6.51}$$

where the $(n \times m)$*-dimensional matrix* $\mathbf{K}$ *is given by the following formula*:

$$\mathbf{K} = \sum_{i=0}^{M} \exp(-\mathbf{A}h_i)\mathbf{B}_i. \tag{4.6.50}$$

Proof:
Corollary 4.6.5 follows directly from Theorems 1.4.1 and 4.6.5 and formula (1.2.10), which describes the transition matrix for stationary dynamical systems. ■

Absolute controllability in $[t_0, t_1]$ of the dynamical system (4.6.1) strongly depends on the values of the functions $v_i(t)$, $i = 0, 1, 2, \ldots, M$ and, in the case of dynamical systems (4.6.8) or (4.6.11), on the values of delays h_i, $i = 1, 2, \ldots, M$. For some special values of the delays, a dynamical system can be absolutely controllable in a given time interval, and for other values of the delays the same dynamical system will not be absolutely controllable in the same time interval. This phenomenon is illustrated in Example 4.6.4. Example 4.6.4 also shows, that relative controllability in $[t_0, t_1]$ is not a sufficient condition for absolute controllability in the same time interval.

Example 4.6.4
Let us consider a dynamical system of form (4.6.11) for which the matrices $\mathbf{A}$, $\mathbf{B}_0$, and $\mathbf{B}_1$ are of the following form:

$$\mathbf{A} = \begin{bmatrix} -1 & 1 \\ 0 & -2 \end{bmatrix}, \quad \mathbf{B}_0 = \begin{bmatrix} 0 \\ -4 \end{bmatrix}, \quad \mathbf{B}_1 = \begin{bmatrix} 1 \\ 1 \end{bmatrix}, \quad M = 1.$$

Hence

$$\exp(-\mathbf{A}h_1) = \begin{bmatrix} \exp h_1 & \exp h_1 - \exp 2h_1 \\ 0 & \exp 2h_1 \end{bmatrix}.$$

In the case where $h_1 = \ln 2$, the above equality gives

$$\mathbf{K} = \mathbf{B}_0 + \exp(-\mathbf{A}\ln 2)\mathbf{B}_1 = \begin{bmatrix} 0 \\ -4 \end{bmatrix} + \begin{bmatrix} 2 & -2 \\ 0 & 4 \end{bmatrix}\begin{bmatrix} 1 \\ 1 \end{bmatrix} = \begin{bmatrix} 0 \\ 0 \end{bmatrix}.$$

Hence

$$\operatorname{rank}[\mathbf{K} \vdots \mathbf{AK}] = \operatorname{rank}\begin{bmatrix} 0 & \vdots & 0 \\ 0 & \vdots & 0 \end{bmatrix} = 0 < 2 = n.$$

Thus by Corollary 4.6.5 the above dynamical system with single delay $h_1 = \ln 2$ is not absolutely controllable in any time interval $[0, t_1]$ for $t_1 > \ln 2$.

In the case where $h_1 = \ln 4$, we have

$$\mathbf{K} = \mathbf{B}_0 + \exp(-\mathbf{A}\ln 4)\mathbf{B}_1 = \begin{bmatrix} 0 \\ -4 \end{bmatrix} + \begin{bmatrix} 4 & -12 \\ 0 & 16 \end{bmatrix}\begin{bmatrix} 1 \\ 1 \end{bmatrix} = \begin{bmatrix} -8 \\ 12 \end{bmatrix}.$$

Hence

$$\operatorname{rank}[\mathbf{K} \vdots \mathbf{AK}] = \operatorname{rank}\begin{bmatrix} -8 & \vdots & 20 \\ 12 & \vdots & -24 \end{bmatrix} = 2 = n.$$

Thus, by Corollary 4.6.5, the above dynamical system with single delay $h_1 = \ln 4$ is absolutely controllable in any time interval $[0, t_1]$ for all $t_1 > \ln 4$.

Using Theorem 4.6.1, we can easily check that the dynamical system in question is relatively controllable in any time interval $[0, t_1]$, for $t_1 > h_1 = \ln 2$. This directly follows from the fact that

$$\operatorname{rank}[\mathbf{B}_0 \vdots \mathbf{B}_1 \vdots \mathbf{AB}_0 \vdots \mathbf{AB}_1] = \operatorname{rank}\begin{bmatrix} 0 & \vdots & 1 & \vdots & -4 & \vdots & 0 \\ -4 & \vdots & 1 & \vdots & 8 & \vdots & -2 \end{bmatrix} = 2 = n.$$

Therefore the relative controllability in $[0, t_1]$ does not imply the absolute controllability in the same time interval.

In the case, where the dynamical system (4.6.1) is relatively controllable in $[t_0, t_1]$, there generally exist many different controls $\mathbf{u}(t)$ $(t_0 \leqslant t \leqslant t_1)$, which can transfer the initial complete state $\mathbf{z}(t_0)$ of the dynamical system to the desired final instantaneous state $\mathbf{x}_1 \in \mathbf{R}^n$ at time t_1. Among the possible controls which achieve the same transfer, we may look for the optimal one according to the following cost function (performance index):

$$J(\mathbf{u}) = \int_{t_0}^{t_1} \|\mathbf{u}(t)\|^2_{\mathbf{R}(t)} dt, \tag{4.6.52}$$

where $\mathbf{R}(t)$ is an $(n \times n)$-dimensional, continuous, symmetric, positive definite weighting matrix for all $t \in [t_0, t_1]$.

Using the same method as in Section 1.7, we can prove a theorem, which gives us an analytic formula for the optimal control function and, moreover, the corresponding minimal value of the performance index (4.6.52).

THEOREM 4.6.6 (Klamka (1977c)). *Suppose the dynamical system* (4.6.1) *is relatively controllable in* $[t_0, t_1]$, *and let* $u' \in L^2([t_0, t_1], \mathbf{R}^m)$ *be any control which steers the dynamical system* (4.6.1) *from the initial complete state* $\mathbf{z}(t_0)$ *to the desired final instantaneous state* $\mathbf{x}_1 \in \mathbf{R}^n$ *at time* t_1. *Then the control* $\mathbf{u}^0 \in L^2([t_0, t_1], \mathbf{R}^m)$ *given by*

$$\mathbf{u}^0(t) = \mathbf{R}^{-1}(t)\mathbf{B}_{t_1}^{\mathrm{T}}(t)\mathbf{F}^{\mathrm{T}}(t_1, t)\mathbf{W}_R^{-1}(t_0, t_1)\big(\mathbf{x}_1 - \mathbf{q}(\mathbf{z}(t_0))\big), \quad t \in [t_0, t_1], \tag{4.6.53}$$

where

$$\mathbf{W}_{\mathbf{R}}(t_0, t_1) = \int_{t_0}^{t_1} \mathbf{F}(t_1, t)\mathbf{B}_{t_1}(t)\mathbf{R}^{-1}(t)\mathbf{B}_{t_1}^{\mathrm{T}}(t)\mathbf{F}^{\mathrm{T}}(t_1, t)\,dt \tag{4.6.54}$$

steers the dynamical system (4.6.1) *from the initial complete state* $\mathbf{z}(t_0)$ *to the desired final instantaneous state* $\mathbf{x}_1 \in \mathbf{R}^n$ *at time* t_1 *and additionally*

$$\int_{t_0}^{t_1} \|\mathbf{u}'(t)\|_{\mathbf{R}(t)}^2\,dt \geqslant \int_{t_0}^{t_1} \|\mathbf{u}^0(t)\|_{\mathbf{R}(t)}^2\,dt. \tag{4.6.55}$$

Moreover, the minimal value of the performance index $J(\mathbf{u}^0)$ *is given by the formula*

$$J(u^0) = \int_{t_0}^{t_1} \|\mathbf{u}^0(t)\|_{\mathbf{R}(t)}^2\,dt = \|\mathbf{x}_1 - \mathbf{q}(\mathbf{z}(t_0))\|_{\mathbf{W}_{\mathbf{R}^{-1}(t_0,t_1)}}^2 . \tag{4.6.56}$$

Proof:

The proof of the theorem is almost the same as the proof of Theorem 1.7.1. The only difference lies in considering the matrix $\mathbf{B}_{t_1}(t)$ given by formula (4.6.16) instead of the matrix $\mathbf{B}(t)$ and in replacing the vector $\mathbf{F}(t_1, t_0)\mathbf{x}(t_0)$ by the vector $\mathbf{q}(\mathbf{z}(t_0))$ given by formula (4.6.15). ∎

A similar problem can also be solved in the case where a dynamical system (4.6.1) is absolutely controllable in $[t_0, t_1]$. Then the minimum energy control problem is formulated as follows: it is necessary to calculate the control $\mathbf{u}(t)$, $t_0 \leqslant t \leqslant v_M(t_1)$, which transfers the dynamical system (4.6.1) from the initial complete state $\mathbf{z}(t_0)$ to the desired final complete state $\mathbf{z}(t_1)$ and minimizes the cost function

$$J(\mathbf{u}) = \int_{t_0}^{v_M(t_1)} \|\mathbf{u}(t)\|_{\mathbf{R}(t)}^2\,dt, \tag{4.6.57}$$

where the $(n \times n)$-dimensional matrix $\mathbf{R}(t)$ satisfies the assumptions given in Section 1.7.

The solution of the problem stated above can be obtained in the same way as in Section 1.7. As a result we derive the analytical formula for the minimum energy control function and, moreover, the corresponding value of the performance index (4.6.57).

THEOREM 4.6.7 (Klamka (1977b)). *Suppose the dynamical system* (4.6.1) *is absolutely controllable in* $[t_0, t_1]$, *and let* $\mathbf{u}' \in L^2([t_0, v_M(t_1)], R^m)$ *be any control which steers the dynamical system* (4.6.1) *from the initial complete state* $\mathbf{z}(t_0)$ *to the desired final complete state* $\mathbf{z}(t_1)$. *Then the control* $\mathbf{u}^0 \in L^2([t_0, v_M(t_1)], R^m)$ *given by*

$$\mathbf{u}^0(t) = \mathbf{R}^{-1}(t)\hat{\mathbf{B}}^{\mathrm{T}}(t)\mathbf{F}^{\mathrm{T}}(t_1, t)\hat{\mathbf{W}}_{\mathbf{R}}^{-1}(t_0, v_M(t_1))\big(\mathbf{q}(\mathbf{z}(t_0)) - \mathbf{q}(\mathbf{z}(t_1))\big),$$
$$t \in [t_0, v_M(t_1)], \qquad (4.6.58)$$

where

$$\hat{\mathbf{W}}_{\mathbf{R}}(t_0, v_M(t_1)) = \int_{t_0}^{v_M(t_1)} \mathbf{F}(t_1, t)\hat{\mathbf{B}}(t)\mathbf{R}^{-1}(t)\hat{\mathbf{B}}^{\mathrm{T}}(t)\mathbf{F}^{\mathrm{T}}(t_1, t)\,dt \qquad (4.6.59)$$

steers the dynamical system (4.6.1) *from the initial complete state* $\mathbf{z}(t_0)$ *to the desired final complete state* $\mathbf{z}(t_1)$ *and additionally satisfied the inequality*

$$\int_{t_0}^{v_M(t_1)} ||\mathbf{u}'(t)||^2_{\mathbf{R}(t)}\,dt \geqslant \int_{t_0}^{v_M(t_0)} ||\mathbf{u}^0(t)||^2_{\mathbf{R}(t)}\,dt. \qquad (4.6.60)$$

Moreover, the minimal value of the performance index $J(\mathbf{u}^0)$ *is given by the formula*

$$J(\mathbf{u}^0) = \int_{t_0}^{t_1} ||\mathbf{u}^0(t)||^2_{\mathbf{R}(t)}\,dt = ||\mathbf{q}(\mathbf{z}(t_1)) - \mathbf{q}(\mathbf{z}(t_0))||_{\hat{\mathbf{W}}_{\mathbf{R}^{-1}}(t_0, v_M(t_1))},$$
$$(4.6.61)$$

Proof:

The proof of Theorem 4.6.7 is analogous to the proof of Theorems 1.7.1 and 4.6.6. It is only necessary to substitute $\hat{\mathbf{B}}(t)$ for the matrix $\mathbf{B}(t)$ and the vectors $\mathbf{q}(\mathbf{z}(t_0))$ and $\mathbf{q}(\mathbf{z}(t_1))$ for the vectors $\mathbf{F}(t_1, t_0)\mathbf{x}(t_0)$ and $\mathbf{x}_1$, respectively. After this substitution the proof proceeds almost automatically. ∎

4.7 Remarks and Comments

In Chapter 4 we have considered only the controllability of the simplest linear time-invariant dynamical systems of neutral type with lumped multiple constant delays (equation (4.2.1)). The results presented in this chapter can be generalized to linear time-invariant dynamical systems of neutral type with distributed delays in state variables and in control, which are described by the following functional-differential equation:

$$\frac{d}{dt}(\mathbf{x}(t) + C\mathbf{x}_t + D\mathbf{u}_t) = A\mathbf{x}_t + B\mathbf{u}_t, \qquad t \geqslant 0, \qquad (4.7.1)$$

where A, B, C, D are linear bounded operators with values in the space $\boldsymbol{R}^n$, defined as follows:

$$A\mathbf{x}_t = \int_{-h}^{0} d\bar{\mathbf{A}}(s)\mathbf{x}_t(s), \qquad \mathbf{x}_t \in C([-h, 0], \boldsymbol{R}^n),$$

$$B\mathbf{u}_t = \int_{-h}^{0} d\bar{\mathbf{B}}(s)\mathbf{u}_t(s), \qquad \mathbf{u}_t \in C([-h, 0], \boldsymbol{R}^m),$$

$$C\mathbf{x}_t = \int_{-h}^{0} d\bar{\mathbf{C}}(s)\mathbf{x}_t(s), \qquad \mathbf{x}_t \in C([-h, 0], \boldsymbol{R}^n),$$

$$D\mathbf{u}_t = \int_{-h}^{0} d\bar{\mathbf{D}}(s)\mathbf{u}_t(s), \qquad \mathbf{u}_t \in C[(-h, 0], \boldsymbol{R}^m),$$

where $\bar{\mathbf{A}}(s)$, $\bar{\mathbf{B}}(s)$, $\bar{\mathbf{C}}(s)$ and $\bar{\mathbf{D}}(s)$ are normalized matrix functions with bounded variation in $[-h, 0]$ and with values in $\boldsymbol{R}^{n\times n}$, $\boldsymbol{R}^{n\times m}$, $\boldsymbol{R}^{n\times n}$, $\boldsymbol{R}^{n\times m}$, respectively.

Numerous results concerning various kinds of controllability of the dynamical system (4.7.1) and its special cases (including, for example, the so-called *F-controllability* can be found in the following papers: Angeli and Galizia (1987), Banks and Jacobs (1973), Banks and Manitius (1974, 1975), Banks, Jacobs and Langenhop (1975), Choudhury (1973), Colonius (1982, 1984), Denham and Yamashita (1979a, 1979b), Kocięcki (1983), Korytowski (1975), Bartosiewicz (1980, 1984), Chukwu (1979, 1980, 1984, 1987), Delfour and Manitius (1974), Delfour, Manitius and Lee (1978), Frost and Storey (1978, 1979), Jacobs and Langenhop (1975, 1976), Klamka (1976a, 1976b, 1976c, 1976d, 1976e, 1976f, 1977a, 1977b, 1978a, 1978b, 1980a, 1980b, 1985c, 1985a, 1984c, 1981, 1982a, 1987, 1986a), Manitius (1972, 1976a, 1976b, 1982), Manitius and Olbrot (1972), Manitius and Triggiani (1978a, 1978b), Marchenko and Merezcha (1981), Miniuk (1983), O'Connor and Tarn (1983a, 1983b), Olbrot (1983, 1977, 1973, 1976, 1981), Olbrot and Jakubczyk (1981), Olbrot and Żak (1980b), Pandolfi (1976b), Rodas and Langenhop (1978), Reinbacher (1987), Sinha (1985, 1986), Sinha and Yamamoto (1980), Somasundaram and Balachandram (1983, 1984a, 1984b, 1985), Tarn and Spong (1981), Thowsen (1976, 1980a, 1980b), Underwood and Young (1979), Underwood and Chukwu (1988), Williams and Zakian (1977), Zmood (1974), Krakhotko, Alsevich and Razmyslovich (1980), Kurcyusz and Olbrot (1977), Lee and Olbrot (1983), Levsen and Nazaroff (1973), Olbrot and Sosnowski (1981b), Salamon (1984), Tadmor (1984).

Generalizations to time-varying linear dynamical systems with distributed or lumped multiple delays in state variables or in control are given in the following publications: Klamka (1976a, 1976b, 1976c, 1976d, 1976e, 1976f), Levsen and Nazaroff (1973), Manitius (1976a, 1976b, 1982), Warga (1972). In these papers several necessary and sufficient conditions for various kinds of controllability are presented. The minimum energy control problem is considered and solved in papers of Klamka (1976b, 1976d).

As in the case of dynamical systems without delays, stabilizability in dynamical systems with delays is strongly connected with various kinds of controllability. The relations between stabilizability and controllability in dynamical systems with delays are extensively investigated in the following papers: Datko (1974, 1976), Delfour (1977), Delfour and Manitius (1974), Kwon and Pearson (1977), Manitius (1976b), Manitius and Triggiani (1978b), Manitius and Olbrot (1972), O'Connor and Tarn (1983b), Olbrot (1976), Pandolfi (1975, 1976b, 1978a), Przyłuski (1977), Sontag (1976a, 1976b), Ignatenko and Janovich (1980), Manitius and Manousiouthakis (1985).

The dynamical systems (4.2.1), (4.6.1) and (4.7.1) have their trajectories in a finite-dimensional space, i.e., $\mathbf{x}(t) \in \boldsymbol{R}^n$, for $t \geqslant 0$. Recently, there have appeared papers in which dynamical systems with delays and with trajectories in infinite-dimensional Banach or Hilbert spaces are considered: Kamen (1976), Klamka (1978b, 1978c, 1981a, 1982c), Pandolfi (1978a), Warga (1972). The criteria for various kinds of controllability for these systems have been obtained by using the methods and results presented in Chapters 3 and 4.

For dynamical systems with delays, we can define many different types of observability and formulate several kinds of the so-called duality principle. The precise form of the duality principle strongly depends on the type of dynamical system and on the kinds of controllability and observability. It also depends on the form of the output equation. The most general time-invariant output equation which contains delays in the state variables and in the control is the equation of the following form:

$$\mathbf{y}(t) = E\mathbf{x}_t + F\dot{\mathbf{x}}_t + G\mathbf{u}_t, \qquad t \geqslant 0, \tag{4.7.2}$$

where E, F, G are linear bounded operators which have values in $\boldsymbol{R}^p$ and are defined as follows:

$$E\mathbf{x}_t = \int_{-h}^{0} d\overline{\mathbf{E}}(s)\mathbf{x}_t(s), \qquad \mathbf{x}_t \in C([-h, 0], \boldsymbol{R}^n),$$

$$F\dot{\mathbf{x}}_t = \int_{-h}^{0} d\overline{\mathbf{F}}(s)\dot{\mathbf{x}}_t(s), \qquad \mathbf{x}_t \in C^1([-h, 0], \boldsymbol{R}^n),$$

$$G\mathbf{u}_t = \int_{-h}^{0} d\overline{\mathbf{G}}(s)\mathbf{u}_t(s), \qquad \mathbf{u}_t \in C([-h, 0], \boldsymbol{R}^m),$$

where $\overline{\mathbf{E}}(s)$, $\overline{\mathbf{F}}(s)$ and $\overline{\mathbf{G}}(s)$ are normalized matrix functions with bounded variation in $[-h, 0]$ and with values in $\boldsymbol{R}^{p\times n}$, $\boldsymbol{R}^{p\times n}$ and $\boldsymbol{R}^{p\times m}$, respectively.

The results concerning the various kinds of observability of the dynamical systems with delays in state equation (4.7.1) and in the output equation (4.7.2) are given in the following publications: Bhat and Koivo (1976), Delfour and Mitter (1972a), Delfour and Manitius (1974), Desoer and Chen (1976), Klamka (1982a), Lee and Olbrot (1981), Lee, Neftci and Olbrot (1982), Manitius (1976a, 1976b, 1982), Marchenko (1981), Olbrot (1976, 1981), Olbrot and Sosnowski (1981a), Olbrot and Żak (1980a, 1980b), Pandolfi (1979), Williams and Zahian (1977), Przyłuski and Sosnowski (1987), Ha (1980). With the use of the duality principle some interesting special cases are considered. It should also be stressed that the problem of the observability of dynamical systems with delays is strongly connected with the problem of detectability and design of observers. This observation directly leads to methods of design of observers, based on functional analysis or algebraic considerations.

Bibliographical Notes on the Controllability of Nonlinear Systems

Controllability theory for nonlinear dynamical systems is not so uniform and connected as in the case of linear dynamical systems. Most of the results obtained and of the controllability criteria have a local character or concern only a very narrow class of dynamical systems. The main difficulty arising in the investigation of controllability for nonlinear dynamical systems is the lack of general methods for solving nonlinear differential or functional-differential equations. Hence we cannot use methods based on the evident form of solution (such methods can be used in linear dynamical systems, where the solution is given in the integral form).

Hence in controllability investigations for nonlinear dynamical systems we use many different methods, which depend on the type of nonlinear equation and the kinds of controllability. The most popular methods are based on:

(1) fixed point theorems for nonlinear maps or implicit function theorems,

(2) theory of vector fields and Lie algebras,

(3) perturbation methods—the controllability of a perturbed nonlinear system is verified by the investigation of controllability for an appropriate linear dynamical system without any perturbations,

(4) maximum principle.

The choice of the appropriate method is based on the type of nonlinearity in the state equation.

Using fixed point theorems and implicit function theorems (Magnusson, Pritchard and Quinn (1985)) we usually obtain sufficient conditions for local controllability as well as for global controllability, but with stronger assumptions about system parameters. However, it should be stressed that this method may be used only for a very narrow class of nonlinear dynamical systems. Namely, we can only consider dynamical systems with the state equation containing the following term: $\mathbf{A}(t)\mathbf{x}(t)$, or $\mathbf{A}(t, \mathbf{x}(t))\mathbf{x}(t)$, or $\mathbf{A}(t, \mathbf{x}(t), \mathbf{u}(t))\mathbf{x}(t)$ (Carmichael and Quinn (1988), Davison and Kunze (1970), Klamka (1975a, 1975b), Mirza and Womack (1972a), Kartsatos and Dannon (1987)). On the other hand, this method allows us to derive sufficient conditions for the relative controllability of the dynamical systems with delays in the state variables (Dauer and Gahl (1977)), in the control (Klamka (1975c, 1976a, 1978a, 1980a, 1980b), Mirza and Womack (1972b), Wei (1976), and even for neutral nonlinear dynamical systems (Chukwu (1980, 1984, 1987), Gahl (1978)). This is a valuable advantage of the fixed point methods.

Moreover, it should be pointed out that by using fixed point theorems for controllability investigations we obtain controllability conditions in the space of continuous functions. This follows directly from the fact that the criteria for checking the compactness of a set in the space of continuous functions are quite easily applicable (Warga (1972)).

The Lie algebras and the theory of vector fields are also extensively used in the investigation of the controllability of nonlinear systems. These methods allow us to obtain numerous interesting results for some class of nonlinear dynamical systems, which are presented in the following publications: Aeyels (1985), Haynes and Hermes (1970), Hermann and Krener (1977), Krener (1987), Hermes (1971, 1974a, 1974b, 1976a, 1976b, 1982), Hirschorn (1975, 1976, 1986), Hunt (1979, 1982, 1984), Jurdjevic (1972, 1978), Jurdjevic and Sallet (1984), Levitt and Sussmann (1975), Lobry (1974), Sussmann (1975, 1976, 1978, 1983, 1987), Sussmann and Jurdjevic (1972), Tokarzewski (1977, 1979).

The comparative methods belong to the oldest methods (Lee and Markus (1968)) and give valuable results in the case of additive perturbations (Barmish and Schmitendorf (1980, 1982), Chukwu (1975, 1976), Chukwu and Gronski (1977), Chukwu and Silliman (1977), Dauer (1971, 1972a, 1972b, 1973a, 1973b, 1974a, 1974b, 1976), Gauthier and Bornard (1982a, 1982b, 1982c), Rubio (1983), Sannuti (1977, 1978), Gabasov, Karpyuk and Kirilova (1980), Furi, Nistri, Pera and Zezza (1985b), Hou (1985), Lukes (1972, 1974a, 1974b, 1985), Sontag (1983). The comparative methods are among the most popular methods in investigating the controllability of nonlinear dynamical systems.

The special version of the maximum principle can also be used in checking the controllability of nonlinear dynamical systems. However, this method requires introducing very complicated mathematical notations and calculations (Warga (1972)) and hence it is not often used (Grantham and Vincent (1975), Warga (1972, 1976, 1978a, 1978b, 1985), Yorke (1972)).

Numerous results concern the controllability of nonlinear dynamical systems defined on the plane, i.e., with two state variables (Albrecht and Wax (1978), Boothby (1982), Mohler (1973), Wei and Pearson (1978c)).

In the investigations of controllability for nonlinear dynamical systems we can also use the method of the Lapunov function (Gershwin and Jacobson (1971), Jurdjevic and Quinn (1978)), some geometrical methods (Aronson (1973a, 1973b), Korobov (1979)) and some other methods of mathematical analysis (Aubin, Frankowska and Olech (1986), Bacciotti (1983), Bacciotti and Stefani (1981, 1983), Elliot (1971), Khanh (1981), Kisielewicz (1980), Kopeykina (1980), Kopeykina and Gurina (1981), Rubio (1983), Schwarzkopf (1975), Skowroński and Leitmann (1977), Skowroński and Vincent (1982), Vincent and Skowroński (1979), Vinter (1980), Walczak (1984), Frankowska (1987), Papageorgiu (1987), Tsinias and Kalouptsidis (1982), Van der Schoft (1982)).

The stochastic controllability of nonlinear dynamical systems has been considered in papers of Sunahara, Aikara and Kishino (1975), Klamka and Socha (1977, 1980), Mariton (1987), Socha and Klamka (1978), Zabczyk (1981), approximate controllability for infinite-dimensional nonlinear dynamical systems in the papers of Chewning (1976), Fujii and Sokawa (1974), Underwood and Young (1979), and the properties of attainable sets for nonlinear systems have been examined in the paper of Hermes (1971) and Rebhuhn (1977).

A special class of nonlinear systems are bilinear dynamical systems. The bibliography concerning continuous and discrete bilinear dynamical systems has been presented in Section 2.7.

The controllability of nonlinear dynamical systems with delays in control or in the state variables has seldom been considered in the literature, and the methods of its investigation are difficult. This immediately follows, on the one hand, from the nonlinearity of the differential-difference state equation and, on the other hand, from the existing delays.

Of the methods presented at the beginning of this section, the one that gives the best results is the method based on the fixed point theorems for nonlinear mappings. Particularly useful is the Schauder fixed point theorem and the Darboux fixed point theorems. However, it should be pointed out that the Darboux theorem requires introducing the notion of the so-called "measure of noncompactness" for noncompact set (Balachandran (1987, 1988), Dacka (1980a, 1980b, 1981, 1982), Magnusson, Pritchard and Quinn (1985)).

Fixed point methods require the investigation of compactness for the sets in Banach spaces. In order to do that we must use some special compactness conditions (i.e., the Arzelà–Ascoli theorem for the case of the space of continuous functions), which generally require laborious and tedious calculations. It should be mentioned that the Darboux theorem allows us to avoid those difficulties and this is its main advantage. Moreover, the Darboux theorem permits considering controllability for neutral nonlinear dynamical systems (Dacka (1981)) and dynamical systems with delays in control (Dacka (1982)).

In the papers of Dauer and Gahl (1977) and Dłotko (1981) sufficient conditions for the relative controllability of nonlinear dynamical systems with delays in the state variable have been formulated. Publications of Chukwu (1980, 1984, 1987), Gahl (1978) concern the problem of the relative and absolute controllability of nonlinear neutral dynamical systems. Moreover, papers of Klamka (1976a, 1976e, 1978a) contain several sufficient conditions for relative controllability of nonlinear dynamical systems with lumped and distributed delays in control.

Finally, it should be mentioned that the theory of controllability for nonlinear dynamical systems is still in the process of developing and still rather far from a satisfactory solution.

References

Aeyels, D. (1985) "Global controllability for smooth nonlinear systems: a geometric approach", *SIAM Journal on Control and Optimization, 23*, 452–465.

Ahmed, N. U. (1985) "Finite-time null controllability for a class of linear evolution equations on a Banach space with control constraints", *Journal of Optimization Theory and Applications, 47*, 129–158.

Albrecht, F. and Wax, N. (1978) "The controllability of certain nonlinear planar systems", *Journal of Mathematical Analysis and Applications, 63*, 762–771.

Albrecht, F., Grasse, K. A. and Wax, N. (1986) "Path controllability of linear input-output systems", *IEEE Transactions on Automatic Control, 31*, 569–571.

Angeli, M. and Galizia, T. (1987) "Minimal controllability for systems with delays", *International Journal of Control, 45*, 1255–1264.

Anderson, B. D. O. and Min Hong Hui (1982) "Structural controllability and matrix nets", *International Journal of Control, 35*, 397–416.

Anichini, G. (1983) "Controllability and controllability with prescribed controls", *Journal of Optimization Theory and Applications, 39*, 35–45.

Armentano, V. A. (1985) "The pencil ($s\mathbf{A}-\mathbf{E}$) and controllability-observability for generalized linear systems: a geometrical approach", *SIAM Journal on Control and Optimization, 23*, 616–638.

Aronson, G. (1973a) "A new approach to nonlinear controllability", *Journal of Mathematical Analysis and Applications, 44*, 763–772.

Aronson, G. (1973b) "Global controllability and bang-bang steering of certain nonlinear systems", *SIAM Journal on Control, 11*, 1–13.

Artstein, Z. (1973) "Trajectories and the attainable set of an abstract linear control systems, *Mathematical Systems Theory, 7*, 165–280.

Artstein, Z. (1980) "Discrete and continuous bang-bang and facial spaces, or: look for the extreme points", *SIAM Journal Review, 22*, 172–185.

Artstein, Z. (1982a) "Uniform controllability via limiting systems", *Applied Mathematical Optimization, 9*, 111–131.

Artstein, Z. (1982b) "Linear systems with delayed controls: a reduction", *IEEE Transactions on Automatic Control, 27*, 869–879.

Artstein, Z. and Tadmor, G. (1982) "Linear systems with indirect controls: the underlying measure", *SIAM Journal on Control and Optimization, 20*, 96–111.

Athans, M. and Falb, P. (1966), *Optimal Control*, New York: McGraw-Hill.

Aubin, J. P., Frankowska, H. and Olech, C. (1986) "Controllability of convex processes", *SIAM Journal on Control and Optimization, 24*, 1192–1211.

Bacciotti, A. (1983) "On the positive orthant controllability of two-dimensional nonlinear systems", *Systems and Control Letters, 3*, 53–55.

Bacciotti, A. and Stefani, G. (1981) "The region of attainability of nonlinear systems with unbounded controls", *Journal of Optimization Theory and Applications, 35*, 57–84.

Bacciotti, A. and Stefani, G. (1983) "On the relationship between global and local controllability, *Mathematical Systems Theory, 16*, 79–91.

Balakrishnan, A. V. (1977) *Introduction to Optimization Theory in Hilbert Spaces*, New York: Springer–Verlag.

Balachandran, K. (1987) "Global relative controllability of nonlinear systems with time-varying multiple delays in control", *International Journal of Control, 46*, 193–200.

Balachandran, K. (1988) "Controllability of a class of perturbed nonlinear systems" *Kybernetika, 24*, 61–64.

Ball, J. M., Marsden, J. E. and Slemrod, M. (1982) "Controllability for distributed bilinear systems", *SIAM Journal on Control and Optimization, 20*, 575–579.

Banaszuk, A., Kocięcki, M. and Przyłuski, K. M. (1988) "Remarks on controllability of implicit linear discrete-time systems", *Systems and Control Letters, 10*, 67–70.

Banks, H. T., Jacobs, M. Q. and Latina, M. R. (1977) "The synthesis of optimal controls for linear problems with retarded controls", *Journal of Optimization Theory and Applications, 8*, 319–366.

Banks, H. T. and Jacobs, M. Q. (1973) "An attainable sets approach to optimal control of functional differential equations with function space terminal conditions", *Journal of Differential Equations, 13*, 127–149.

Banks, H. T. and Manitius, A. (1974) "Application of abstract variational theory of hereditary systems—a survey", *IEEE Transactions on Automatic Control, 19*, 524–533.

Banks, H. T., Jacobs, M. Q. and Langenhop, C. E. (1975) "Characterization of the controlled states in W_2^1 of linear hereditary systems", *SIAM Journal on Control, 13*, 611–649.

Banks, H. T. and Manitius, A. (1975) "Projection series for retarded functional differential equations with applications to optimal control problems", *Journal of Differential Equations, 18*, 296–332.

Baras, J. S., Brockett, R. W. and Fuhrmann, P. A. (1974) "State-space models for infinite-dimensional systems", *IEEE Transactions on Automatic Control, 19*, 693–700.

Baras, J. S., and Brockett, R. W. (1975) "H^2-functions and infinite-dimensional realisation theory", *SIAM Journal on Control, 13*, 221–241.

Barmish, B. R. and Schmitendorf, W. E. (1980) "New results on controllability of system of the form $\dot{\mathbf{x}}(t) = \mathbf{A}(t)\mathbf{x}(t)+\mathbf{f}(t, \mathbf{u}(t))$", *IEEE Transactions on Automatic Control, 25*, 540–547.

Barmish, B. R. and Schmitendorf, W. E. (1982) "The associated disturbance free system: a means for investigating the controllability of disturbed system", *Journal of Optimization Theory and Applications, 38*, 525–540.

Bar-Ness, Y. and Langholz, G. (1975) "Preservation of controllability under sampling", *International Journal of Control, 22*, 39–48.

Bartosiewicz, Z. (1980) "Density of image of semigroup operators for linear neutral functional differential equations", *Journal of Differential Equations, 38*, 161–175.

Bartosiewicz, Z. (1984) "Approximate controllability of neutral systems with delays in control", *Journal of Differential Equations, 51*, 295–325.

Basile, G. and Marro, G. (1973) "A new characterization of some structural properties of linear systems: unknown-input observability, invertibility and functional controllability", *International Journal of Control, 17*, 931–943.

Benchimol, C. D. (1978a) "A note on weak stabilizability of contraction semigroup", *SIAM Journal on Control and Optimization, 16*, 373–379.

Benchimol, C. D. (1978b) "Feedback stabilizability in Hilbert spaces", *Applied Mathematical Optimization, 4*, 225–248.

Bernhard, P. (1980) "Exact controllability of perturbed continuous-time linear systems", *IEEE Transactions on Automatic Control, 25*, 89–96.

Bhat, K. P. and Koivo, H. N. (1976) "Modal characterization of controllability and observability for time delay systems", *IEEE Transactions on Automatic Control, 21*, 292–293.

Bhattacharyya, S. and Gomes, A. (1973) "Output controllability and eigenvalues assignability in single-output linear systems", *IEEE Transactions on Automatic Control, 18*, 540–541.

Bianchini, R. M. (1983a) "Local controllability, rest states and cyclic points", *SIAM Journal on Control and Optimization, 21*, 714–720.

Bianchini, R. M. (1983b) "Instant controllability of linear autonomous systems", *Journal of Optimization Theory and Applications, 39*, 237–250.

Bittanti, S., Guardabassi, G., Maffezoni, C. and Silverman, L. (1978) "Controllability and the matrix Riccati equation", *SIAM Journal on Control and Optimization, 16*, 37–40.

Bittanti, S., Colaneri, P. and Guardabassi, G. (1984) "*H*-controllability and observability of linear periodic systems", *SIAM Journal on Control and Optimization, 22*, 889–893.

Boothby, W. M. (1982) "Some comments on positive orthant controllability of bilinear systems", *SIAM Journal on Control and Optimization, 20*, 634–644.

Brammer, R. F. (1972) "Controllability in linear autonomous systems with positive controllers", *SIAM Journal on Control, 10*, 339–353.

Brammer, R. F. (1975) "Differential controllability and the solution of linear inequalities", *IEEE Transactions on Automatic Control, 20*, 128–131.

Brasch, F., Howze, J. and Pearson, J. (1971) "On the controllability of composite systems", *IEEE Transactions on Automatic Control, 16*, 210–220.

Brogan, L. W. (1968) "Optimal control theory applied to systems described by partial differential equations", *Advances in Control Systems, 6*, 222–316.

Brockett, R. W. (1972) "On the reachable set for bilinear systems", in: *Theory and Applications of Variable Structure Systems*, New York; Academic Press, 54–63.

Brockett, R. W. (1976) "Finite and infinite dimensional bilinear realization", *Journal of the Franklin Institute, 301*, 509–520.

Brockett, R. W. and Fuhrmann, P. A. (1976) "Normal symmetric dynamical systems", *SIAM Journal on Control and Optimization, 14,* 117–119.

Brunovsky, P. (1969) "Controllability and linear closed-loop controls in linear periodic systems", *Journal of Differential Equations, 6,* 296–313.

Brunovsky, P. (1970) "A classification of linear controllable systems", *Kybernetika, 6,* 173–188.

Burrows, C. R. and Sahinkaya, M. N. (1981) "A new algorithm for determining structural controllability", *International Journal of Control, 33,* 379–392.

Burrows, C. R. and Sahinkaya, M. N. (1983) "Modified algorithm for determining structural controllability", *International Journal of Control, 37,* 1417–1431.

Busłowicz, M. (1981) "Controllability of linear discrete-delay systems", *Proceedings of the Symposium on Functional Differential Systems and Related Topics" II,* Zielona Góra, 47–51.

Butkovski, A. G. (1975) *Control Methods for Systems with Distributed Parameters* (in Russian), Moscow: Nauka.

Butkovski, A. G. (1977) *Structure Theory of Distributed Systems* (in Russian), Moscow: Nauka.

Carja, O. (1988) "On constraint controllability of linear systems in Banach spaces", *Journal of Optimization Theory and Applications, 56,* 215–225.

Carmichael, N. and Quinn, M. D. (1988) "Fixed-point methods in nonlinear control", *IMA Journal of Matmematical Control and Information, 5,* 41–67.

Casti, J. L. (1984) *Nonlinear System Theory,* New York: Academic Press.

Casti, J. L. and Wood, E. F. (1984) "Some question of reachability in natural-resource management", *Applied Mathematical Optimization, 15,* 185–207.

Chan, W. L. and Lau, C. K. (1982) "Constrained stochastic controllability of infinite dimensional linear systems", *Journal of Mathmeatical Analysis and Applications, 85,* 46–78.

Chan, W. L. and Li, C. W. (1981) "Controllability, observability and duality in infinite dimensional linear systems, with multiple norm-bounded controllers", *International Journal of Control, 33,* 1039–1058.

Chan, W. L. and Li, C. W. (1982) "Max-min controllability in pursuit games with norm bounded controls", *Journal of Optimization Theory and Applications, 37,* 89–113.

Chang, C. S., Tarn, T. J. and Elliot, D. L. (1975) "Controllability of bilinear systems", in: *Variable Systems with Application to Economics and Biology,* New York: Springer-Verlag.

Chen, C. T. (1967) "Output controllability of composite systems", *IEEE Transactions on Automatic Control, 12,* 201–202.

Chen, C. T. (1970) *Introduction to Linear System Theory,* New York: Holt, Rinehart and Winston Inc.

Chen, C. T., Desoer, C. A. and Niederliński, A. (1966) "Simplified conditions for controllability and observability of linear time-invariant systems", *IEEE Transactions on Automatic Control, 11,* 613–614.

Chen, C. T. and Desoer, C. A. (1967) "Controllability and observability of composite systems", *IEEE Transactions on Automatic Control, 12,* 402–409.

Chen, C. T. and Desoer, C. A. (1968) "A proof of controllability of Jordan form state equations" *IEEE Transactions on Automatic Control, 13,* 195–196.

Chen, G. (1979a) "Control and stabilization for the wave equation in a bounded domain", *SIAM Journal on Control and Optimization*, *17*, 66–81.

Chen, G. (1979b) "Energy decay estimates and exact boundary value controllability for the wave equation in a bounded domain", *Journal Mathematiques Pures et Appliquées*, *58*, 249–273.

Chewning, W. C. (1976) "Controllability of the nonlinear wave equation in several space variables", *SIAM Journal on Control and Optimization*, *14*, 19–25.

Choudhury, A. K. (1973) "A contribution to the controllability of time-lag systems", *International Journal of Control*, *17*, 365–373.

Chu, H. (1966) "Some topological properties of complete controllability", *IEEE Transactions on Automatic Control*, *11*, 612–613.

Chu, K. W. E. (1988) "A controllability condensed form and a state feedback pole assignment algorithm for descriptor systems", *IEEE Transactions on Automatic Control*, *33*, 336–370.

Chukwu, E. N. (1975) "Finite time controllability of nonlinear control processes", *SIAM Journal on Control*, *13*, 807–816.

Chukwu, E. N. (1976) "On the controllability to a closed sets of nonlinear and related systems", *Nonlinear Analysis Theory, Methods and Applications*, *4*, 429–441.

Chukwu, E. N. (1979) "Euclidean controllability of linear delay systems with limited controls", *IEEE Transactions on Automactic Control*, *24*, 798–800.

Chukwu, E. N. (1980) "On the null controllability of nonlinear delay systems with restrained controls", *Journal of Mathematical Analysis and Applications*, *78*, 283–296.

Chukwu, E. N. (1984) "Null controllability in function space of nonlinear retarded systems with limited control", *Journal of Mathematical Analysis and applications*, *103*, 198–210.

Chukwu, E. N. (1987) "Function space null controllability of linear delay systems with limited power", *Journal of Mathematical Analysis and Applications*, *124*, 392–304.

Chukwu, E. N. and Gronski, J. M. (1977) "Approximate and complete controllability of nonlinear systems to a convex target set", *Journal of Mathematical Analysis and Applications*, *61*, 97–112.

Chukwu, E. N. and Silliman, S. D. (1977) "Complete controllability to a closed target set", *Journal of Optimization Theory and Applications*, *21*, 369–383.

Ciftibasi, T. and Yuksel, C. (1983) "Sufficient or necessary conditions for modal controllability and observability of Roesser's 2*D* system model", *IEEE Transactions on Automatic Cotntrol*, *28*, 527–529.

Clarke, F. H. and Loewen, P. D. (1986) "The value function in optimal control: sensitivity, controllability and time-optimality", *SIAM Journal on Control and Optimization*, *24*, 243–263.

Clayton, P. R. and Kuo, Y. L. (1971) "Controllability and observability of multivariable systems", *IEEE Transactions on Automatic Control*, *16*, 207–209.

Colonius, F. (1982) "Stable and regular reachability for relaxed hereditary differential systems", *SIAM Journal on Control and Optimization*, *20*, 675–694.

Colonius, F. (1984) "On approximate and exact null controllability of delay systems", *Systems and Control Letters*, *5*, 209–211.

Corfmat, J. P. and Morse, A. S. (1976) "Structurally controllable and structurally canonical systems", *IEEE Transactions on Automatic Control*, *21*, 130–131.

Coxson, P. G. and Shapiro, H. (1987) "Positive input reachability and controllability of positive systems", *Linear Algebra and its Applications, 94*, 35–53.

Cullen, D. J. (1986) "An algebraic condition for controllability at infinity", *Systems and Control Letters, 6*, 321–324.

Dacka, C. (1980a) "On the controllability of a class of nonlinear systems", *IEEE Transactions on Automatic Control, 24*, 263–266.

Dacka, C. (1980b) "On the controllability of certain nonlinear systems", *Archiwum Automatyki i Telemechaniki, 25*, 483–494.

Dacka, C. (1981) "On the controllability of nonlinear systems with time variable delays", *IEEE Transactions on Automatic Control, 26*, 956–959.

Dacka, C. (1982) "Relative controllability of perturbed nonlinear systems with delay in control", *IEEE Transactions on Automatic Control, 27*, 268–270.

Datko, R. (1974) "Neutral autonomous functional equations with quadratic cost", *SIAM Journal on Control, 12*, 70–82.

Datko, R. (1976) "The stabilization of linear functional differential equations", in: *Calculus of Variations and Control Theory*, New York: Academic Press, 353–369.

Dauer, J. P. (1971) "Perturbation of linear control systems", *SIAM Journal on Control, 9*, 393–400.

Dauer, J. P. (1972a) "Controllability of nonlinear systems using a growth condition", *Journal of Optimization Theory and Applications, 9*, 90–98.

Dauer J. P. (1972b) "A controllability technique for nonlinear systems", *Journal of Mathematical Analysis and Applications, 37*, 442–451.

Dauer, J. P. (1973a) "A note on bounded perturbations of controllable systems", *Journal of Mathematical Analysis and Applications, 42*, 221–225.

Dauer, J. P. (1973b) "On controllability of systems of the form $\dot{\mathbf{x}}(t) = \mathbf{A}(t)\mathbf{x}(t)+\mathbf{g}(t, \mathbf{u})$", *Journal of Optimization Theory and Applications, 11*, 132–138.

Dauer, J. P. (1974a) "Controllability of nonlinear systems with restrained controls" *Journal of Optimization Theory and Applications, 14*, 251–262.

Dauer, J. P. (1974b) "Approximate controllability of nonlinear systems with restrained controls", *Journal of Mathematical Analysis and Applications, 46*, 126–131.

Dauer, J. P. (1976) "Nonlinear perturbations of quasi-linear control systems", *Journal of Mathematical Analysis and Applications, 54*, 717–725.

Dauer, J. P. and Gahl, R. D. (1977) "Controllability of nonlinear delay systems", *Journal of Optimization Theory and Applications, 21*, 59–70.

Davison, E. J. (1977) "Connectability and structural controllability of composite systems", *Automatica, 13*, 109–124.

Davison, E. J. and Kunze, E. G. (1970) "Some sufficient conditions for the global and local controllability of nonlinear time-varying systems", *SIAM Journal on Control and Optimization, 8*, 489–498.

Davison, E. J. and Wang, S. H. (1975) "New results on the controllability and observability of general composite systems", *IEEE Transactions on Automatic Control, 20*, 123–128.

Delfour, M. C. (1977) "The linear quadratic optimal control problem for hereditary differential systems: theory and numerical solution", *Applied Mathematical Optimization, 3*, 101–162.

Delfour, M. C. and Mitter, S. K. (1972) "Controllability, observability and optimal feedback control of affine hereditary differential systems", *SIAM Journal on Control, 10*, 298–328.

Delfour, M. C. and Mitter, S. K. (1972b) "Controllability and observability for infinite dimensional systems", *SIAM Journal on Control, 10*, 329–333.

Delfour, M. C. and Manitius, A. (1971) "Control systems with delays: areas of applications and present status of the linear theory", in: *New Trends in Systems Analysis*, New York: Springer-Verlag, 2, 421–437.

Delfour, M. C., Manitius, A. and Lee, E. B. (1978) "*F*-reduction of the operator Riccati equation for hereditary differential systems", *Automatica, 14*, 385–395.

Denham, M. J. and Yamashita, K. (19791) "Dead-beat output zeroing by input and output feedback in multivariable delay-differential systems", *International Journal of Control, 30*, 803–812.

Denham, M. J. and Yamashita, K. (1979b) "On the controllability of delay-differential systems", *International Journal of Control, 30*, 813–822.

Desoer, C. A. and Chen, C. T. (1976) "Controllability and observability of feedback systems", *IEEE Transactions on Automatic Control, 12*, 474–475.

Dickinson, B.W. (1973) "Matrical indices and controllability of linear differential systems", *SIAM Journal on Applied Mathematics, 25*, 613–617.

Dłotko, T. (1981) "On the controllability of certain nonlinear systems of differential equations with deviating argument", *Proceedings of the Symposium on Functional Differential Systems and Related Topics II*, Zielona Góra, 96–101.

Dłotko, T., Klamka, J. and Ligęza, J. (1976) "Controllability and minimum energy control of linear systems", *Podstawy Sterowania, 6*, 211–218.

Dolecki, S. (1976) "A classifications of controllability concepts for infinite dimensional systems", *Control and Cybernetics, 5*, 33–44.

Dolecki, S. and Russell, D. L. (1977) "A general theory of observation and control", *SIAM Journal on Control and Optimization, 15*, 185–220.

Dunford, N. and Schwartz, J. T. (1958) *Linear Operators. General Theory*, New York: Inetrscience Publishers, Inc.

Dunford, N. and Schwartz, J. T. (1963) *Linear Operatos. Spectral Theory*, New York: Interscience Publishers, Inc.

Eising, R. (1978) "Realization and stabilization of 2-*D* systems", *IEEE Transactions on Automatic Control, 23*, 793–799.

Eising, R. (1979a) "Controllability and observability of 2-*D* systems", *IEEE Transactions on Automatic Control, 24*, 132–133.

Eising, R. (1979b) "Low-order realizations for 2-*D* transfer functions", *Proceedings of the IEEE, 67*, 866–868.

Eising, R. (1979c) "Separability of 2-*D* transfer matrices", *IEEE Transactions on Automatic Control 24*, 508–510.

Eising, R. and Hautus, M. J. L. (1981) "Realization algorithm for systems over a principal ideal domain", *Mathematical Systems Theory, 14*, 353–366.

Elliot, D. L. (1971) "A consequence of controllability", *Journal of Differential Equations, 10*, 364–370.

Elliot, D. L., Tarn, T. J. and Goka, T. (1973) "Controllability of discrete bilinear systems with bounded controls", *IEEE Transactions on Automatic Control, 18*, 298–301.

Emre, E. (1980) "A polynomial characterization of (A, B)-invariant and reachability subspaces", *SIAM Journal on Control and Optimization, 18*, 420–436.

Evans, M. E. and Murthy, D. N. (1977a) "Controllability of a class of discrete-time bilinear systems" *IEEE Transactions on Automatic Control, 22*, 78–83.

Evans, M. E. and Murthy, D. N. (1977b) "Controllability of discrete-time systems with positive controls", *IEEE Transactions on Automatic Control, 22*, 942–945.

Evans, M. E. and Murthy, D. N. (1978) "Controllability of discrete-time inhomogeneous bilinear systems", *Automatica, 14*, 147–151.

Evans, M. E. (1986) "Boundary control and discrete-time controllability", *International Journal of Systems Science, 17*, 943–951.

Farlov, S. J. (1986) "Exact controllability of partial differential equations by piecewise-constant controls", *Inernational Journal of Systems Science, 17*, 151–159.

Fattorini, H. O. (1966a) "Some remarks on complete controllability", *SIAM Journal on Control, 4*, 686–694.

Fattorini, H. O. (1966b) "Controllability of higher order linear systems", in: *Functional Analysis and Optimization*, New York: Academic Press, 301–311.

Fattorini, H. O. (1967) "On complete controllability of linear systems", *Journal of Differential Equations, 3*, 391–402.

Fattorini, H. O. (1974) "The time optimal control problem in Banach spaces", *Applied Mathematical Optimization, 1*, 163–188.

Fattorini, H. O. (1975a) "Boundary control of temperature distributions in a parallelepipedon" *SIAM Journal on Control, 13*, 1–13.

Fattorini, H. O. (1975b) "Local controllability of nonlinear wave equation", *Mathematical Systems Theory, 9*, 30–45.

Fattorini, H. O. (1983) *The Cauchy Problem*, Reading, U. S. A: Addison-Wesley Company.

Fattorini, H. O. and Russell, D. L. (1971) "Exact controllability theorem for linear parabolic equations in one space dimension", *Archive for Rational Mechanics and Analysis, 43*, 272–292.

Fattorini, H. O. and Russell, D. L. (1974) "Uniform bounds on biorthogonal functions for real exponentials with an application to the control theory of parabolic equations", *Quarterly of Applied Mathematics, 2*, 45–60.

Feintuch, A. (1978) "On single input controllability for infinite dimensional linear systems", *Journal of Mathematical Analysis and Applications, 62*, 538–546.

Fessas, P. (1979) "An analytic determination of the (A, B)-invariant and controllability subspaces", *International Journal of Control, 30*, 491–512.

Fessas, P. (1981) "Controllability of interconnected systems; local feedbacks and decentralized control", *International Journal of Control, 33*, 997–1004.

Ford, D. A. and Johnson, C. D. (1968) "Invariant subspaces and the controllability and observability of linear dynamical systems", *SIAM Journal on Control, 6*, 553–558.

Fornasini, E. and Marchesini, G. (1976) "State-space realization theory of two-dimensional filters", *IEEE Transactions on Automatic Control, 21*, 484–491.

Fornasini, E. and Marchesini, G. (1978) "Doubly-indexed dynamical systems: state-space models and structural properties", *Mathematical System Theory, 12*, 59–72.

Fornasini, E. and Marchesini, G. (1979) "On the internal stability of two-dimensional filters", *IEEE Transactions on Automatic Control, 24*, 129–130.

Fornasini, E. and Marchesini, G. (1982) "Global properties and duality in 2-*D* systems", *Systems and Control Letters, 2*, 30–38.

Frankowska, H. (1987) "Local controllability and infinitesimal generators of semigroups of set-valued maps", *SIAM Journal on Control and Optimization, 25*, 412–432.

Frazho, A. E. (1980) "A shift operator approach to bilinear system theory", *SIAM Journal on Control and Optimization, 18*, 640–658.

Frazho, A. E. (1981) "Abstract bilinear systems: the forward shift approach", *Mathematical Systems Theory, 14*, 83–94.

Frost, M. G. and Storey, C. (1978) "A note on the controllability of linear constant delay-differential systems", *International Journal of Control, 28*, 673–679.

Frost, M. G. and Storey, C. (1979) "Further remarks on the controllability of linear constant delay-differential systems", *International Journal of Control, 30*, 863–870.

Fuhrmann, P. (1972) "On weak and strong reachability and controllability of infinite-dimensional inear systems", *Journal of Optimization Theory and Applications, 9*, 77–89.

Fuhrmann, P. (1973) "On observability and stability in infinite-dimensional linear systems", *Journal of Optimization Theory and Applications, 12*, 173–181.

Fuhrmann, P. (1974a) "On realization of linear systems and applications to some question of stability", *Mathematical Systems Theory, 8*, 132–141.

Fuhrmann, P. (1974b) "Some remarks on controllability", *Richerche di Automatica, 5*, 1–5.

Fuhrmann, P. (1975) "Realization theory in Hilbert space for a class of transfer functions", *Journal of Functional Analysis, 18*, 338–349.

Fuhrmann, P. (1976) "Exact controllability and observability and realization theory in Hilbert spaces", *Journal of Mathematical Analysis and Applications, 53*, 337–392.

Fujii, N. (1980) "Feedback stabilization of distributed parameter systems by a functional observer", *SIAM Journal on Control and Optimization, 18*, 108–120.

Fujii, N. and Sakawa, Y. (1974) "Controllability for nonlinear differential equations in Banach space", *Automatic Control Theory and Applications, 2*, 44–46.

Furi, N., Nistri, P., Pera, M. P. and Zezza, P. L. (1985a) "Linear controllability by piecewise constant controls with assigned switching times". *Journal of Optimization Theory and Applications, 45*, 219–229.

Furi, N., Nistri, P., Pera, M. P. and Zezza, P. L. (1985b) "Topological methods for the global controllability of nonlinear systems", *Journal of Optimization Theory and Applications, 45*, 231–256.

Gabasov, R. and Kirilova, F. M. (1971) *Qualitative Theory of Optimal Processes* (in Russian), Moscow: Nauka.

Gabasov, R., Karpyuk, V. V. and Kirilova, F. M. (1980) "Some control problems for hereditary systems", *Proceedings of the Seminar on Functional Differential Systems and Related Topics I*, Zielona Góra, 105–114.

Gahl, R. D. (1978) "Controllability of nonlinear systems of neutral type", *Journal of Mathematical Analysiss and Applications, 63*, 33–42.

Gauthier, J. P. and Bornard, G. (1982a) "Controllabilite des systems bilineaires", *SIAM Journal on Control and Optimization, 20*, 377–384.

Gauthier, J. P. and Bornard, G. (1982b) "An openness conditions for the controllability of nonlinear systems", *SIAM Journal on Control and Optimization, 20*, 808–814.

Gauthier, J. P. and Bornard, G. (1982c) "An openness condition for controllability", *SIAM Journal on Control and Optimization, 20*, 1010–1030.

Germani, A. and Monaco, S. (1981) "Some results on the controllability of perturbed linear systems in Hilbert spaces", *Systems and Control Letters, 1*, 140–147.

Germani, A. and Monaco, S. (1983) "Functional output controllability for linear systems on Hilbert spaces", *Systems and Control Letters, 2*, 313–320.

Gershwin, S. B. and Jacobson, D. H. (1971) "A controllability theory for nonlinear systems", *IEEE Transactions on Automatic Control, 16*, 37–46.

Gibson, J. S. (1980) "A note on stabilization of infinite-dimensional linear oscillators by compact linear feedback", *SIAM Journal on Control and Optimization, 18*, 311–316.

Glower, K. and Silverman, L. M. (1976) "Characterization of structural controllability", *IEEE Transactions on Automatic Control, 21*, 534–537.

Goncalves, J. B. (1987) "Controllability on codimension one", *Journal of Differential Equations, 68*, 1–9.

Górecki, H., Fuksa, S., Korytowski, A. and Mitkowski, W. (1983) *Optimal Control in Linear Systems with Quadratic Performance Index* (in Polish), Warszawa: PWN.

Graham, K. D. and Russell, D. L. (1975) "Boundary value control of the wave equation in a spherical region", *SIAM Journal on Control, 13*, 174–196.

Grantham, W. J. and Vincent, T. L. (1975) "A controllability minimum principle", *Journal of Optimization Theory and Applications, 17*, 93–114.

Goka, T., Tarn, T. J. and Zaborszky, J. (1973) "On the controllability of a class of discrete bilinear systems", *Automatica, 9*, 615–622.

Guardabassi, G., Locatelli, A. and Rinaldi, S. (1974) "The role of controllability in some sensitivity problem", *International Journal of Control, 19*, 57–64.

Ha, T. T. (1980) "Further results on the pointwise controllability of delay-differential systems", *IEEE Transactions on Automatic Control, 25*, 981–983.

Harris, S. E. (1978) "Stochastic controllability of linear discrete systems with multiplicative noise", *International Journal of Control, 27*, 213–227.

Haynes, G. W. and Hermes, H. (1970) "Nonlinear controllability via Lie theory", *SIAM Journal on Control, 8*, 450–460.

Helton, J. W. (1974) "Discrete time systems, operators, models and scattering theory", *Journal of Functional Analysis, 16*, 15–38.

Helton, J. W. (1976) "Systems with infinite-dimensional state space: the Hilbert space approach", *Proceedings of the IEEE, 64*, 145–160.

Hermann, R. and Krener, A. (1977) "Nonlinear controllability and observability", *IEEE Transacions on Automatic Control, 22*, 728–740.

Hermes, H. (1971) "On the structure of attainable sets for generalized differential equations and control systems", *Journal of Differential Equations, 9*, 141–154.

Hermes, H. (1974a) "On local and global controllability", *SIAM Journal on Control and Optimization, 12*, 252–261.

Hermes, H. (1974b) "On the necessary and suffiicient conditions for local controllability along a reference trajectory", in: *Geometric Methods in Systems Theory*, Dordrecht: Reidel, 165–173.

Hermes, H. (1976a) "Local controllability and sufficient conditions in singular problems, I", *Journal of Differential Equations, 20*, 213–232.

Hermes, H. (1976b) "Local controllability and sufficient conditions in singular problems, II", *SIAM Journal on Control, 14*, 1049–1062.

Hermes, H. (1982) "On local controllability", *SIAM Journal on Control and Optimization, 20*, 211–220.

Heymann, M. (1976) "Controllability subspaces and feedback simulation", *SIAM Journal on Control and Optimization, 14*, 769–789.

Heymann, M. and Stern, P. (1975) "Controllability of linear systems with positive controls: geometric considerations", *Journal of Mathematical Analysis and Applications, 52*, 36–41.

Hiris, V. and Topuzu, P. (1974) "On complete controllability and cyclic vectors", *Analele Universitati Timişoara, Mathematics, 12*, 111–114.

Hiris, V. and Megan, M. (1977a) "Controllability, observability and invariant subspaces", *Analele Universitati Timişoara, Mathematics, 15*, 103–112.

Hiris, V. and Megan, M. (1977b) "Controllability of linear systems and boundary value problems", *Aequationes Mathematicae, 16*, 81–91.

Hirschorn, R. (1975) "Controllability in nonlinear systems", *Journal of Differential Equations, 19*, 46–61.

Hirschorn, R. (1976) "Global controllability of nonlinear systems", *SIAM Journal on Control and Optimization, 14*, 700–711.

Hirschorn, R. (1986) "Strong controllability of nonlinear systems", *SIAM Journal on Control and Optimization, 24*, 264–275.

Hollis, P. and Murthy, D. (1982) "Study of uncontrollable discrete bilinear systems", *IEEE Transactions on Automatic Control, 27*, 184–186.

Hosoe, S. (1980) "Determination of generic dimensions of controllable subspaces and its applications", *IEEE Transactions on Automatic Control, 25*, 1192–1196.

Hosoe, S. and Matsumoto, K. (1979) "On the irreducibility condition in the structural controllability theorems", *IEEE Transactions on Automatic Control, 24*, 963–966.

Hou, S. H. (1985) "Controllability and feedback systems", *Nonlinear Analysis, 9*, 1487–1493.

Hunt, L. (1979) "Controllability of general nonlinear systems", *Mathematical Systems Theory, 12*, 361–370.

Hunt, L. (1982) "N-dimensional controllability with $(N-1)$ controls", *IEEE Transactions on Automatic Control, 27*, 113–117.

Hunt, L. (1984) "Controllability of nonlinear hypersurface systems", in: *Algebraic and Geometric Methods in Linear Systems Theory*, Lectures in Applied Mathematics, Providence; American Mathematical Society, *18*, 225–237.

Ignatenko, V. and Janovich, V. (1980) "Controllability and reconstruction of systems with delays", *Proceedings of the Seminar on Functional Differential Systems and Related Topics*, I, Zielona Góra, 115–122.

Inouye, Y. (1982) "Notes on controllability and constructibility of linear discrete-time systems", *International Journal of Control, 35*, 1081–1084.

Jacobs, M. and Langenhop, C. (1975) "Controllable two-dimensional neutral systems", *Banach Center Publications*, Warsaw, *1*, 107–113.

Jacobs, M. and Langenhop, C. (1976) "Criteria for function space controllability of linear neutral systems", *SIAM Journal on Control and Optimization, 14*, 1009–1048.

Jakubczyk, B. and Olbrot, A. W. (1978) "Dynamic feedback stabilization of linear time lag systems", *Proceedings of the 7-th IFAC Congress*, Helsinki, 2083–2098.

Johnson, C. (1974) "State over description and uncontrollability of dynamical systems", *International Journal of Control, 10*, 1097–1100.

Jurdjevic, V. (1970) "Abstract control systems: controllability and observability", *SIAM Journal on Control, 8*, 424–439.

Jurdjevic, V. (1972) "Certain controllability properties of analytic control systems", *SIAM Journal on Control, 10*, 354–360.

Jurdjevic, V. and Quinn, J. (1978) "Controllability and stability", *Journal of Differential Equations, 28*, 381–389.

Jurdjevic, V. and Sallet, G. (1984) "Controllability properties of affine systems", *SIAM Journal on Control and Optimization, 22*, 501–508.

Kaczorek, T. (1971) "Controllability of the linear nonstationary systems with stochastic inputs", *Bulletin Polish Academy of Sciences, Technical Sciences, 19*, 44–50.

Kaczorek, T. (1974) *Theory of Automatic Control Systems* (in Polish), Warsaw: WNT.

Kaczorek, T. (1977) *Control Theory*, Part I (in Polish), Warsaw: PWN.

Kaczorek, T. (1980), *Control Theory*, Part II (in Polish), Warsaw: PWN.

Kaczorek, T. (1981) "Controllability and observability of multivariable linear systems with proportional-multiple derivative output feedbacks", *Bulletin Polish Academy of Sciences, Technical Sciences, 29*, 427–430.

Kaczorek, T. (1983a) *Theory of Multivariable Linear Dynamical Systems* (in Polish), Warsaw: WNT.

Kaczorek, T. (1983b) "Minimum energy control of 3-*D* linear systems", *Proceedings of the Seminar on Real Time Process Control, Jablonna*, 137–152.

Kaczorek, T. (1985) *Two-Dimensional Linear Systems*, Berlin, Springer-Verlag.

Kaczorek, T. (1986a) "General response formula for two-dimensional linear systems with variable coefficients", *IEEE Transactions on Automatic Control, 31*, 278–280.

Kaczorek, T. (1986b) "Deadbeat servoproblem for multivariable *n-D* systems", *IEEE Transactions on Automatic Control, 31*, 360–362.

Kaczorek, T. (1986c) "Coefficient assignment of 2-*D* systems by periodic feedback", *IEEE Transactions on Automatic Control, 31*, 870–872.

Kaczorek, T. (1987a) "Straight line reachability of Roesser model", *IEEE Transactions on Automatic Control*, *32*, 637–639.

Kaczorek, T. (1987b) "Straight line reachability of two-dimensional linear systems", *Foundations of Control Engineering*, *12*, 143–149.

Kaczorek, T. and Klamka, J. (1987) "Local controllability and minimum energy control of *n-D* linear systems", *Bulletin Polish Academy of Sciences, Technical Sciences*, *35*, 679–685.

Kaczorek, T. and Klamka, J. (1988) "Minimum energy control for general model of 2-*D* linear systems", *International Journal of Control*, *67*, 1555–1562.

Kalman, R. E. (1960) "On the general theory of control systems", *Proceedings of the 1-st IFAC Congress, London*, 481–493.

Kalman, R. E. (1963) "Mathematical description of linear dynamical systems", *SIAM Journal on Control*, *1*, 152–192.

Kalman, R. E., Falb, P. and Arbib, M. (1969) *Topics in Mathematical System Theory*, New York: McGraw-Hill.

Kamen, E. W. (1975a) "On an algebraic theory of systems defined by convolution operators", *Mathematical Systems Theory*, *9*, 57–74.

Kamen, E. W. (1975b) "On the operator theory of linear systems with pure and distributed delays", *Proceedings of the IEEE Conference on Decision and Control*, 1–4.

Kamen, E. W. (1976) "Module structure of infinite-dimensional systems with applications to controllability", *SIAM Journal on Control and Optimization*, *14*, 389–408.

Karcianas, N. (1979a) "On the basic matrix characterization of controllability", *International Journal of Control*, *29*, 767–786.

Karcianas, N. (1979b) "Zero time adjustment of initial conditions and its relationship to controllability subspaces", *International Journal of Control*, *29*, 749–765.

Kartsatos, A. G. and Dannon, V. C. (1987) "The controllability of a quasilinear functional differential systems", *Annales Polonici Mathematici*, *67*, 371–380.

Kern, G. (1978) "On the controllability properties of bilinear control systems with delays", *International Journal of Control*, *28*, 557–569.

Khanh, V. Q. (1981) "Global controllability of general nonlinear systems", *Proceedings of the Seminar on Functional Differential Systems and Related Topics II*, Zielona Góra, 156–163.

Kisielewicz, M. (1980) "Controllability of generalized systems to a closed target set", *Proceedings of the Seminar on Functional Differential Systems and Related Topics, I*", Zielona Góra, 161–167.

Klamka, J. (1972) "Uncontrollability and unobservability of multivariable systems", *IEEE Transactions on Automatic Control*, *17*, 725–726.

Klamka, J. (1973) "Uncontrollability and unobservability of composite systems", *IEEE Transactions on Automatic Control*, *18*, 539–540.

Klamka, J. (1974a) "Uncontrollability of composite systems", *IEEE Transactions on Automatic Control*, *19*, 280–281.

Klamka, J. (1974b) "Controllability and observability conditions via Jordan canonical form" (in Polish), *Podstawy Sterowania*, *4*, 349–370.

Klamka, J. (1975a) "On the global controllability of perturbed nonlinear systems", *IEEE Transactions on Automatic Control*, *20*, 170–172.

Klamka, J. (1975b) "On the local controllability of perturbed nonlinear systems", *IEEE Transactions on Automatic Control, 20*, 289–291.

Klamka, J. (1975c) "Controllability of nonlinear systems with delay in control", *IEEE Transactions on Automatic Control, 20*, 702–704.

Klamka, J. (1975d) "Controllability conditions for composite systems via Jordan canonical form" (in Polish), *Podstawy Sterowania, 5*, 43–61.

Klamka, J. (1976a) "Controllability of systems with delays" (in Polish), *Podstawy Sterowania, 6*, 179–194.

Klamka, J. (1976b) "Relative controllability and minimum energy control of linear systems with distributed delays in control", *IEEE Transactions on Automatic Control, 21*, 594–595.

Klamka, J. (1976c) "Relative controllability of linear systems with varying delay", *Systems Science, 2*, 17–24.

Klamka, J. (1976d) "Controllability and minimum energy control for systems with distributed delays in control" (in Polish), *Archiwum Automatyki i Telemechaniki, 21*, 437–446.

Klamka, J. (1976e) "Relative controllability of nonlinear systems with delays in control", *Automatica, 12*, 633–634.

Klamka, J. (1976f) "Controllability of linear systems with time-variable delays in control", *International Journal of Control, 24*, 869–878.

Klamka, J. (1977a) "On the controllability of linear systems with delays in the control", *International Journal of Control, 25*, 875–883.

Klamka, J. (1977b) "Absolute controllability of linear systems with time-variable delays in control", *International Journal of Control, 26*, 57–63.

Klamka, J. (1977c) "Relative and absolute controllability of discrete systems with delays in control", *International Journal of Control, 26*, 65–74.

Klamka, J. (1977d) "Minimum energy control of discrete systems with delays in control", *International Journal of Control, 26*, 737–744.

Klamka, J. (1978a) "Relative controllability of nonlinear systems with distributed delays in control", *International Journal of Control, 29*, 307–312.

Klamka, J. (1978b) "Relative controllability of infinite-dimensional systems with delays in control", *Systems Science, 4*, 43–52.

Klamka, J. (1978c) "Controllability of linear infinite-dimensional discrete systems with delays in control", *Analele Universitati Timişoara, Mathematics, 16*, 49–61.

Klamka, J. (1980a) "Controllability of nonlinear systems with delays in control", *Proceedings of the Seminar on Functional Differential Systems and Related Topics*, Zielona Góra, 168–178.

Klamka, J. (1980b) "Controllability of nonlinear systems with distributed delays in control", *International Journal of Control, 31*, 811–819.

Klamka, J. (1980c) "Controllability for distributed parameter systems with delays in control" *Systems Science, 6*, 269–277.

Klamka, J. (1981a) "Controllability of partial differential systems with delays", *Proceedings of the Seminar on Functional Differential Systems and Related Topics, II*, Zielona Góra, 174–182.

Klamka, J. (1981b) "Controllability of dynamical system —a survey" (in Polish), *Archiwum Automatyki i Telemechaniki, 26*, 279–309.

Klamka, J. (1982a) "Observer for linear feedback control of systems with distributed delays in controls and outputs", *Systems and Control Letters*, *1*, 326–331.

Klamka, J. (1982b) "Controllability, observability and stabilizability of distributed systems with infinite domain", *Proceedings of the 3-rd IFAC Symposium on Control of Distributed Parameter Systems VI*, Toulouse, 7–11.

Klamka, J. (1982c) "Controllability of systems with delays", *Systems Science*, *8*, 205–212.

Klamka, J. (1983a) "Controllability of *M*-dimensional linear systems", *Foundations of Control Engineering*, *8*, 65–74.

Klamka, J. (1983b) "Minimum energy control of *M-D* systems", *Proceedings of the Symposium IMECO*, Patras, Greece, session 19, 1–5.

Klamka, J. (1983c) "Minimum energy control of 2-*D* systems in Hilbert spaces", *Systems Science*, *9*, 33–42.

Klamka, J. (1984a) "Controllability and optimal control of 2-*D* linear systems", *Foundations o Control Engineering*, *9*, 15–24.

Klamka, J. (1984b) "Function of 2-*D* matrix", *Foundations of Control Engineering*, *9*, 71–82.

Klamka, J. (1984c) "Relative controllability of nonlinear systems with lumped and distributed delays in control", *Proceedings of the Seminar on Functional Differential Systems and Related Topics III*, Zielona Góra, 139–148.

Klamka, J. (1985a) "Controllability of infinite-dimensional systems with delays in control", *Acta Mathematicae Silesianae*, Katowice, *1*, 51–65.

Klamka, J. (1985b) "Controllability of linear infinite-dimensional discrete systems with delays", *Foundations of Control Engineering*, *10*, 123–133.

Klamka, J. (1985c) "A note on controllability of nonlinear systems with delays in control", *Proceedings of the Seminar on Functional Differential Systems and Related Topics, IV*, Zielona Góra, 79–88.

Klamka, J. (1986a) "Local controllability of nonlinear systems with delays in control", *Proceedings of the IMACS-IFAC Symposium on Modelling and Simulation for Control of Lumped and Distributed Parameter Systems*, Lille, France, 97–98.

Klamka, J. (1986b) "Controllability of *M*-dimensional linear discrete systems in Banach spaces", *Proceedings of the V Seminar on Real Time Process Control*, Radziejowice, 217–226.

Klamka, J. (1986c) "Relative controllability of nonlinear systems with delays", *Proceedings of the V International Conference on Systems, Modelling, Control*, Zakopane, 87–92.

Klamka, J. (1987) "Controllability of dynamical systems with constrained controls—a survey" (in Polish), *Archiwum Automatyki i Telemechaniki*, *32*, 21–34.

Klamka, J. (1988a) "Constrained controllability of 2-*D* linear systems", *Proceedings of the 12 World Congress IMACS*, *2*, Paris, 166–169.

Klamka, J. (1988b) "*M*-dimensional linear discrete systems in Banach spaces", *Proceedings of the 12 World Congress IMACS*, *4*, Paris, 31–33.

Klamka, J. and Socha, L. (1977) "Some remarks about stochastic controllability", *IEEE Transactions on Automatic Control*, *22*, 880–881.

Klamka, J. and Socha, L. (1980) "Some remarks about stochastic controllability for delayed linear systems", *International Journal of Control*, *32*, 561–566.

Klamka, J. and Kaczorek T. (1986) "Minimum energy control of 2-*D* linear systems with variable coefficients", *International Journal of Control, 44*, 645–650.

Kocięcki, M. (1983) "Genericity of some properties of systems with delays", *Proceedings of the Seminar on Functional Differential Systems and Related Topics III*, Zielona Góra, 149–154.

Kobayashi, T. (1978) "Some remarks on controllability for distributed parameter systems", *SIAM Journal on Control and Optimization, 16*, 733–742.

Kobayashi, T. (1980a) "Discrete-time observability for distributed parameter systems", *International Journal of Control, 31*, 181–193.

Kobayashi, T. (1980b) "Discrete-time controllability for distributed parameter systems", *International Journal of Systems Science, 11*, 1063–1074.

Kobayashi, T. (1982) "Controllability and stabilizability of sensitivity combined systems for distributed parameter systems", *International Journal of Control, 35*, 309–321.

Kokotovic, P. and Haddad, A. (1975) "Controllability and time-optimal control of systems with slow and fast modes", *IEEE Transactions on Automatic Control, 20*, 111–113.

Kopeykina, T. W. (1980) "A controllability nonlinear delay systems", *Proceedings of the Seminar on Functional Differential Systems and Related Topics, I*, Zielona Góra, 197–206.

Kopeykina, T. W. and Gurina, T. M. (1981) "On controllability of nonlinear systems in a critical case", *Proceedings of the Seminar on Functional Differential Systems and Related Topics", II*, Zielona Góra, 183–189.

Korobov, V. (1979) "Geometric criteria for local controllability of dynamical systems with constrained controls" (in Russian), *Differential Equations, 15*, 1592–1599.

Korobov, V., Marinich, A. and Podolski, E. (1975) "Controllability of linear, autonomous systems with constrained controls" (in Russian), *Differential Equations, 11*, 1967–1979.

Korobov, V. and Rabakh, R. (1979) "Exact controllability in Banach spaces" (in Russian), *Differential Equations, 15*, 2142–2150.

Korobov, V. and Shon, N. H. (1980a) "ε-controllability of linear autonomous systems with constrained controls" (in Russian), *Differential Equations, 16*, 395–404.

Korobov, V. and Shon, N. H. (1980b) "Controllability of linear systems in Banach spaces with constrained controls", Part I (in Russian), *Differential Equations, 16*, 806–817.

Korobov, V. and Shon, N. H. (1980c) "Controllability of linear systems in Banach spaces with constrained controls", Part II (in Russian), *Differential Equations, 16*, 1010–1022.

Korytowski, A. (1975) "Function controllability of delayed systems" (in Polish), *Archiwum Automatyki i Telemechaniki, 20*, 19–28.

Koussiouris, T. (1981) "Controllability indices of a system, minimal indices of its transfer function matrix and their relations", *International Journal of Control 34*, 613–622.

Koussiouris, T. (1982) "The module structure of controllable systems", *International Journal of Control, 35*, 665–675.

Krakhotko, V., Alsevich, V. and Razmyslovich, G. (1980) "On the control theory of dynamical systems with delays", *Proceedings of the Seminar on Functional Differential Systems and Related Topics, I*, Zielona Góra, 207–216.

Krabs, W., Leugering., G. and Seidman, T. (1985) "On boundary controllability of a vibrating plate", *Applied Mathematical Optimization, 13*, 205–229.

Kreindler, A. and Sarachik, P. (1964) "On the concepts of controllability and observability of linear systems" *IEEE Transactions on Automatic Control, 9,* 129–136.

Krener, A. J. (1987) "A causal realization theory, I. Linear deterministic system", *SIAM Journal on Control and Optimization, 25,* 499–525.

Kucera, V. (1980) "Testing controllability and constructibility in discrete linear systems", *IEEE Transactions on Automatic Control, 25,* 297–298.

Kung, S., Levy, B., Morf, M. and Kailath, T. (1977) "New results in 2-*D* systems theory: 2-*D* state space models-realization and the notions of controllability, observability and minimality", *Proceedings IEEE, 65,* 945–961.

Kurcyusz, S. and Olbrot, A. W. (1977) "On the closure in W_1^q of attainable subspace of linear time lag systems", *Journal of Differential Equations, 24,* 29–50.

Kurek, J. E. (1987) "Reachability of a system described by the multi-dimensional Roesser model", *International Journal of Control, 45,* 1559–1563.

Kwon, W. and Pearson, A. (1977) "A note on feedback stabilization of a differential-difference systems", *IEEE Transactions on Automatic Control, 22,* 468–470.

Kwon, W. and Pearson, A. (1980) "Feedback stabilization of linear systems with delayed control", *IEEE Transactions on Automatic Control, 25,* 266–269.

Lagnese, J. (1977) "Boundary value control of a class of hyperbolic equation in a general region", *SIAM Journal on Control and Optimization, 15,* 973–983.

Lagnese, J. (1978) "Exact boundary value controllability of a class of hyperbolic equations", *SIAM Journal on Control and Optimizaton, 16,* 1000–1017.

Lagnese, J. (1979) "On the support of solutions of the wave equation with applications to exact boundary controllability", *Journal of Mathematical Analysis and Applications, 58,* 121–135.

Lasiecka, I. and Triggiani, R. (1983a) "Feedback semigroups and cosine operators for boundary feedback parabolic and hyperbolic equation", *Journal of Differential Equations, 47,* 246–272.

Lasiecka, I. and Triggiani, R. (1983b) "Stabilization and structural assignment of Dirichlet boundary feedback parabolic equations", *SIAM Journal on Control and Optimization, 21,* 766–803.

Lee, E. and Markus, L. (1968) *Foundations of Optimal Control Theory,* New York: Academic Press.

Lee, E. and Olbrot, A. W. (1981) "Observability and related structural results for linear hereditary systems", *International Journal of Control, 34,* 1061–1078.

Lee, E. and Olbrot, A. W. (1983) "On reachability over polynomial rings and a related genericity problem", *Proceedings of the 19-th Conference on Information Science and Systems,* Princeton, U.S.A., 1–3.

Lee, E., Neftci, S. and Olbrot, A. W. (1982) "Canonical forms for time delay systems", *IEEE Transactions on Automatic Control, 27,* 128–132.

Lee, K. K. and Arapostathis, A. (1987) "On the controllability of piecewise linear hypersurface systems", *Systems and Control Letters, 9,* 89–96.

Leugering, G. (1985) "Boundary controllability in one-dimensional linear thermoviscoelasticity", *Journal of Integral Equations, 10,* 157–173.

Levan, N. (1980) "Controllability, *-controllability and stabilizability", *Journal of Differential Equations, 38,* 61–79.

Levitt, N. and Sussmann, H. (1975) "On the controllability by means of two vector fields", *SIAM Journal on Control, 13*, 1271–1281.

Levis, R. (1979) "Control-delayed system properties via an ordinary model", *International Journal of Control, 30*, 474–490.

Levsen, L. and Nazaroff, G. (1973) "A note on the controllability of linear time variable delay systems", *IEEE Transactions on Automatic Control, 18*, 188–189.

Lin, C. T. (1974) "Structural controllability", *IEEE Transactions on Automatic Control, 19*, 201–208.

Lin, C. T. (1977) "System structure and minimal structure controllability", *IEEE Transactions on Automatic Control, 22*, 855–362.

Lin, T., Kawamata, M. and Higuchi, T. (1987) "New necessary and sufficient conditions for local controllability and local observability of 2-*D* separable dominator systems", *IEEE Transactions on Automatic Control, 32*, 254–256.

Lion, A. (1987) "Controllability of generalized linear time-invariant systems", *IFEE Transactions on Automatic Control, 32*, 429–432.

Littman, W. and Markus, L. (1988) "Exact boundary controllability of a hybrid system of elasticity", *Archive for Rational Mechanics and Analysis, 103*, 193–236.

Lobry, C. (1974) "Controllability of nonlinear systems on compact manifolds", *SIAM Journal on Control and Optimization, 12*, 1–4.

Luckel, J. and Müller, P. (1975) "Analyse von Steuerbarkeits, Beobachtbarkeits, und Störbarkeits-strukturen linearer zeitinvarianter Systeme", *Regelungstechnik, 5*, 163–171.

Lukes, D. (1972) "Global controllability of nonlinear systems", *SIAM Journal on Control, 10*, 112–126.

Lukes, D. (1974) "Global controllability of distributed nonlinear equation", *SIAM Journal on Control, 12*, 695–704.

Lukes, D. (1974b) "A class of globally controllable nonlinear differential equations", *Proceedings of the Joint Automatic Control Conference*, Texas, 50–56.

Lukes, D. (1985) "Affine feedback controllability of constant coefficient differential equations", *SIAM Journal on Control and Optimization, 23*, 952–972.

MacCamy, R., Mizel, V. and Seidman, T. (1968) "Approximate boundary controllability for the heat equation, I", *Journal of Mathematical Analysis and Applications, 23*, 699–703.

MacCamy, R., Mizel, V. and Seidman, T. (1969) "Approxinmate boundary controllability for the heat equation, II", *Journal of Mathematical Analysis and Applications, 28*, 482–492.

Magnusson, K., Pritchard, A. J. and Quinn, M. D. (1985) "The application of fixed point theorems to global nonlinear controllability problems", *Banach Center Publications, Mathematical Control Theory, 14*, Warsaw, 319–344.

Mallela, P. (1982) "State controllability and identifiability of discrete stationary linear systems with arbitrary lag", *Mathematical Modelling, 3*, 59–67.

Manitius, A. (1972) "On the controllability conditions for systems with distributed delays in state and control", *Archiwum Automatyki i Telemechaniki, 18*, 119–131.

Manitius, A. (1976a) "Optimal control of hereditary systems", *Control Theory and Topics in Functional Analysis, 3*, 43–172.

Manitius, A. (1976b) "Controllability, observability and stabilizability of retarded systems", *Proceedings of the Conference on Decision and Control*, Piscataway U.S.A., 752-758.

Manitius, A. (1982) "*F*-controllability and observability of linear retarded systems, *Applied Mathematical Optimization*, *9*, 73–79.

Manitius, A. and Olbrot, A. W. (1972) "Controllability conditions for linear systems with delayed state and control", *Archiwum Automatyki i Telemechaniki*, *17*, 119–131.

Manitius, A. and Triggiani, R. (1978a) "Function space controllability of retarded systems: a derivation from abstract operator conditions", *SIAM Journal on Control and Optimization*, *16*, 599–645.

Manitius, A. and Triggiani, R. (1978b) "Sufficient conditions for function space controllability and feedback stabilizability of linear retarded systems", *IEEE Transactions on Automatic Control*, *23*, 659–665.

Manitius, A. and Manousiouthakis, V. (1985) "On spectral controllability of multi-input time-delay systems", *Systems and Control Letters*, *6*, 199–205.

Marchenko, V. M. (1981) "New approach to controllability and observability of dynamical systems", *Proceedings of the Seminar on Functional Differential Systems and Related Topics*, *II*, Zielona Góra, 214–218.

Marchenko, V. M. and Merezscha, V. L. (1981) "Concerning controllability of linear systems", *Proceedings of the Seminar on Functional Differential Systems and Related Topics*, *II*, Zielona Góra, 219–224.

Mariton, M. (1987) "Stochastic controllability of linear systems with Markovian jumps", *Automatica*, *23*, 783–785.

Martin, J. (1976) "On the controllability of parabolic systems by scanning control", *IEEE Transactions on Automatic Control*, *21*, 616–618.

Martin, J. (1977) "Controllability and observability of parabolic systems and addendum to two recent papers of Y. Sakawa", *SIAM Journal on Control and Optimization*, *15*, 363–366.

McGlothin, G. (1974) "A modal control model for distributed systems with application to boundary controllability", *International Journal of Control*, *20*, 417–432.

McGlothin, G. (1978) "Controllability, observability and duality in a distributed parameter systems with continuous and point spectrum", *IEEE Transactions on Automatic Control*, *23*, 687–690.

Megan, M. (1974) "Similarity, controllability and spectral mappings", *Analele Universitati Timişoara, Mathematics*, *12*, 71–78.

Megan, M. (1976) "On exponential stability of linear control systems in Hilbert spaces", *Analele Universitati Timişoara, Mathematics*, *14*, 125–130.

Megan, M. (1978a) "On controllability, stability and the Lapunov differential equation for linear control systems", *Glasnik Matematicki*, *13*, 183–198.

Megan, M. (1978b) "On the input-output stability of linear controllable systems", *Canadian Mathematical Bulletin*, *21*, 187–195.

Megan, M. and Hiris, V. (1975) "On the space of linear controllable systems in Hilbert spaces", *Glasnik Mathematicki*, *10*, 161–167.

Mesarovic, M. (1971) "Controllability of general systems", *Mathematical Systems Theory*, *5*, 353–364.

Miniuk, C. (1983) "Controllability theory for linear systems with delays" (in Russian), *Differential Equations, 19*, 575–584.

Mirza, K. and Womack, B. (1972a) "On the controllability of a class of nonlinear systems", *IEEE Transactions on Automatic Control, 17*, 531–535.

Mirza, K. and Womack, B. (1972b) "On the controllability of nonlinear time delay systems", *IEEE Transactions on Automatic Control, 17*, 812–814.

Mitter, S. K. and Foulkes, R. (1971) "Controllability and pole assignment for discrete time linear systems defined over arbitrary fields", *SIAM Journal on Control, 9*, 1–7.

Mohler, R. (1973) *Bilinear Control Processes*, New York: Academic Press.

Molinari, B. (1976a) "Extended controllability and observability for linear systems", *IEEE Transactions on Automatic Control, 21*, 136–137.

Molinari, B. (1976b) "A strong controllability and observability in linear multivariable control", *IEEE Transactions on Automatic Control, 21*, 761–764.

Moore, B. (1981) "Principial component analysis in linear systems: controllability, observability and model reduction", *IEEE Transactions on Automatic Control, 26*, 17–32.

Moore, B. and Laub. A. (1978) "Computation of supremal (**A**, **B**)-invariant and controllability subspaces", *IEEE Transactions on Automatic Control, 23*, 783–792.

Morf, M., Levy, B. and Kung, S. (1977) "New results in 2-*D* systems theory: 2-*D* polynomial matrices, factorization and coprimes", *Proceedings IEEE, 65*, 861–872.

Morse, A. (1971) "Output controllability and system synthesis", *SIAM Journal on Control, 9*, 143–148.

Morse, A. (1973) "Structural invariants of linear multivariable systems", *SIAM Journal on Control, 11*, 446–465.

Müller, P. and Weber, H. (1972) "Analysis and optimization of certain qualities of controllability and observability for linear dynamical systems", *Automatica, 8*, 237–246.

Mullis, C. (1973) "On the controllability of discrete linear systems with output feedback", *IEEE Transactions on Automatic Control, 18*, 608–615.

Murota, K. (1987) "Refined study on structural controllability of descriptor systems by means of matroids", *SIAM Journal on Control and Optimization, 25*, 967–989.

Murthy, D. N. P. (1986) "Controllability of linear positive dynamical systems", *International Journal of Systems Science, 17*, 49–54.

Naito, K. (1987) "Controllability of semilinear control systems dominated by the linear part", *SIAM Journal on Control and Optimization, 25*, 715–722.

Nambu, T. (1979a) "Feedback stabilization for distributed parameter system of parabolic type", *Journal of Differential Equations, 33*, 167–188.

Nambu, T. (1979b) "Remarks on approximate boundary controllability for distributed parameter systems of parabolic type: supremum norm problem", *Journal of Mathematical Analysis and Applications, 69*, 194–204.

Narukawa, K. (1981) "Admissible null controllability and optimal time control", *Hiroshima Mathematical Journal, 11*, 533–551.

Narukawa, K. (1982) "Admissible controllability of vibrating systems with constrained controls", *SIAM Journal on Control and Optimization, 20*, 770–782.

Narukawa, K. (1983) "Boundary value control of isotropic elastodynamic system with constrained controls", *Journal of Mathematical Analysis and Applications, 93*, 250–272.

Narukawa, K. (1984) "Complete controllability of one-dimensional vibrating systems with bang-bang controls", *SIAM Journal on Control and Optimization, 22*, 788–804.

Nguyen, K. S. (1986) "On the null-controllability of linear discrete-time systems with restrained controls", *Journal of Optimization Theory and Applications, 50*, 313–329.

Niederliński, A. (1974) *Multivariable Control Systems* (in Polish), Warszawa: WNT.

O'Brien, R. (1978) "Contraction semigroups, stabilization and the mean ergodic theorem", *Proceedings of the American Mathematical Society, 71*, 82–94.

O'Brien, R. (1979) "Perturbation of controllable systems", *SIAM Journal on Control and Optimization, 17*, 175–179.

O'Connor, D. A. and Tarn, T. J. (1983a) "On the function space controllability of linear neutral systems", *SIAM Journal on Control and Optimization, 21*, 306–329.

O'Connor, D. A. and Tarn, T. J. (1983b) "On stabilization by state feedback for neutral differential-difference equations", *IEEE Transactions on Automatic Control, 28*, 615–618.

Ogata, K. (1967) *State Space Analysis in Control Systems*, New York: Prentice-Hall.

Olbrot, A. W. (1972) "On the controllability of linear systems with time delays in control", *IEEE Transactions on Automatic Control, 17*, 664–666.

Olbrot, A. W. (1973) "Algebraic criteria of controllability to zero function for linear constant time-lag systems", *Control and Cybernetics, 2*, 59–77.

Olbrot, A. W. (1976) "Stabilizability, detectability and spectrum assignment for linear autonomous systems with general time delays", *IEEE Transactions on Automatic Control, 23*, 887–890.

Olbrot, A. W. (1977) "Control of retarded systems with function space constraints: approximate controllability", *Control and Cybernetics, 6*, 17–71.

Olbrot, A. W. (1980) "Genericity and nongenericity of mathematical models" (in Polish), *Archiwum Automatyki i Telemechaniki, 25*, 437–481.

Olbrot, A. W. (1981) "Observability and observers for a class of linear systems with delays", *IEEE Transactions on Automatic Control, 26*, 513–517.

Olbrot, A. W. (1983) "Control to equilibrium of linear delay-differential systems", *IEEE Transactions on Automatic Control, 28*, 521–523.

Olbrot, A. W. and Jakubczyk, B. (1981) "Canonical delays, $R^n(s)$-controllability and function space controllability of linear differential-difference systems", *International Journal of Control, 33*, 1–29.

Olbrot, A. W. and Sosnowski, A. (1981a) "Duality theorems on control and observation of discrete-time infinite-dimensional systems", *Mathematical Systems Theory, 14*, 137–187.

Olbrot, A. W. and Sosnowski, A. (1981b) "Approximate null controllability of linear hereditary systems", *Proceedings of the Seminar on Functional Differential Systems and Related Topics, II*, Zielona Góra, 259–264.

Olbrot, A. W. and Żak, S. H. (1980a) "Controllability and observability problem for linear functional differential systems", *Foundations of Control Engineering, 5*, 79–89.

Olbrot, A. W. and Żak, S. H. (1980b) "Sufficient and necessary conditions for spectral controllability and observability of linear retarded systems", *Proceedings of the Seminar on Functional Differential Systems and Related Topics, I*, Zielona Góra, 267–273.

Paige, C. (1981) "Properties of numerical algorithm related to computing controllability", *IEEE Transactions on Automatic Control*, *26*, 130–138.

Panda, S. (1970) "Controllability of discrete composite systems", *Automatica*, *6*, 309–313.

Panda, S., Chen, C. T. and Desoer, C. A. (1970) "Comments on controllability and observability of composite systems", *IEEE Transactions on Automatic Control*, *15*, 280–281.

Pandolfi, L. (1975) "On feedback stabilization of functional differential equation", *Bolletino della Unione Matematica Italiana*, *12*, 1–10.

Pandolfi, L. (1976a) "Linear control systems: controllability with constrained controls", *Journal of Optimization Theory and Applications*, *19*, 577–585.

Pandolfi, L. (1976b) "Stabilization of neutral functional equations", *Journal of Optimization Theory and Applications*, *20*, 191–204.

Pandolfi, L. (1978a) "Stabilization of control processes in Hilbert spaces", *Proceedings of the Royal Society of Edinburgh*, *81*, 247–258.

Pandolfi, L. (1978b) "Stabilization and controllability for a class of control systems", *Atti Academia Nazionale dei Lincei*, *24*, 130–136.

Pandolfi, L. (1979) "Canonical realizations of systems with delayed controls", *Richerche di Automatica*, *10*, 27–37.

Papageorgiu, N. S. (1987) "On the attainable set of differential inclusions and control systems", *Journal of Mathematical Analysis and Applications*, *125*, 305–322.

Pearson, A. and Kwon, W. (1976) "A minimum energy feedback regulator for linear systems subject to an average power constraint", *IEEE Transactions on Automatic Control*, *21*, 757–761.

Pecsvaradi, T. and Narenda, K. (1971) "Reachable sets for linear dynamical systems", *International Journal of Control*, *19*, 319–345.

Peichl, G. and Schappacher, W. (1986) "Constrained controllability in Banach spaces", *SIAM Journal on Control and Optimization*, *24*, 1261–1275.

Petersen, I. (1987) "Notions of stabilizability and controllability for a class of uncertain linear systems", *International Journal of Control*, *46*, 409–322.

Petersen, I. and Barmish, B. (1983a) "On certain controllability gap", *SIAM Journal on Control and Optimization*, *21*, 86–94.

Petersen, I. and Barmish, B. (1983b) "A conjecture concerning certain notion of controllability for a class of uncertain systems", *Systems and Control Letters*, *2*, 359–362.

Popov, V. M. (1972) "Invariant description of linear time-invariant controllable systems", *SIAM Journal on Control*, *10*, 252–264.

Porter, B. and Bradsham, A. (1972) "Mode-controllability characteristic of multivariable time-invariant systems with input derivative", *International Journal of Control*, *16*, 105–107.

Prabhu, S. and McCausland, I. (1976) "Methods of moments and controllability of certain distributed parameter systems", *International Journal of Control*, *23*, 89–95.

Przyłuski, K. (1977) "Stabilizability of the system $\dot{\mathbf{x}}(t) = \mathbf{F}\mathbf{x}(t)+\mathbf{G}\mathbf{u}(t-h)$ by a discrete feedback control", *IEEE Transactions on Automatic Control*, *22*, 269–270.

Przyłuski, K. and Sosnowski, A. (1987) "Observability and detectability of linear time-delay systems with unknown disturbances", *International Journal of Control*, *45*, 1115–1129.

Rastovic, D. (1979) "On the space of linear controllable systems, operator algebras and controllability", *Analele Universitati Timişoara, Mathematics*, *17*, 159–162.

Rebhuhn, D. (1977) "On the set of attainability of nonlinear nonautonomous control systems", *SIAM Journal on Control and Optimization*, *15*, 803–812.

Reghis, M. (1976) "On Kalman's canonical decomposition of linear control systems", *Analele Universitati Timişoara, Mathematics*, *14*, 65–84.

Reghis, M. and Megan, M. (1977) "On the controllability and stabilizability in Hilbert spaces", *Analele Universitati Timişoara, Mathematics*, *15*, 37–54.

Reinbacher, H. (1987) "New algebraic conditions for controllability of neutral differential equations", *Journal of Optimization Theory and Applications*, *54*, 93–111.

Rink, R. and Mohler, R. (1968) "Completely controllable bilinear systems", *SIAM Journal on Control*, *6*, 477–486.

Rodas, H. R. and Langenhop, C. E. (1978) "A sufficient condition for function space controllability of a linear neutral systems", *SIAM Journal on Control and Optimization*, *16*, 429–435.

Roesser, R. (1975) "A discrete state-space model for linear image processing", *IEEE Transactions on Automatic Control*, *20*, 1–10.

Rolewicz, S. (1987) *Functional Analysis and Control Theory*, Dordrecht—Boston—Lancaster—Tokyo—Warszawa, Reidel — PWN.

Rubio, J. (1983) "Weak global controllability of nonlinear systems", *Journal of Optimization Theory and Applications*, *39*, 251–259.

Russell, D. (1971) "Boundary value control of a higher-dimensional wave equation", *SIAM Journal on Control*, *9*, 29–42.

Russell, D. (1973) "A unified boundary controllability theory for hyperbolic and parabolic differential equation", *Studies in Applied Mathematics*, *3*, 189–211.

Sakawa, Y. (1974) "Controllability for partial differential equations of parabolic type", *SIAM Journal on Control*, *12*, 389–400.

Sakawa, Y. (1975) "Observability and related problems for partial differential equation of parabolic type", *SIAM Journal on Control and Optimization*, *13*, 14–27.

Sakawa, Y. (1983) "Feedback stabilization of linear diffusion systems", *SIAM Journal on Control and Optimization*, *21*, 667–676.

Sakawa, Y. (1985) "Feedback control of second-order evolution equations with unbounded observation", *International Journal of Control*, *41*, 717–731.

Sakawa, Y. and Matsushita, T. (1975) "Feedback stabilization of a class of distributed systems and construction of a state estimator", *IEEE Transactions on Automatic Control*, *20*, 748–753.

Salamon, D. (1984) "On controllability and observability of time delay systems", *IEEE Transactions on Automatic Control*, *29*, 432–439.

Sannuti, P. (1977) "On the controllability of singularly perturbed systems", *IEEE Transactions on Automatic Control*, *22*, 622–624.

Sannuti, P. (1978) "On the controllability of some singularly perturbed nonlinear systems", *Journal of Mathematical Analysis and Applications*, *64*, 579–591.

Saperstone, S. H. (1973) "Global controllability of linear systems with positive controls", *SIAM Journal on Control*, *11*, 417–423.

Saperstone, S. H. and Yorke, J. A. (1971) "Controllability of linear oscillatory systems using positive controls", *SIAM Journal on Control, 9*, 253–262.

Sastry, S. S. and Desoer, C. A. (1982) "The robustness of controllability and observability of linear time-varying systems", *IEEE Transactions on Automatic Control, 27*, 933–939.

Schmidt, G. (1980a) "The bang-bang principle for the time-optimal problem in boundary control of the heat equation", *SIAM Journal on Control and Optimization, 18*, 101–107.

Schmidt, G. (1980b) "Boundary control for the heat equation with steady-state targets", *SIAM Journal on Control and Optimization, 18*, 145–154.

Schmitendorf, W. and Barmish, B. (1980) "Null controllability of linear system with constrained controls", *SIAM Journal on Control and Optimization, 18*, 327–345.

Schmitendorf, W. and Barmish, B. (1981) "Controlling a constrained linear system to an affine target", *IEEE Transactions on Automatic Control, 26*, 761–763.

Schmitendorf, W. and Elenbogen, B. (1982) "Constrained max-min controllability", *IEEE Transactions on Automatic Control, 27*, 731–733.

Schwarzkopf, A. (1975) "Controllability and tenability of nonsingular systems with state equality constraints", *SIAM Journal on Control, 13*, 695–705.

Sebakhy, O. and Bayoumi, M. (1973) "Controllability of linear time-varying systems with delays in control", *International Journal of Control, 17*, 127–135.

Sebakhy, O. and Bayoumi, M. (1976) "A simplified criterion for the controllability of linear systems with delay in control", *IEEE Transactions on Automatic Control, 16*, 364–365.

Sebek, M., Bisiacco, H. and Fornasini, E. (1988) "Controllability and reconstructibility conditions for 2-*D* systems", *IEEE Transactions on Automatic Control, 33*, 496–499.

Seidman, T. (1977) "Observation and prediction for the heat equation. Path observability and controllability", *SIAM Journal on Control and Optimization, 15*, 412–427.

Seidman, T. (1978) "Exact boundary control for some evolution equations", *SIAM Journal on Control and Optimization, 16*, 979–999.

Seidman, T. (1979) "Time invariance of the reachable set for linear control problems", *Journal of Mathematical Analysis and Applications, 72*, 17–20.

Seidman, T. (1987) "Invariance of the reachable set under nonlinear perturbations", *SIAM Journal on Control and Optimization, 25*, 1173–1191.

Sezer, E. and Huseyin, O. (1979) "On the controllability of composite systems", *IEEE Transactions on Automatic Control, 24*, 327–329.

Sharma, P. K. (1986) "Some results on pole-placement and reachability", *Systems and Control Letters, 6*, 325–328.

Shields, R. and Pearson, J. (1976) "Structural controllability of multi-input linear systems", *IEEE Transactions on Automatic Control, 21*, 203–212.

Silvermann, L. M. (1969) "Asymptotic controllability", *IEEE Transactions on Automatic Control, 14*, 78–80.

Sinha, A. S. C. (1986) "Controllability of nonlinear delay systems", *International Journal of Control, 43*, 1305–1315.

Sinha, A. S. C. (1985) "Null controllability of nonlinear infinite delay systems with restrained controls", *International Journal of Control, 42*, 735–741.

Sinha, A. S. C. and Yokomoto, C. (1980) "Null controllability of nonlinear system with variable time delay", *IEEE Transactions on Automatic Control, 25*, 1234–1236.

Skowroński, J. H. and Leitmann, G. (1977) "Avoidance control", *Journal of Optimization Theory and Applications, 23*, 581–591.

Skowroński, J. M. and Vincent, T. L. (1982) "Playability with and without capture", *Journal of Optimization Theory and Applications, 36*, 111–128.

Slemrod, M. (1974) "A note on complete controllability and stabilizability for linear control systems in Hilbert space", *SIAM Journal on Control, 12*, 500–508.

Socha, L. and Klamka, J. (1978) "Stochastic controllability of dynamical systems" (in Polish), *Podstawy Sterowania, 8*, 191–200.

Somasundaram, D. and Balachandran, K. (1983) "Relative controllability of a class of nonlinear systems with delay in control", *Indian Journal of Pure and Applied Mathematics, 14*, 1327–1334.

Somasundaram, D. and Balachandran, K. (1984a) "Controllability of nonlinear systems consisting of a bilinear mode with distributed delays in control", *IEEE Transactions on Automatic Control", 29*, 573–575.

Somasundaram, D. and Balachandran, K. (1984b) "Controllability of nonlinear systems consisting of a bilinear mode with time-varying delays in control", *Automatica, 20*, 257–258.

Somasundaram, D. and Balachandran, K. (1985) "Relative controllability of nonlinear systems with time varying delays in control", *Kybernetika, 21*, 65–72.

Sontag, E. (1976a) "Linear systems over commutative rings: a survey", *Richerche di Automatica, 7*, 1–34.

Sontag, E. (1976b) "On linear systems and noncommutative rings", *Mathematical Systems Theory, 9*, 327–344.

Sontag, E. (1976c) "On finitely accessible and finitely observable rings", *Journal of Pure and Applied Algebra, 8*, 97–104.

Sontag, E. (1978) "On first order equations for multidimensional filters", *IEEE Transactions on Automatic, Control, 26*, 480–482.

Sontag, E. (1983) "A Lapunov-like characterization of asymptotic controllability", *SIAM Journal on Control and Optimization, 21*, 462–471.

Sorenson, E. (1968) "Controllability and observability of linear stochastic time discrete control systems", *Advances in Control Theory, 6*, 95–158.

Sourour, A. (1982) "On strong controllability of infinite-dimensional linear systems", *Journal of Mathematical Analysis and Applications, 87*, 460–462.

Stanford, D. and Connor, L. (1980) "Controllability and stabilizability in multi-pair systems", *SIAM Journal on Control and Optimization, 18*, 488–497.

Sunahara, Y., Aikara, S. and Kishino, K. (1975) "On the stochastic observability and controllability for nonlinear systems", *International Journal of Control, 22*, 65–82.

Sundarashan, M. (1979) "Decentral and multilevel controllability of large-scale systems", *International Journal of Control, 30*, 71–80.

Sussmann, H. (1975) "On the number of directions needed to achieve controllability", *SIAM Journal on Control, 13*, 414–419.

Sussmann, H. (1976) "Some properties of vector fields systems that are not altered by small perturbations", *Journal of Differential Equations, 20*, 292–315.

Sussmann, H. (1978) "A sufficient condition for local controllability", *SIAM Journal on Control and Optimization, 16*, 790–802.

Sussmann, H. (1983) "Lie brackets and local controllability: a sufficient condition for scalar-input systems", *SIAM Journal on Control and Optimization, 21*, 685–713.

Sussmann, H. (1987) "A general theorem on local controllability", *SIAM Journal on Control and Optimization, 25*, 158–194.

Sussmann, H. and Jurdjevic, V. (1972) "Controllability of nonlinear systems", *Journal of Differential Equations, 12*, 95–116.

Tadmor, G. (1984) "Functional differential equations of retarded and neutral type: analytic solutions and piecewise continuous controls", *Journal of Differential Equations, 51*, 151–181.

Takahashi, K. (1984) "Exact controllability and spectrum assignment", *Journal of Mathematical Analysis and Applications, 104*, 537–545.

Tarn, T. J., Elliot, D. L. and Goka, T. (1973) "Controllability of discrete bilinear systems with bounded control", *IEEE Transactions on Automatic Control, 18*, 298–301.

Tarn, T. J. and Spong, M. W. (1981) "On the spectral controllability of delay-differential equations", *IEEE Transactions on Automatic Control, 26*, 527–528.

Theodorou, N. J. and Tzafestas, S. G. (1984) "A canonical state-space model for three-dimensional systems", *International Journal of Systems Science, 15*, 1353–1379.

Thowsen, A. (1976) "Function space null controllability by augmented delay feedback control", *IEEE Transactions on Automatic Control, 21*, 298–299.

Thowsen, A. (1980a) "Characterization of state controllable time-delay systems with piecewise constant inputs", Part I, *International Journal of Control, 31*, 31–42.

Thowsen, A. (1980b) "Characterization of state controllable time-delay systems with piecewise constant inputs", Part II, *International Journal of Control, 31*, 43–49.

Tokarzewski, J. (1977) "Local controllability properties of nonlinear dynamical systems", *Archiwum Automatyki i Telemechaniki, 22*, 175–199.

Tokarzewski, J. (1979) "Controllability of a class of locally symmetrical nonlinear dynamical systems", *Archiwum Automatyki i Telemechaniki, 25*, 29–41.

Tokashi, Y. and Hirokazu, M. (1979) "Strong structural controllability", *SIAM Journal on Control and Optimization, 17*, 123–138.

Triggiani, R. (1975a) "Controllability and observability in Banach space with bounded operators", *SIAM Journal on Control, 13*, 462–491.

Triggiani, R. (1975b) "Pathological asymptotic behaviour of control systems in Banach space", *Journal of Mathematical Analysis and Applications, 43*, 411–429.

Triggiani, R. (1975c) "On the lack of exact controllability for mild solutions in Banach spaces", *Journal of Mathematical Analysis and Applications, 50*, 438–446.

Triggiani, R. (1975d) "On the stabilizability problem in Banach space", *Journal of Mathematical Analysis and Applications, 52*, 383–403.

Triggiani, R. (1976) "Extensions of rank conditions for controllability and observability to Banach space and unbounded operators", *SIAM Journal on Control and Optimization, 14*, 313–338.

Triggiani, R. (1977) "A note on the lack of exact controllability for mild solutions in Banach spaces", *SIAM Journal on Control and Optimization, 15*, 407–441.

Triggiani, R. (1978) "On the relationship between first and second order controllable systems in Banach spaces", *SIAM Journal on Control and Optimization, 16*, 847–859.

Triggiani, R. (1979) "On Nambu's boundary controllability problem for diffusion processes", *Journal of Differential Equations, 33*, 189–200.

Troch, I. (1971) "A note on complete controllability", *SIAM Journal on Control, 9*, 543–546.

Tsinias, J. and Kalouptsidis, N. (1982) "Output feedback design and controllable cascade connections of nonlinear systems", *Systems and Control Letters, 2*, 230–236.

Underwood, R. and Young, D. (1979) "Null controllability of nonlinear functional differential equations", *SIAM Journal on Control and Optimization, 17*, 753–772.

Underwood, R. and Chukwu, E. N. (1988) "Null controllability of nonlinear neutral differential equations", *Journal of Mathematical Analysis and Applications, 129*, 326–345.

Van der Schaft, A. J. (1982) "Observability and controllability for smooth nonlinear systems", *SIAM Journal on Control and Optimization, 20*, 338–354.

Veliov, V. M. and Krastanov, M. I. (1986) "Controllability of piecewise linear systems", *Systems and Control Letters, 7*, 335–341.

Vincent, T. L. and Skowroński, J. M. (1979) "Controllability with capture", *Journal of Optimization Theory and Applications, 29*, 77–86.

Vinter, R. (1980) "A characterization of the reachable set for nonlinear control systems", *SIAM Journal on Control and Optimization, 18*, 599–610.

Walczak, S. (1984) "A note on the controllability of nonlinear systems", *Mathematical Systems Theory, 17*, 351–356.

Wang, P. K. C. (1964) "Control of distributed parameter systems", *Advances in Control Theory, 1*, 75–172.

Wang, P. K. C. (1972) "Modal feedback stabilization of a linear distributed system", *IEEE Transactions on Automatic Control, 17*, 552–553.

Warga, J. (1972) *Optimal Control of Differential and Functional Equations*, New York: Academic Press.

Warga, J. (1976) "Controllability and necessary conditions in unilateral problem without differentiability assumptions", *SIAM Journal on Control and Optimization, 14*, 546–573.

Warga, J. (1978a) "Controllability and a multiplier for nondifferentiable optimization problems", *SIAM Journal on Control and Optimization, 16*, 803–812.

Warga, J. (1978b) "Controllability of nondifferentiable hereditary processes", *SIAM Journal on Control and Optimization, 16*, 813–831.

Warga, J. (1985) "Second order controllability and optimization with ordinary controls", *SIAM Journal on Control and Optimization, 23*, 49–60.

Warren, M. and Eckberg, A. (1975) "On the dimensions of controllability subspaces: a characterization via polynomial matrices and Kronecker invariants", *SIAM Journal on Control, 13*, 434–445.

Weck, N. (1982) "A remark on controllability for symmetric hyperbolic systems in one space dimension", *SIAM Journal on Control and Optimization, 20*, 1–8.

Weck, N. (1984) "More states reachable by boundary control of the heat equation", *SIAM Journal on Control and Optimization, 22*, 699–710.

Wei, K. C. (1976) "A class of controllable nonlinear systems", *IEEE Transactions on Automatic Control, 21*, 787–789.

Wei, K. C. and Pearson, A. (1978a) "Global controllability for a class of bilinear systems", *IEEE Transactions on Automatic Control, 23*, 486–488.

Wei, K. C. and Pearson, A. (1978b) "On minimum energy control of commutative bilinear systems", *IEEE Transactions on Automatic Control, 23*, 1020–1023.

Wei, K. C. and Pearson, A. (1978c) "Control law for an intercept system", *Journal of Guidance and Control, 1*, 298–304.

Weiss, L. (1972) "Controllability, realization and stability of discrete-time systems", *SIAM Journal on Control, 10*, 230–251.

Weiss, L. (1973) "Use of semigroups for derivation of controllablity and observability conditions in time-invariant linear systems", *International Journal of Control, 18*, 475–479.

White, L. W. (1980) "Controllability properties of pseudo-parabolic boundary control problems", *SIAM Journal on Control and Optimization, 18*, 534–539.

Willems, J. L. (1986) "Structural controllability and observability", *Systems and Control Letters, 8*, 5–12.

Williams, N. and Zakian, V. (1977) "A ring of delay operators with applications to delays differential systems", *SIAM Journal on Control Optimization, 15*, 247–255.

Wolovich, W. A. (1968) "On the stabilization of controllable systems", *IEEE Transactions on Automatic Control, 13*, 569–572.

Wolovich, W. A. (1974) *Linear Multivariable Systems*, New York: Springer-Verlag.

Wolovich, W. A. and Hwang, H. I. (1974) "Composite system controllability and observability", *Automatica, 10*, 209–212.

Wonham, W. M. (1978) *Linear Multivariable Control, a Geometric Approach*, New York: Springer-Verlag.

Voronov, A. A. (1979) *Stability, Controllability, Observability* (in Russian), Moscow: Nauka.

Yamamoto, Y. (1980) "Some remarks on reachability of infinite-dimensional linear systems", *Journal of Mathematical Analysis and Optimization, 74*, 568–577.

Yamamoto, Y. (1981) "Module structure of constant linear systems and its applications to controllability", *Journal of Mathematical Analysis and Applications, 83*, 411–437.

Yip, E. L. and Sincovec, R. F. (1981) "Solvability, controllability and observability of continuous descriptor systems", *IEEE Transactions on Automatic Control, 26*, 702–707.

Yonemura, Y. and Ito, M. (1972) "Controllability of composite systems of tandem connection", *IEEE Transactions on Automatic Control, 17*, 722–724.

Yorke, J. A. (1972) "The maximum principle and controllability of nonlinear equations", *SIAM Journal on Control, 10*, 334–338.

Yousif S. M. (1975) "On approximate controllability of linear dynamic systems", *Proceedings of the Midwest Symposium on Circuits and Systems*, Montreal, 225–230.

Youlong, Y., Laixiang, S. and Huisheng, Z. (1982) "Controllability and observability of discrete systems", *Chinese Annales of Mathematics, 3*, 273–278.

Zabczyk, J. (1981) "Controllability of stochastic linear systems", *Systems and Control Letters, 1*, 25–31.

Zadeh, L. A. and Desoer, C. A. (1963) *Linear System Theory. The State Space Approach*, New York: McGraw-Hill.

Zhou, H. X. (1982) "A note on approximate controllability for semilinear one-dimensional heat equation", *Applied Mathematical Optimization*, *8*, 275–285.

Zhou, H. X. (1983) "Approximate controllability for a class of semilinear abstract equations" *SIAM Journal on Control and Optimization*, *21*, 551–565.

Zhou, H. X. (1984) "Controllability properties of linear and semilinear absrtact control systems", *22*, 405–422.

Zmood, R. B. (1974) "The Euclidean space controllability of control systems with delay", *SIAM Journal on Control*, *12*, 609–623.

Zwart, H. (1988) "Characterization of all controlled invariant subspaces for spectral systems", *SIAM Journal on Control and Optimization*, *26*, 369–386.

Index

The bold-faced characters indicate definitions